Methods in Molecular Biology

For further volumes:
http://www.springer.com/series/7651

For over 35 years, biological scientists have come to rely on the research protocols and methodologies in the critically acclaimed *Methods in Molecular Biology* series. The series was the first to introduce the step-by-step protocols approach that has become the standard in all biomedical protocol publishing. Each protocol is provided in readily-reproducible step-by-step fashion, opening with an introductory overview, a list of the materials and reagents needed to complete the experiment, and followed by a detailed procedure that is supported with a helpful notes section offering tips and tricks of the trade as well as troubleshooting advice. These hallmark features were introduced by series editor Dr. John Walker and constitute the key ingredient in each and every volume of the *Methods in Molecular Biology* series. Tested and trusted, comprehensive and reliable, all protocols from the series are indexed in PubMed.

Innate Lymphoid Cells

Methods and Protocols

Edited by

Shoba Amarnath

Laboratory of T cell Regulation, Institute of Cellular Medicine, Newcastle University, Newcastle Upon Tyne, UK

Humana Press

Editor
Shoba Amarnath
Laboratory of T cell Regulation
Institute of Cellular Medicine
Newcastle University
Newcastle Upon Tyne, UK

ISSN 1064-3745 ISSN 1940-6029 (electronic)
Methods in Molecular Biology
ISBN 978-1-0716-0337-6 ISBN 978-1-0716-0338-3 (eBook)
https://doi.org/10.1007/978-1-0716-0338-3

This Humana imprint is published by the registered company Springer Science+Business Media, LLC, part of Springer Nature.
The registered company address is: 1 New York Plaza, New York, NY 10004, U.S.A.

Preface

The immune system is broadly divided into the innate and the adaptive immune system. In recent years, a new population of immune cells called Innate Lymphoid Cells (ILCs) has been discovered. ILCs have been classified as first responders to environmental insults within the tissue micro-environment. An explosion of research in this area has identified the pleiotropic nature of these cells in driving innate immune responses, initiating adaptive immune responses, imparting tissue repair, and regulating thermogenesis. Furthermore, dysregulation of these subsets has been reported in human diseases. The purpose of this book is to provide relevant background and methodologies currently available for characterizing murine and human ILC subsets. It has been written with the aim to provide the reader with useful phenotypical and functional assays that may be relevant for studying ILC function in mice and humans. Moreover, since ILCs occupy tissue niches and are known to be important in barrier protection, methodologies to study the microbiome and enteroid cultures, which will be essential for studying human ILC function, have also been provided within the book. Detailed chapters on characterizing ILC precursors, various ILC subsets, and cellular engineering of ILCs for therapeutic use are described within the book. This book will be useful for all the readers interested in ILC biology and for those interested in fundamental immunology, cellular engineering, cell signaling, and clinician scientists.

Newcastle Upon Tyne, UK *Shoba Amarnath*

Acknowledgements

The editor thanks all contributing authors for their collaborations, which has made this project possible; the series editor Prof. John Walker whose help and guidance have been instrumental in the success of this project; Mr. Patrick Marton and Ms. Anna Rakovsky for making the production of this book possible. I would also like to thank Dr. Arian Laurence for his help and encouragement throughout this project.

Contents

Preface . . . . *v*
Acknowledgements . . . . *vii*
Contributors . . . . *xi*

1 Protocols for Innate Lymphoid Cell Phenotypic and Functional Characterization: An Overview 1
Shoba Amarnath

2 Protocols for Studying Murine ILC Development 7
Matthew T. Stier and R. Stokes Peebles Jr

3 Identification of Murine Innate Lymphoid Cell Subsets in Barrier Tissues and Their Draining Lymph Nodes Using Flow Cytometry 23
Emma E. Dutton and David R. Withers

4 Identification and Functional Characterization of Murine Group 3 Innate Lymphoid Cell (ILC3) Subsets in the Intestinal Tract and Associated Lymphoid Tissues 37
Felipe Melo-Gonzalez and Matthew R. Hepworth

5 Identification of Murine and Human Innate Lymphoid Cells in Frozen Tissue Sections Using Immunofluorescence 51
Emma E. Dutton and David R. Withers

6 Assessing Phosphorylation of STAT Transcription Factors in Mouse Innate Lymphoid Cells 59
Gianluca Scarno, Giuseppe Pietropaolo, Chiara Di Censo, Giovanna Peruzzi, and Giuseppe Sciumè

7 Determining Coreceptor Expression and Function in Murine ILC2 Through Flow Cytometry Characterization and Coculture Techniques 71
Christian Schwartz and Padraic G. Fallon

8 Protocol for Determining the Effect of Neuroendocrine Hormones on Murine ILC Function 83
Linda Quatrini and Sophie Ugolini

9 Role of G-Protein Coupled Receptors in Chemotaxis of Innate Lymphoid Cells 93
Brinda Bhatt, Huabin Zhu, Nikhil Patel, and Nagendra Singh

10 Group 2 ILC Functional Assays in Allergic Airway Inflammation 99
Lewis J. Entwistle, Franz Puttur, Lisa G. Gregory, and Clare M. Lloyd

11 Isolation, Culture, and Induction of Plasticity in ILC2s 115
Jonathan Silver, Alison A. Humbles, and Yoichiro Ohne

12 Bacterial Infection Allows for Functional Examination of Adoptively Transferred Mouse Innate Lymphoid Cell Subsets 129
Goo-Young Seo, Daniel A. Giles, and Mitchell Kronenberg

13 Isolation, Characterization, and Culture of Intestinal Intraepithelial Lymphocytes 141
Olivia Jane James, Maud Vandereyken, and Mahima Swamy

14 Isolation and Characterization of Innate Lymphoid Cells within the Murine Tumor Microenvironment 153
Grace Mallett, Warren Patterson, Megan Payne, and Shoba Amarnath

15 Microbiota Analysis Using Sequencing by Synthesis: From Library Preparation to Sequencing 165
Andrew Nelson and Christopher J. Stewart

16 Establishing Human Intestinal Enteroid/Organoid Lines from Preterm Infant and Adult Tissue 185
Christopher J. Stewart, Mary K. Estes, and Sasirekha Ramani

17 Characterizing Innate Lymphoid Cell Phenotype and Function in Human Inflammatory Bowel Disease 199
Omer S. Omer, Nick Powell, and Graham M. Lord

18 CRISPR/Cas9-Based Gene Engineering of Human Natural Killer Cells: Protocols for Knockout and Readouts to Evaluate Their Efficacy 213
Mélanie Lambert, Caroline Leijonhufvud, Filip Segerberg, J. Joseph Melenhorst, and Mattias Carlsten

Index *241*

Contributors

SHOBA AMARNATH • *Laboratory of T Cell Regulation, Institute of Cellular Medicine, Newcastle University, Newcastle Upon Tyne, UK*

BRINDA BHATT • *Department of Biochemistry and Molecular Biology, Georgia Cancer Center, Augusta University, Augusta, GA, USA*

MATTIAS CARLSTEN • *Department of Medicine, Huddinge, Center for Hematology and Regenerative Medicine, Karolinska Institutet, Stockholm, Sweden*

CHIARA DI CENSO • *Department of Molecular Medicine, Laboratory Affiliated to Istituto Pasteur Italia—Fondazione Cenci Bolognetti, Sapienza University of Rome, Rome, Italy*

EMMA E. DUTTON • *Institute of Immunology and Immunotherapy, College of Medical and Dental Sciences, University of Birmingham, Birmingham, UK*

LEWIS J. ENTWISTLE • *Inflammation, Repair and Development, National Heart and Lung Institute, Imperial College London, London, UK*

MARY K. ESTES • *Department of Molecular Virology and Microbiology, Baylor College of Medicine, Houston, TX, USA; Department of Medicine, Baylor College of Medicine, Houston, TX, USA*

PADRAIC G. FALLON • *Trinity Biomedical Sciences Institute, School of Medicine, Trinity College Dublin, Dublin 2, Ireland; National Children's Research Centre, Our Lady's Children's Hospital, Dublin, Ireland; Trinity Translational Medicine Institute, St James's Hospital, Trinity College Dublin, Dublin 2, Ireland*

DANIEL A. GILES • *Division of Developmental Immunology, La Jolla Institute for Immunology, La Jolla, CA, USA*

LISA G. GREGORY • *Inflammation, Repair and Development, National Heart and Lung Institute, Imperial College London, London, UK*

MATTHEW R. HEPWORTH • *Lydia Becker Institute of Immunology and Inflammation, University of Manchester, Manchester, UK; Division of Infection, Immunity and Respiratory Medicine, Faculty of Biology, Medicine and Health, Manchester Collaborative Centre for Inflammation Research (MCCIR), Manchester Academic Health Science Centre, School of Biological Sciences, University of Manchester, Manchester, UK*

ALISON A. HUMBLES • *AstraZeneca, Gaithersburg, MD, USA*

OLIVIA JANE JAMES • *Medical Research Council Protein Phosphorylation and Ubiquitylation Unit (MRC-PPU), School of Life Sciences, University of Dundee, Dundee, UK*

MITCHELL KRONENBERG • *Division of Developmental Immunology, La Jolla Institute for Immunology, La Jolla, CA, USA; Division of Biology, University of California, San Diego, La Jolla, CA, USA*

MÉLANIE LAMBERT • *Center for Hematology and Regenerative Medicine, Department of Medicine, Huddinge, Karolinska Institutet, Stockholm, Sweden*

CAROLINE LEIJONHUFVUD • *Center for Hematology and Regenerative Medicine, Department of Medicine, Huddinge, Karolinska Institutet, Stockholm, Sweden*

CLARE M. LLOYD • *Inflammation, Repair and Development, National Heart and Lung Institute, Imperial College London, London, UK*

GRAHAM M. LORD • *Peter Gorer Department of Immunobiology, School of Immunology and Microbial Sciences, King's College London, London, UK; Faculty of Biology, Medicine and Health, University of Manchester, Manchester, UK*

Grace Mallett • *Laboratory of T cell Regulation, Institute of Cellular Medicine, Newcastle University, Newcastle upon Tyne, UK*
J. Joseph Melenhorst • *Center for Cellular Immunotherapies, Perelman School of Medicine at the University of Pennsylvania, Philadelphia, PA, USA; Institute for Immunology, Perelman School of Medicine at the University of Pennsylvania, Philadelphia, PA, USA; Department of Pathology and Laboratory Medicine, Perelman School of Medicine at the University of Pennsylvania, Philadelphia, PA, USA; Abramson Cancer Center, University of Pennsylvania, Philadelphia, PA, USA*
Felipe Melo-Gonzalez • *Lydia Becker Institute of Immunology and Inflammation, University of Manchester, Manchester, UK; Division of Infection, Immunity and Respiratory Medicine, Faculty of Biology, Medicine and Health, Manchester Collaborative Centre for Inflammation Research (MCCIR), Manchester Academic Health Science Centre, School of Biological Sciences, University of Manchester, Manchester, UK*
Andrew Nelson • *Faculty of Health and Life Sciences, Northumbria University, Newcastle upon Tyne, UK*
Yoichiro Ohne • *AstraZeneca, Gaithersburg, MD, USA*
Omer S. Omer • *Peter Gorer Department of Immunobiology, School of Immunology and Microbial Sciences, King's College London, London, UK*
Nikhil Patel • *Department of Pathology, Augusta University, Augusta, GA, USA*
Warren Patterson • *Laboratory of T cell Regulation, Institute of Cellular Medicine, Newcastle University, Newcastle upon Tyne, UK*
Megan Payne • *Laboratory of T cell Regulation, Institute of Cellular Medicine, Newcastle University, Newcastle upon Tyne, UK*
R. Stokes Peebles Jr • *Division of Allergy, Pulmonary and Critical Care Medicine, Department of Medicine, Vanderbilt University Medical Center, Nashville, TN, USA*
Giovanna Peruzzi • *Center for Life Nano Science@Sapienza, Istituto Italiano di Tecnologia, Rome, Italy*
Giuseppe Pietropaolo • *Department of Molecular Medicine, Laboratory Affiliated to Istituto Pasteur Italia—Fondazione Cenci Bolognetti, Sapienza University of Rome, Rome, Italy*
Nick Powell • *Department of Inflammation Biology, School of Immunology and Microbial Sciences, King's College London, London, UK*
Franz Puttur • *Inflammation, Repair and Development, National Heart and Lung Institute, Imperial College London, London, UK*
Linda Quatrini • *Department of Immunology, IRCSS Bambino Gesù Children's Hospital, Rome, Italy*
Sasirekha Ramani • *Department of Molecular Virology and Microbiology, Baylor College of Medicine, Houston, TX, USA*
Gianluca Scarno • *Department of Molecular Medicine, Laboratory Affiliated to Istituto Pasteur Italia—Fondazione Cenci Bolognetti, Sapienza University of Rome, Rome, Italy*
Christian Schwartz • *Mikrobiologisches Institut—Klinische Mikrobiologie, Immunologie und Hygiene, Universitätsklinikum Erlangen and Friedrich-Alexander Universität (FAU) Erlangen-Nürnberg, Erlangen, Germany; Trinity Biomedical Sciences Institute, School of Medicine, Trinity College Dublin, Dublin 2, Ireland*
Giuseppe Sciumè • *Department of Molecular Medicine, Laboratory Affiliated to Istituto Pasteur Italia—Fondazione Cenci Bolognetti, Sapienza University of Rome, Rome, Italy*
Filip Segerberg • *Center for Hematology and Regenerative Medicine, Department of Medicine, Huddinge, Karolinska Institutet, Stockholm, Sweden*

GOO-YOUNG SEO • *Division of Developmental Immunology, La Jolla Institute for Immunology, La Jolla, CA, USA*
JONATHAN SILVER • *AstraZeneca, Gaithersburg, MD, USA*
NAGENDRA SINGH • *Department of Biochemistry and Molecular Biology, Georgia Cancer Center, Augusta University, Augusta, GA, USA*
CHRISTOPHER J. STEWART • *Institute of Cellular Medicine, Newcastle University, Newcastle upon Tyne, UK*
MATTHEW T. STIER • *Division of Allergy, Pulmonary and Critical Care Medicine, Department of Medicine, Vanderbilt University Medical Center, Nashville, TN, USA*
MAHIMA SWAMY • *Medical Research Council Protein Phosphorylation and Ubiquitylation Unit (MRC-PPU), School of Life Sciences, University of Dundee, Dundee, UK*
SOPHIE UGOLINI • *Aix Marseille University, CNRS, INSERM, Centre d'Immunologie de Marseille-Luminy, Marseille, France*
MAUD VANDEREYKEN • *Medical Research Council Protein Phosphorylation and Ubiquitylation Unit (MRC-PPU), School of Life Sciences, University of Dundee, Dundee, UK*
DAVID R. WITHERS • *Institute of Immunology and Immunotherapy, College of Medical and Dental Sciences, University of Birmingham, Birmingham, UK*
HUABIN ZHU • *Department of Biochemistry and Molecular Biology, Georgia Cancer Center, Augusta University, Augusta, GA, USA*

Chapter 1

Protocols for Innate Lymphoid Cell Phenotypic and Functional Characterization: An Overview

Shoba Amarnath

Abstract

The innate immune system forms a first line of defense against infections and is an essential component in immune defense prior to the generation of sustained adaptive immune responses. Until recently this innate response was believed to be predominantly elicited by cells of myeloid origin. The last 10 years have seen the discovery of an extensive family of innate lymphoid cells (ILCs) characterized by the ability to express cytokines, traditionally associated with activated T effector cells, and lack of expression of antigen receptors.

ILCs are enriched at barrier surfaces and rapidly respond to alarmins within the tissue microenvironment. ILCs lack recombinant activating gene (RAG)-dependent rearranged receptors, lack myeloid and dendritic cell lineage defining factors, and possess lymphoid morphology. The first characterized innate lymphoid cells were natural killer cells (NK) cells and lymphoid tissue inducer (LTi) cells. NK cells, defined in 1975), mediate important early antigen independent immune responses. By contrast, LTi cells, defined in 1997, are vital for formation of lymph nodes during embryogenesis. In the past decade, research on ILCs has defined them as important regulators of barrier immunity. However, ILC phenotype and functional characterization is complex and requires highly specific protocols for delineating ILC specific function from other well-characterized immune cells. Within this book, each chapter provides an in-depth protocol that will expand on techniques used by laboratories to study ILC development, characterization and function in mice and humans.

Key words ILC protocols, ILC characterization

1 ILC Development

All three groups of ILCs are hematopoietic cells that develop within the bone marrow and depend on key transcription factors such as NFIL3, ID2, TOX, ETS1, and GATA3 [1–3]. All ILC subsets including LTi and NK cells develop from the common lymphoid progenitor or CLP. The cytokine IL-7, by activating the transcription factor STAT5 in CLP, induces the expression of Nfil3, a basic leucine zipper transcription factor. Nfil3 in turn regulates ID2 expression within the CLP [3, 4]. ID2 is a key checkpoint in determining if CLPs differentiate into CHILP (common helper innate lymphoid cell precursor) or into NK progenitors. This

Shoba Amarnath (ed.), *Innate Lymphoid Cells: Methods and Protocols*, Methods in Molecular Biology, vol. 2121,
https://doi.org/10.1007/978-1-0716-0338-3_1,

essential fork in the development pathway determines the differentiation of NK cells versus other ILC lineages. The CHILPs then upregulate α4β7 integrin which enable migration of ILC precursors (ILCp) to their respective tissue sites. Within the tissue, ILC precursors develop into various ILC subsets which is dictated by the master transcription factors associated with the three principal T effector cell lineages, namely, Tbet, GATA3, and RORγt. Of note, GATA3 remains highly expressed in all subsets until ILCp differentiate into their individual subsets. On differentiation, GATA3 is exclusively expressed by ILC2 [5].

In addition to transcription factors associated with T effector cells, ILCp also express coreceptors such as programmed death-1 (PD-1), which marks *bcl11b*$^+$ and *Il17rb*$^+$ ILC precursors. Therefore, in order to thoroughly investigate ILC development, a number of transcription factors (T-bet, RORγt), cytokine receptors (IL7R), and coreceptors (PD-1) should be incorporated within the characterization experiments. A comprehensive characterization protocol for analyzing ILC precursors and studying ILC development is outlined within this book.

2 Types of ILCs

ILCs are grouped under three subtypes based on their phenotype and function. Their close similarity to T cells and antigen presenting cells highlights the necessity for accurate protocols for characterizing their phenotype and function without bystander effects from T cells. In recent years, the leap in research on this immune cell compartment has resulted in numerous characterization protocols which are summarized by a number of laboratories within this book. An overview of the various ILC subsets and their phenotypic and functional characterization is summarized below.

2.1 Group-1 ILCs

Group 1 ILCs comprise of NK cells, Eomes$^+$ ILC1, and Eomes$^-$ ILC-1, and can elicit robust antiviral responses [4, 6, 7]. NK cell differentiation is dependent on Eomes, whereas the bulk of ILC1 is dependent on the expression of the related transcription factor T-bet [7, 8]. Mature splenic NK cells can be defined as NK1.1$^+$Nkp46$^+$CD49a$^-$CD49b$^+$, whereas Eomes$^+$ ILC1s can be defined as NKp46$^+$CD49a$^+$CD49b$^+$TRAIL$^+$Tbet$^+$CD73^{+-} TRAIL$^+$CD69$^+$CD103$^+$ (mostly found within the salivary glands), and the traditional ILC1s that occupy tissue niches such as liver, siLPL, siIEL, skin, and peritoneum are characterized as NK1.1$^+$NKp46$^+$CD49a$^+$ CD127$^+$TRAIL$^+$Tbet$^+$ [9]. Furthermore, ILC1s can be classified as cells that respond to cytokines IL-12, IL-18, and IL-15 [5, 8, 10] by producing IFNγ and TNFα [11–13]. The major differences between NK cells and ILC-1s include cytotoxicity, tissue residency, and IL7R expression. NK cells are known to elicit cytotoxic defense via production of perforin [14] whereas ILC1s are characterized as IL7R$^+$ and are tissue resident. However, recent work has demonstrated that ILC1s can migrate and traffic between lymphoid and nonlymphoid tissues [15]. Therefore, emerging literature suggests that ILC1s are a defined population that can carry out specific immune functions and can be characterized by specific surface receptors, cytokines, and transcription factor expression. In addition to these differences, the expression of NKp46 and T-bet can also represent ILC1s in mice. While it is ambiguous whether to classify this population as "ex-ILC3s" or ILC1s, it can be postulated that the cytokine phenotype of these cells can dictate their function and hence classification as either Group-1 (IFNγ) or Group-3 (IL-17 or IL-22) ILCs.

2.2 Group-2 ILCs

A substantial amount of research has been performed on ILC2 characterization due to their well-defined type-2 function within the immune system. These cells were first described by multiple groups as "natural helper cells," "nuocytes", "innate helper cells" but were broadly reclassified as group 2 ILCs [16]. Group-2 ILCs are stimulated by alarmins IL-33, TSLP, and IL-25 and respond by producing effector cytokines IL-13, IL-4, and IL-5. ILC2s are implicated in anti-helminth immunity, in autoimmune diseases such as allergy and atopic dermatitis, in tissue repair through Amphiregulin (Areg), and in regulating thermogenesis by converting white fat into beige fat. Group-2 ILCs are further defined as either naïve ILC2s (ST2$^+$KLRG1$^-$), inflammatory ILC2s (KLRG1$^+$ST2$^-$) or mature ILC2s (KLRG1$^+$ST2$^+$) in mice [2, 17–20]. All three ILC2 subsets express the master transcription factor GATA3. In addition to these surface receptors, cytokines and transcription factors, the expression of coreceptors and their respective ligands on ILC2s has also been reported. A number of coreceptors such as ICOS, ICOS-L [21], PD-1 [22], PDL-1 [23], and GITR [24] are expressed on ILC2s and can be used as ideal surface receptor markers for characterization purposes. In addition to coreceptors, the function of ILC2s as antigen presenting cells has been illustrated by the expression of MHC-class II, CD80 and to a lesser extent by CD86 expression. ILC2s can also present lipid antigens through CD1A expression. Therefore, the pleiotropic function of ILC2s not only in maintaining immunity but also in restoring tissue homeostasis and thermogenesis has highlighted these immune cells as regulators of both the immune system and tissue integrity.

2.3 Group-3 ILCs

Group 3 ILCs are classified as a group of innate lymphocytes that are dependent on the transcription factor RORγt. Lymphoid tissue inducer cells (LTi) were the first to be described within this subset. LTi cells are vital for the formation of lymphoid structure during fetal development and are essential in promoting secondary lymphoid tissue organogenesis. In adults, LTis are predominantly found within Peyer's patches, draining lymph nodes and tertiary lymphoid structures. Over the years, LTi cells have been extensively characterized and in adults depict a heterogeneous expression of the CD4 receptor, along with expressing CXCR5, GPR183, and CCR6. In addition to LTi cells, the characterization of IL-22 producing ILC3s has undergone substantial reclassification over the years. These were identified in mucosal tissues and were defined as NK22 cells, NKR-LTi cells, NCR22 cells, and ILC22s [16]. However, this nomenclature has been simplified whereby ILC-3s are subdivided by the expression of NK receptor NKp46 (in mice) or NKp44 (in humans). NCR^{+} ILC3s are positive for NKp46/NKp44 [7, 25, 26] whereas NCR^{-} do not express Nkp46/Nkp44 [27, 28]. Group-3 ILCs are activated by cytokines such as IL-23 and are thus efficient at inducing antibacterial responses through the production of cytokines such as IL-17 and IL-22. In addition to their helper like function, similar to Group-2 ILCs, Group-3 cells can also mimic antigen presenting cell [APC] function through expression of MHC class II, OX-40L, CD30L, CD40L, Notch ligands, HVEM, and CD80 [24] which, in turn, can modulate adaptive immunity and control T cell responses [29, 30]. Therefore, understanding the function of ILC3s in modulating the adaptive immune system is an active area of research that requires a number of optimized protocols which are discussed in detail within this book.

2.4 Regulatory ILCs

Regulatory ILCs have been recently reported and are characterized by the expression of the cytokine IL-10. Unlike the other ILC subsets, ILCregs are not defined by any transcription factors; however, they require ID3 for development and are regulated by TGFβ [31]. ILCregs can efficiently suppress ILC1 and ILC-3 function but a comprehensive characterization profile for these cells is yet to be fully defined.

3 Innate Lymphoid Cell Plasticity

Although ILCs are classified into various subgroups, there exists a dynamic functional plasticity between the various subsets whereby ILC-3s can become "ex-ILC3" [7] and obtain an ILC-1 phenotype. Similarly, ILC2s can be converted to ILC3s or ILC-1 under appropriate pathogenic stimulation [32–34]. Understanding the functional plasticity of ILCs in disease states will enable determining the environmental cues that dictate ILC phenotype and function thereby providing insight into modulating ILC subsets for translational benefit. A comprehensive protocol for studying mouse and human ILCs is outlined in the book.

4 Conclusion

In summary, we have attempted to provide a comprehensive number of protocols within this book which will enable the readers to efficiently characterize various ILC subset phenotype and function in both murine and human immune system. The assays outlined comprise detailed protocols for in vitro characterization assays, cell expansion cultures, coculture assays, and functional plasticity assays that are essential for studying the function of ILCs. In addition, the use of new cellular engineering technology that utilizes ILC as cell therapy products is also discussed. Therefore, the chapters of this book not only provide an in-depth fundamental immunological assays for ILC research but also offer translationally relevant protocols that can be of clinical use. In addition to ILC focused chapters, we have also expanded the protocols to include studying the microbiome, enteroid cultures, and intraepithelial lymphocyte characterization protocols. These chapters provide an example of the varied protocols required for studying ILC communication with the microenvironment and the adaptive immune system.

Acknowledgments

We would like to thank Dr. Arian Laurence for his critical reading and comments on the manuscript. We would like to thank the Academy of Medical Sciences, Springboard Award [SBF003 \1129] sponsored by the Wellcome Trust and Newcastle University Research Fellowship for supporting this work.

References

1. Monticelli LA, Sonnenberg GF, Abt MC et al (2011) Innate lymphoid cells promote lung-tissue homeostasis after infection with influenza virus. Nat Immunol 12(11):1045–1054
2. Hoyler T, Klose CS, Souabni A et al (2012) The transcription factor GATA-3 controls cell fate and maintenance of type 2 innate lymphoid cells. Immunity 37(4):634–648
3. Xu W, Domingues RG et al (2012) Nonlinear dendritic integration of sensory and motor input during an active sensing task. Nature 492(7428):247–251
4. Serafini N, Vosshenrich CA, Di Santo JP (2015) Transcriptional regulation of innate lymphoid cell fate. Nat Rev Immunol 15 (7):415–428
5. Hoyler T, Connor CA, Kiss EA et al (2013) T-bet and Gata3 in controlling type 1 and type 2 immunity mediated by innate lymphoid cells. Curr Opin Immunol 25(2):139–147
6. Kiessling R, Klein E, Pross H et al (1975) "Natural" killer cells in the mouse. II. Cytotoxic cells with specificity for mouse Moloney leukemia cells. Characteristics of the killer cell. Eur J Immunol 5(2):117–121
7. Vonarbourg C, Mortha A, Bui VL et al (2010) Regulated expression of nuclear receptor RORgammat confers distinct functional fates to NK cell receptor-expressing RORgammat (+) innate lymphocytes. Immunity 33 (5):736–751
8. Bernink JH, Peters CP, Munneke M et al (2013) Human type 1 innate lymphoid cells accumulate in inflamed mucosal tissues. Nat Immunol 14(3):221–229

9. Cortez VS, Colonna M (2016) Diversity and function of group 1 innate lymphoid cells. Immunol Lett 179:19–24
10. Mebius RE, Rennert P, Weissman IL (1997) Developing lymph nodes collect CD4+CD3− LTbeta+ cells that can differentiate to APC, NK cells, and follicular cells but not T or B cells. Immunity 7(4):493–504
11. Gil-Cruz C, Perez-Shibayama C, Onder L et al (2016) Fibroblastic reticular cells regulate intestinal inflammation via IL-15-mediated control of group 1 ILCs. Nat Immunol 17 (12):1388–1396
12. Weizman OE, Adams NM, Schuster IS et al (2017) ILC1 confer early host protection at initial sites of viral infection. Cell 171 (4):795–808.e12
13. Klose CSN, Flach M, Mohle L et al (2014) Differentiation of type 1 ILCs from a common progenitor to all helper-like innate lymphoid cell lineages. Cell 157(2):340–356
14. Loh J, Chu DT, O'Guin AK et al (2005) Natural killer cells utilize both perforin and gamma interferon to regulate murine cytomegalovirus infection in the spleen and liver. J Virol 79 (1):661–667
15. Dutton EE, Gajdasik DW, Willis C et al (2019) Peripheral lymph nodes contain migratory and resident innate lymphoid cell populations. Sci Immunol 4(35):eaau8082
16. Spits H, Artis D, Colonna M et al (2013) Innate lymphoid cells—a proposal for uniform nomenclature. Nat Rev Immunol 13 (2):145–149
17. Moro K, Yamada T, Tanabe M et al (2010) Innate production of T(H)2 cytokines by adipose tissue-associated c-Kit(+)Sca-1(+) lymphoid cells. Nature 463(7280):540–544
18. Halim TY, MacLaren A, Romanish MT et al (2012) Retinoic-acid-receptor-related orphan nuclear receptor alpha is required for natural helper cell development and allergic inflammation. Immunity 37(3):463–474
19. Mjosberg JM, Trifari S, Crellin NK et al (2011) Human IL-25- and IL-33-responsive type 2 innate lymphoid cells are defined by expression of CRTH2 and CD161. Nat Immunol 12 (11):1055–1062
20. Huang Y, Guo L, Qiu J et al (2015) IL-25-responsive, lineage-negative KLRG1(hi) cells are multipotential 'inflammatory' type 2 innate lymphoid cells. Nat Immunol 16(2):161–169
21. Maazi H, Patel N, Sankaranarayanan I et al (2015) ICOS:ICOS-ligand interaction is required for type 2 innate lymphoid cell function, homeostasis, and induction of airway hyperreactivity. Immunity 42(3):538–551
22. Taylor S, Huang Y, Mallett G et al (2017) PD-1 regulates KLRG1(+) group 2 innate lymphoid cells. J Exp Med 214(6):1663–1678
23. Schwartz C, Khan AR, Floudas A et al (2017) ILC2s regulate adaptive Th2 cell functions via PD-L1 checkpoint control. J Exp Med 214 (9):2507–2521
24. Sonnenberg GF, Hepworth MR (2019) Functional interactions between innate lymphoid cells and adaptive immunity. Nat Rev Immunol 19:599. https://doi.org/10.1038/s41577-019-0194-8
25. Cella M, Fuchs A, Vermi W et al (2009) A human natural killer cell subset provides an innate source of IL-22 for mucosal immunity. Nature 457(7230):722–725
26. Satoh-Takayama N, Lesjean-Pottier S, Vieira P et al (2010) IL-7 and IL-15 independently program the differentiation of intestinal CD3-NKp46+ cell subsets from Id2-dependent precursors. J Exp Med 207 (2):273–280
27. Buonocore S, Ahem PP, Uhlig HH et al (2010) Innate lymphoid cells drive interleukin-23-dependent innate intestinal pathology. Nature 464(7293):1371–1375
28. Sonnenberg GF, Monticelli LA, Elloso MM et al (2011) CD4(+) lymphoid tissue-inducer cells promote innate immunity in the gut. Immunity 34(1):122–134
29. Hepworth MR, Fung TC, Masur SH et al (2015) Immune tolerance. Group 3 innate lymphoid cells mediate intestinal selection of commensal bacteria-specific CD4(+) T cells. Science 348(6238):1031–1035
30. Hepworth MR, Monticelli LA, Fung TC et al (2013) Innate lymphoid cells regulate CD4+ T-cell responses to intestinal commensal bacteria. Nature 498(7452):113–117
31. Wang S, Xia P, Chen Y et al (2017) Regulatory innate lymphoid cells control innate intestinal inflammation. Cell 171(1):201–216. e18
32. Bernink JH, Ohne Y, Teunissen MBM et al (2019) c-Kit-positive ILC2s exhibit an ILC3-like signature that may contribute to IL-17-mediated pathologies. Nat Immunol 20 (8):992–1003
33. Silver JS, Kearley J, Copenhaver AM et al (2016) Inflammatory triggers associated with exacerbations of COPD orchestrate plasticity of group 2 innate lymphoid cells in the lungs. Nat Immunol 17(6):626–635
34. Zhang K, Xu X, Pasha MA et al (2017) Cutting edge: notch signaling promotes the plasticity of group-2 innate lymphoid cells. J Immunol 198 (5):1798–1803

Chapter 2

Protocols for Studying Murine ILC Development

Matthew T. Stier and R. Stokes Peebles Jr

Abstract

Understanding the origins and developmental trajectory of innate lymphoid cell (ILC) progenitors has been of substantial interest to the fields of ILC biology and immunology. While mature ILC are rare lymphocytes, ILC progenitors represent an even smaller fraction of cells, providing additional challenges in studying them. Moreover, though the approaches to studying these cells are conceptually straightforward, the technical nuances that underlie them can substantially affect the quality of the data. Herein, we provide a detailed protocol for assessing the frequency of ILC progenitors in the bone marrow, their phenotype, and their potential to develop into mature ILC. These methods make up the foundation of in vivo investigations into ILC development, and we hope these thorough protocols and associated notes facilitate additional, high-quality inquiries into this fascinating field.

Key words Innate lymphoid cells (ILC), Natural killer (NK) cells, Common lymphoid progenitors, Flow cytometry, Adoptive transfer

1 Introduction

Innate lymphoid cells (ILCs) are a heterogeneous group of effector lymphocytes that play a critical role in tissue homeostasis and defense. ILCs are categorized by their developmental relationship and effector functions with the major lineages including group 1, 2, and 3 ILCs (ILC1, ILC2, and ILC3, respectively), natural killer (NK) cells, and lymphoid tissue inducer (LTi) cells. Progression from hematopoietic stem cell to terminally differentiated ILC involves several stepwise progressions, each with defining transcription factors and/or surface markers and with increasingly restricted developmental potential [1]. A simplified version of this developmental cascade is shown in Fig. 1 [2]. All subsets of innate and adaptive lymphocytes derive from lymphoid-primed multipotent progenitors (LMPP) and common lymphoid progenitors (CLP) [3–5]. In mice, expression of Id2 is associated with ILC lineage commitment and the ordered cascade of alpha-integrin expressing lymphoid precursors (αLP), early innate lymphoid progenitors (EILP), common helper innate lymphoid progenitors (CHILP),

Shoba Amarnath (ed.), *Innate Lymphoid Cells: Methods and Protocols*, Methods in Molecular Biology, vol. 2121,
https://doi.org/10.1007/978-1-0716-0338-3_2, © Springer Science+Business Media, LLC, part of Springer Nature 2020

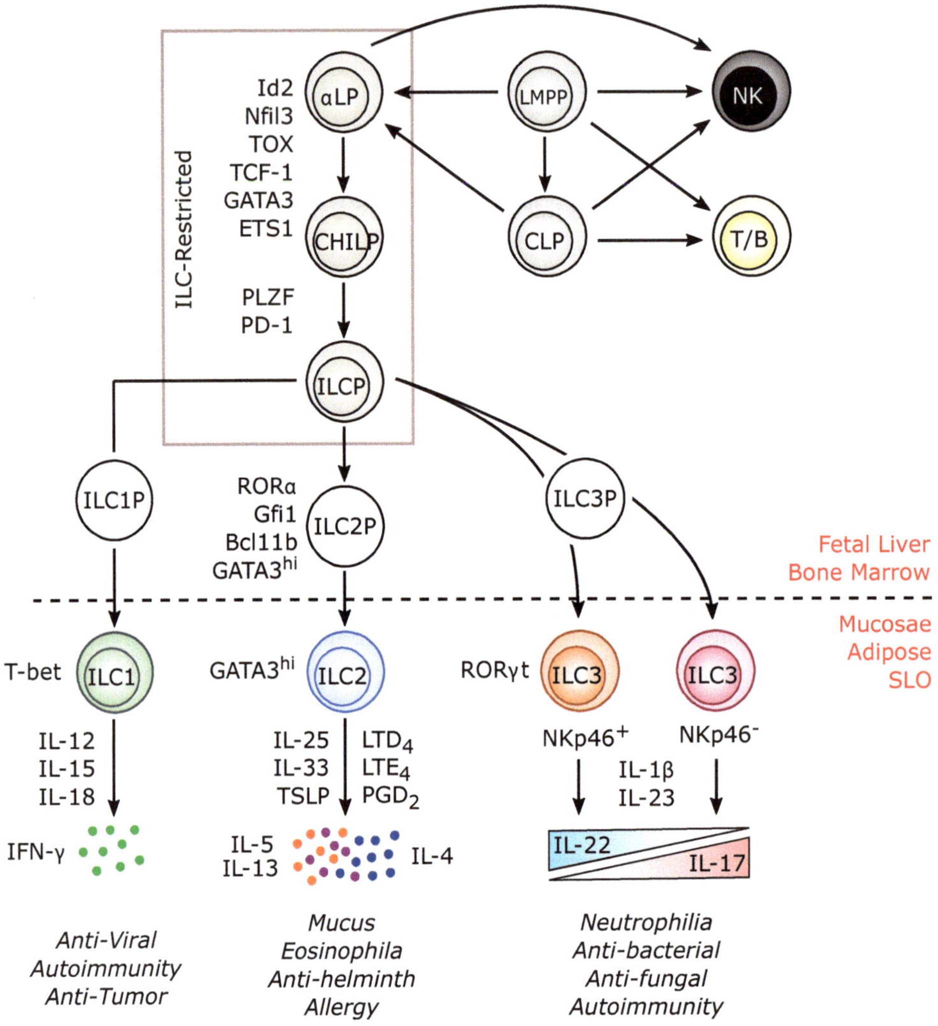

Fig. 1 ILC development. Innate lymphoid cell (ILC) development and effector functions. ILCs develop from lymphoid-primed multipotent progenitor (LMPP) or common lymphoid progenitor (CLP) cells similar to T and NK (NK) cells in the bone marrow and fetal liver. ILC development proceeds through several stepwise, ILC-committed progenitors and requires a number of transcription factors that are expressed at unique points throughout development. Mature ILC1s, ILC2s, and ILC3s represent the major lineages of ILCs with unique transcriptional governance, stimuli, and cytokine production. These are primarily found at mucosal sites, in adipose tissue, and in limited quantities in secondary lymphoid organs (SLO). *αLP* α-integrin-expressing lymphoid precursor, *CHILP* common helper innate lymphoid progenitor, *IFN* interferon, *IL* interleukin, *ILCP* innate lymphoid cell progenitor, *LTD_4* leukotriene D_4, *LTE_4* leukotriene E_4, *PGD_2* prostaglandin D_2, *TSLP* thymic stromal lymphopoietin. (Image adapted and reprinted from *Annals of Allergy, Asthma & Immunology*, 119(6), Stier et al., "Innate lymphoid cells and allergic disease," 480–488, Dec. 2017, with permission from Elsevier [2])

and innate lymphoid cell progenitors (ILCP), each with further restricted lineage potential [6–9]. Briefly, αLP can produce all ILC lineages including NK cells whereas the later stage CHILP can only generate ILC1, ILC2, ILC3, and LTi cells [7, 10]. ILCP are further restricted, lacking capacity to generate LTi cells [11]. Major regulators of these stages are shown in Fig. 1 and include TCF-1 (αLP/EILP), ETS1 (CHILP), and PLZF (ILCP) [10–13]. This chapter will focus predominantly on the evaluation of the "helper-like" ILC lineages—ILC1, ILC2, and ILC3—in mice, though the principles can largely be applied to other known and yet to be described ILC populations.

Assessment of ILC development in vivo relies heavily on flow cytometric analysis of the major precursor populations, with special attention to both their frequency and phenotype. Herein, we describe our approach to performing flow cytometry for hematopoietic stem cells and ILC precursors in murine bone marrow. A substantial volume of work defining ILC development has been performed in the bone marrow, and evidence suggest that ILC or ILC progenitor trafficking from developmental sites such as the bone marrow may be influential in the perinatal period or during processes that disrupt the tissue homeostasis [14–20]. However, at steady-state in adult mice, ILC are a tissue-resident population and ILC progenitors may be embedded in tissues, as is observed in fetal gut, and support ILC populations in situ [21–25]. Thus, a comprehensive evaluation of ILC development may require evaluation at multiple sites and developmental stages unique to a given study.

Beyond evaluation of frequency and phenotype, lineage potential is a benchmark criterion for establishing the degree of lineage restriction of specific precursor populations and inferring developmental trajectory. In this chapter, we discuss an in vivo approach to assessing lineage potential. In vitro parallels exist for evaluating lineage potential [6, 7, 10, 11, 26–28] and represent an important methodology for studying ILC development but are beyond the scope of this section (though such methods rely principally on the same flow cytometry and fluorescence activated cell sorting (FACS), as detailed in this chapter).

Finally, our knowledgebase about ILCs is rapidly expanding and improved techniques continue to refine our understanding of specific ILC precursor populations, their transcriptional regulation, and their lineage potential. In this chapter, we aim to present a set of methods that are fundamental to the study of ILC development yet are adaptable even in light of new discoveries in the field.

2 Materials

2.1 Supplies for Bone Marrow Harvest

1. Surgical tools including sharp scissors, tweezers, and scalpel.
2. 70% ethanol in sprayer.
3. Delicate task wipes (i.e., Kimwipes).
4. Petri dishes (polystyrene, 60 × 15 mm).
5. RPMI, stored at 4 °C.

6. 1× phosphate buffered saline (PBS), stored at 4 °C.
7. 3 and 10 mL disposable syringes.
8. 18 G/1.5 in. and 25 G/1 in. needles.
9. 15 and 50 mL conical tubes.
10. Centrifuge compatible with 15 and 50 mL conical tubes.
11. 70 μm cell strainer for 50 mL conical tubes.
12. RBC lysis buffer (Tonbo Biosciences).
13. Trypan blue.
14. Hemocytometer.

2.2 Supplies for Flow Cytometry

1. Supplies listed in Subheading 2.1, as appropriate.
2. 5 mL FACS tubes (i.e., nonsterile or sterile, Falcon).
3. Flow staining buffer: 1× PBS with 3% fetal bovine serum, sterile filtered.
4. Fc blocking reagent: anti-CD16/32 antibody (i.e., Purified Anti-Mouse CD16/CD32 [clone 2.4G2], Tonbo Biosciences).
5. Flow cytometry-compatible antibodies to identify the desired populations. Fluorophore-conjugated antibodies include c-Kit (clones ACK2 and 2B8), Sca-1 (clone D7), Flt3 (clone A2F10), CD127 (clone SB/199), α4β7 (clone DATK32), PD-1 (clone J43), and CD25 (clone PC61). Biotin-conjugated antibodies contained in the Miltenyi Lineage Cocktail target CD5, CD11b, CD45R, 7-4, Gr-1 (Ly6G/C), and Ter-119 (*see* **Note 1**). We supplement the Lineage Cocktail with an antibody targeting CD3 (clone 17A2) to ensure the exclusion of T cells. A careful review of the most current literature should be performed regularly to select suitable reagents, as contemporary gating strategies may call for additional or alternative antibodies.
6. If analyzing intracellular markers using fluorescently tagged antibodies, fixation and permeabilization reagents (i.e., Foxp3/Transcription Factor Staining Buffer Kit, Tonbo Biosciences).
7. Viability marker: DAPI, propidium iodide, or 7-AAD are good options for nonfixed cells, each with distinct excitation/emission properties. Fixation-compatible reagents are required if cells are fixed and/or permeabilized (i.e., Ghost Dye Fixable Viability Dyes, Tonbo Biosciences).
8. Multilaser flow cytometer. In this protocol, we use a 5-laser BD LSR II with the following setup: 355 nm ultraviolet (2 PMT sensors), 405 nm violet (7), 488 nm blue (2), 545 nm green (5), and 635 nm red (3) lasers.
9. Flow cytometry analysis software (i.e., FlowJo, as used in this protocol).

2.3 Supplies for Adoptive Transfer

1. Supplies listed in Subheadings 2.1 and 2.2, as appropriate.
2. MACS buffer: 1× PBS with 0.5% bovine serum albumin (BSA) and 2 mM EDTA, mixed and sterile filtered.
3. Direct Lineage Cell Depletion Cocktail (Miltenyi) containing the following magnetic microbead-conjugated antibodies against the following targets: CD5, CD11b, CD45R (B220), 7-4, Gr-1 (Ly-6G/C), and Ter-119.
4. LS Columns.
5. Magnetic column stand (i.e., QuadroMACS Separator, Miltenyi).
6. FACS tubes with 35 μm cell strainer caps (Falcon).
7. 1 mL disposable syringes.
8. 30 G/0.5 in. needles.

3 Methods

3.1 Isolation of Murine Bone Marrow Cells

1. To collect tibia and femur from mice, euthanize per institutionally approved animal protocol (*see* **Note 2**). Wet and sterilize abdomen and legs with 70% ethanol. Make skin incision in the lower abdomen and cut from this point down each leg to expose muscle. Use the blunt aspect of the scissors or forceps to spread the tissue to find the femoroacetabular joint. Cut just above this joint, removing a small piece of the pelvis. Cut along the underside of the leg removing the skin and some bulk muscle. There is no need to excise all of the muscle; it will be removed in a later step. Cut at the bottom of the tibia to remove the leg. There is very little marrow at the distal end of the tibia so you can cut straight through the bone without appreciable marrow loss. Place bones in a 60 mm plastic petri dish on ice.
2. Using a scalpel and a forceps, briefly scrape along the knee joint to loosen the connective tissue and ligaments. Hyperflex the knee joint to separate the tibia and femur and then cut through the remaining soft issue to fully separate the bones.
3. Using a delicate task wipe (i.e., Kimwipes) for increased grip, pull away the muscle from each of the leg bones. A small amount of muscle can be left on the ends of the bones, as these will be excised in **step 4**.
4. Cut bones at the metaphysis to expose the marrow. You do not need to modify the distal end of the tibia as this has already been exposed in **step 1**.

5. Fill a 50 mL conical tube with 10 mL of ice-cold RPMI (*see* **Note 3**). Using a 3 mL syringe with a 25 G needle, draw up 3 mL of RPMI from the conical tube, insert syringe needle into the end of the bone, and flush until the bone turns a translucent white color (*see* **Notes 4** and **5**).
6. Using a 10 mL syringe and an 18 G needle, aspirate all of the RPMI in the conical tube, now containing flushed bone marrow, and expel it through the syringe twice for each sample to break up the clumps and generate a single cell suspension.
7. Spin samples in 50 mL conical tubes at 350 × *g* for 5 min at 4 °C in a centrifuge. Sample should collect firmly at the bottom of the 50 mL conical tube. Discard supernatant by pouring off into an appropriate biohazard disposal container. This will leave ~250 μL of solution at the bottom of the tube. Resuspend cells by vortex mixing at medium to medium-high speed (*see* **Note 6**).
8. Add 3 mL of cold 1× RBC lysis buffer per sample, mix lightly by swirling, and incubate on ice for 3 min. Quench by pouring ice-cold 1× PBS into the tubes up to the 25 mL mark.
9. Centrifuge samples at 350 × *g*, discard supernatant, and resuspend cells as in **step 7**.
10. Adjust volume by adding 3 mL of flow staining buffer (*see* **Note 7** if intracellular staining) per one tibia and femur (6 mL for both tibias and femurs).
11. Mix resuspended cells with a brief low speed vortex and pour through a 70 μm cell strainer into new 50 mL conical tubes.
12. Mix samples by lightly vortexing and take an aliquot (10–20 μL) to count the cells. Generally, we dilute 1:10 in trypan blue and manually count on a standard hemocytometer.

3.2 Flow Cytometric Assessment of ILC Precursors

1. From Subheading 3.1, add five million cells per sample to FACS tubes. In addition to each sample, you need several control tubes—one unstained tube, one tube for single staining for each fluorophore and marker (i.e., viability marker), and any control tubes desired such as isotype controls, FMOs, etc. (*see* **Note 8**). Spin sample in 5 mL FACS tube at 350 × *g* for 5 min at 4 °C in a centrifuge. Cells should collect firmly at the bottom of the 5 mL FACS tube. Discard supernatant by pouring off into an appropriate biohazard disposal container. This will leave ~100 μL of solution at the bottom of the tube. Resuspend cells by vortexing for 5 s on moderate speed (*see* **Note 9**).
2. Wash cells by adding 2 mL of flow staining buffer (*see* **Note 7** if intracellular staining) to each sample. Centrifuge samples at 350 × *g*, discard supernatant, and resuspend cells as in **step 1**.

3. For up to five million cells (*see* **Note 10**), add Fc blocking agent (anti-CD16/32 antibody). We use Purified Anti-Mouse CD16/CD32 (2.4G2) by mixing 3 μL antibody into 47 μL flow staining buffer (*see* **Note** 7 if intracellular staining) and adding to the sample. If you plan to stain for intracellular antigens, a viability dye needs to be added at this step (including to the single stain control for this viability dye). We use the fixable live/dead dyes from Tonbo Biosciences diluted 1:200 in the Fc block mixture (i.e. add 0.25 μL fixable live/dead dye and 3 μL of Fc blocking agent to 46.75 μL 1× PBS). Mix by vortexing for 1–2 s on moderate speed and place samples at 4 °C for 15 min protected from light.
4. Do not wash between **steps 3** and **4**. Prepare surface antibody cocktail by diluting antibodies into flow staining buffer (*see* **Note** 7 if intracellular staining) for a total volume of 50 μL per sample. Our gating strategy for major ILC precursor populations is listed in Table 1 and shown in Figs. 2 and 3, recognizing that different labs evaluate for additional cell populations (i.e. αLP and EILP) or may use variations on these cellular markers. For analysis of multiple samples, make one master mix of antibodies to be applied to all samples. Add 50 μL of

Table 1
Gating strategy for ILC progenitors

Population	Gating strategy
LMPP	Lin^- $CD25^-$ $Flt3^+$ $CD127^-$ $Sca\text{-}1^{hi}$ $c\text{-}Kit^{hi}$
CLP	Lin^- $CD25^-$ $Flt3^+$ $CD127^+$ $Sca\text{-}1^{int}$ $c\text{-}Kit^{int}$
CHILP	Lin^- $CD25^-$ $Flt3^-$ $CD127^+$ $\alpha_4\beta_7{}^{hi}$
ILCP	Lin^- $CD25^-$ $Flt3^-$ $CD127^+$ $c\text{-}Kit^+$ $\alpha_4\beta_7{}^+$ $PD\text{-}1^+$
ILC2P	Lin^- $IL\text{-}25R^+$ $CD25^+$ $CD127^+$

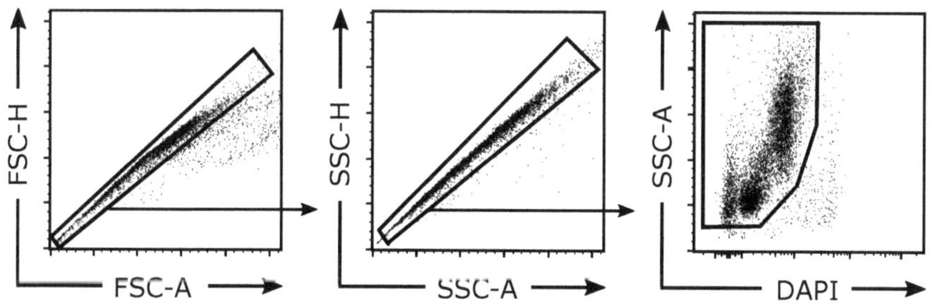

Fig. 2 Flow cytometric exclusion of "doublets" and dead cells. Bone marrow was harvested from 8–12-week-old BALB/c female mice and processed for analytical flow cytometry as described in Subheadings 3.1 and 3.2. Doublets were excluded using 1:1 gates for FSC-A/FSC-H and SSC-A/SSC-H. Dead cells were excluded as $DAPI^+$

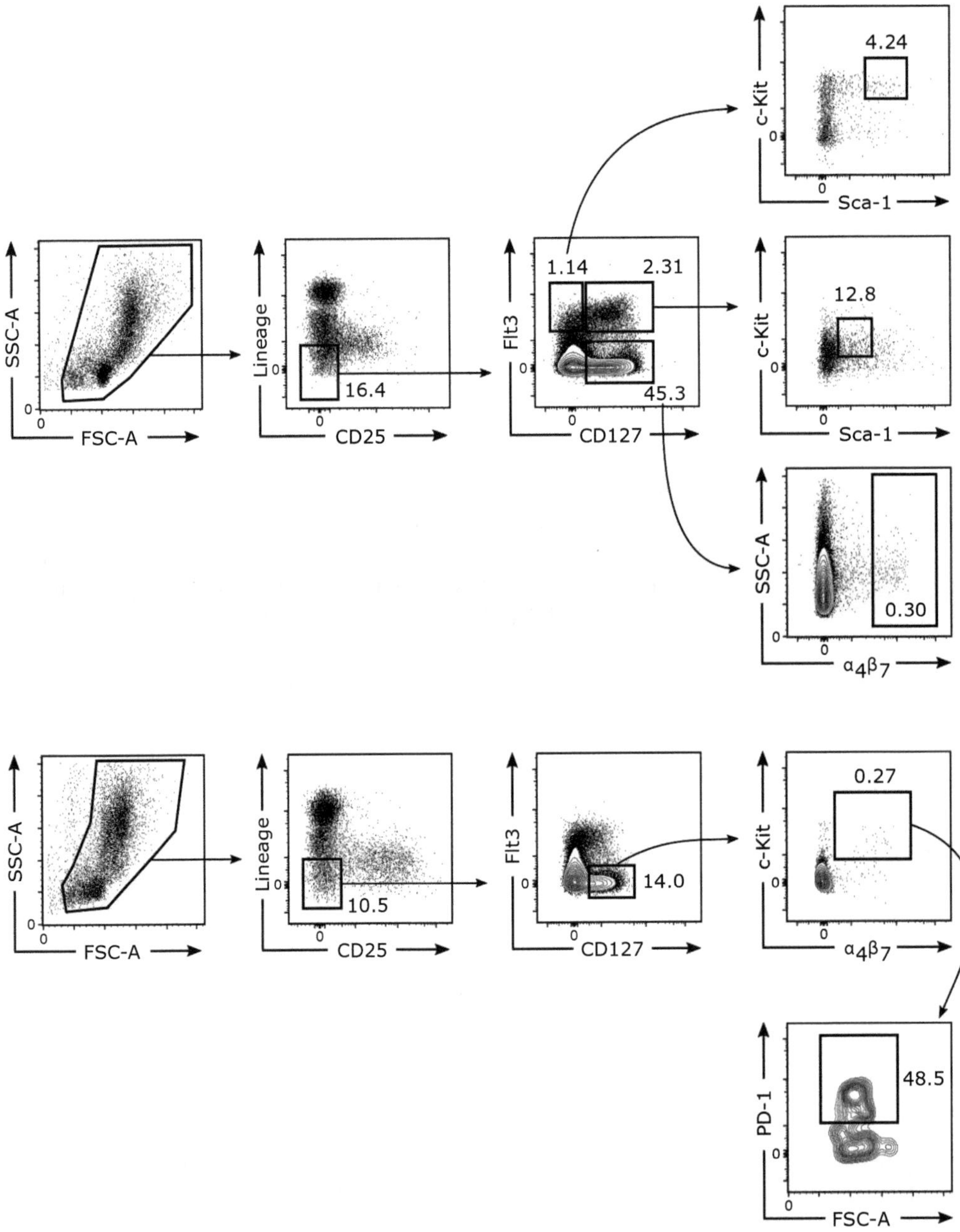

Fig. 3 Gating strategy for LMPPs, CLPs, CHILPs, and ILCPs in the bone marrow. Bone marrow was harvested from 8–12-week-old BALB/c female mice and processed for analytical flow cytometry as described in Subheadings 3.1 and 3.2. Doublets and dead cells excluded as indicated in Fig. 2. Values within flow plots indicate percentage of the parent population. (Image reproduced from Stier et al., *Journal of Experimental Medicine*, 2018 January 2;215 (1):263–281 with permission from Rockefeller University Press [14])

this antibody cocktail to each sample and incubate for 30 min at 4 °C protected from light (*see* **Note 11**). Add each antibody individually in 50 μL to each single stain control tubes as well (one antibody per tube).

5. Wash cells by adding 2 mL of flow staining buffer to each sample. Centrifuge samples at 350 × *g*, discard supernatant, and resuspend cells as in **step 1**.
6. If using biotinylated antibodies (i.e. lineage markers, as in our antibody cocktail), add a streptavidin-conjugated fluorophore to the samples by diluting in 100 μL flow staining buffer. Incubate for 30 min at 4 °C protected from light.
7. Wash cells by adding 2 mL of flow staining buffer to each sample. Centrifuge samples at 350 × *g*, discard supernatant, and resuspend cells as in **step 1**. If staining for intracellular antigens such as PLZF, go to **step 8**. If only staining for cell surface markers, proceed to **step 13**.
8. If performing analysis for intracellular antigens such as PLZF, fix and permeabilize the samples (*see* **Note 12**). We use the Tonbo Foxp3/Transcription Factor Staining Buffer Kit and apply 500 μL of diluted reagent to each sample for 30 min at room temperature protected from light.
9. Wash cells by adding 2 mL of permeabilization buffer to each sample. Centrifuge samples at 350 × *g*, discard supernatant, and resuspend cells as in **step 1**.
10. Prepare intracellular antibody cocktail by diluting antibodies into permeabilization buffer for a total volume of 100 μL per sample. For analysis of multiple samples, make one master mix to be applied to all samples. Add 100 μL of this antibody cocktail to each sample and incubate for 30 min at 4 °C protected from light (*see* **Note 11**). Add each antibody individually in 50 μL to each single stain control tubes as well (one antibody per tube).
11. Wash cells by adding 2 mL of permeabilization buffer to each sample. Centrifuge samples at 350 × *g*, discard supernatant, and resuspend cells as in **step 1**.
12. Wash cells by adding 2 mL of flow staining buffer to each sample. Centrifuge samples at 350 × *g*, discard supernatant, and resuspend cells as in **step 1**.
13. If unfixed cells (surface staining only), add 250 μL of flow staining buffer with a viability stain (including to the single stain control for this viability dye). We use DAPI at a concentration of 25 nM. If cells are fixed to assess for intracellular antigens, a viability dye was used in **step 3** and cells can simply be resuspended in 250 μL of flow staining buffer.
14. Analyze cells on a flow cytometer (*see* **Note 13**). We primarily use a BD LSR II with a five laser configuration as outlined in Subheading 2.2 (*see* **Note 14**). Export data from the flow cytometer data collection interface and evaluate on your preferred flow cytometric analysis software (i.e. FlowJo, as shown

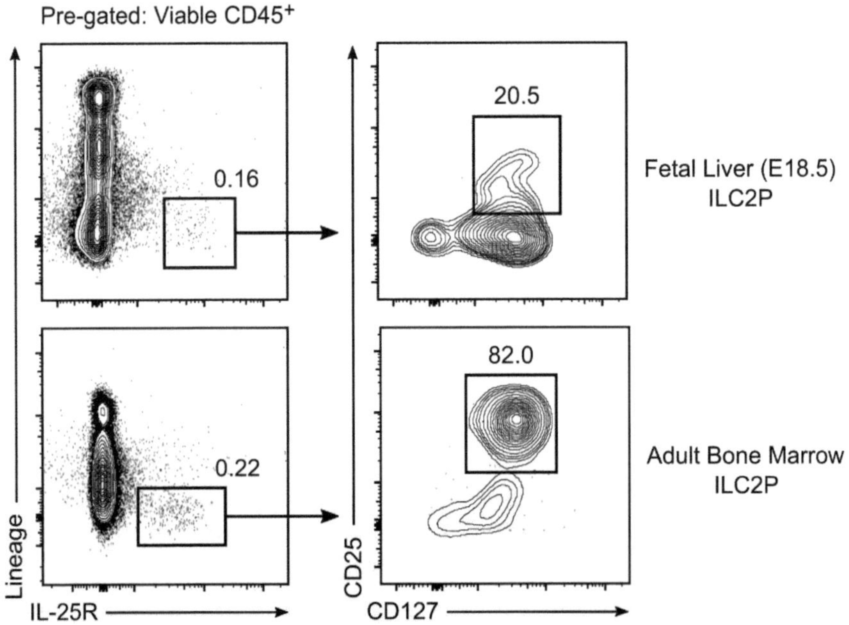

Fig. 4 Analysis of ILC2P in fetal liver and adult bone marrow. Fetal liver leukocytes were isolated by a brief collagenase/DNase digestion, manual disruption, and RBC lysis. Flow cytometric analysis for ILC2P was performed on these fetal liver leukocytes and compared to ILC2P in adult bone marrow. ILC2P were pregated for viable $CD45^+$ singlets (not shown). Values within flow plots indicate percentage of the parent population

in this protocol). The gating strategies outlined in Table 1 and Figs. 2 and 3 provide a template for this analysis (*see* **Note 15**). Additionally, it may be of interest to study ILC precursor populations in sites outside of the bone marrow. This approach can be applied with appropriate modifications in that setting as well. For instance, we have analyzed ILC2P in the fetal liver (Fig. 4).

3.3 Adoptive Transfer of ILC Precursors

1. Isolate bone marrow for both the left and right legs (*see* **Note 16**) of a mouse or multiple mice (*see* **Note 17**) as described in Subheading 3.1 through **step 9**. Perform this entire protocol under sterile or near-sterile conditions.
2. Adjust volume by adding 10 mL of MACS buffer.
3. Mix resuspended cells with a brief low speed vortex and pour through a 70 μm cell strainer into a new 50 mL conical tube.
4. Mix samples by lightly vortexing and take a small aliquot (10–20 μL) to count the cells. For manual counting with a standard hemocytometer, dilution will be dependent upon the number of mice harvested.
5. Spin samples in 50 mL conical tubes at 350 × *g* for 5 min at 4 °C in a centrifuge. Sample should collect firmly at the bottom of the 50 mL conical tube. Discard supernatant by pouring off

into an appropriate biohazard disposal container. This will leave ~250 μL of solution at the bottom of the tube. Resuspend cells by using a vortex mixer on medium to medium-high speed (*see* **Note 18**).

6. Add 90 μL of MACS buffer per ten million cells. Then add 10 μL of Direct Lineage Cell Depletion Cocktail per ten million cells and mix by using a vortex mixer on low speed.
7. Incubate sample for 15 min at 4 °C. Then add 1–2 mL of MACS buffer per ten million cells. Place cap on tube and invert several times to wash the sample.
8. Centrifuge samples and discard supernatant as in **step 5**. This will leave ~250 μL of solution at the bottom of the tube. Resuspend cells by pipetting up and down until the cell pellet is no longer visible.
9. Add 500 μL of MACS buffer per 100 million cells and store at 4 °C.
10. Rinse LS Column by placing in magnetic stand and applying 3 mL of buffer. Discard the flow through. Place a new 15 mL conical tube under the tip of the LS Column.
11. Mix sample by pipetting and add up to 100 million cells (500 μL) to the LS Column (*see* **Note 19**). Allow entire sample to flow into the column. Then add 500 μL of MACS buffer to the LS Column and allow to flow into the column.
12. Add 3 mL of MACS buffer to the LS Column and allow to flow through entirely. Repeat this step two additional times for a total of three flushes.
13. Collect the 15 mL conical tube with flow through; this now contains a population enriched for lineage marker negative cells including lymphoid and ILC progenitors. Spin sample in the 15 mL conical tube at 350 × *g* for 5 min at 4 °C in a centrifuge. Cells should collect firmly at the bottom of the 15 mL conical tube. Discard supernatant by pouring off into an appropriate biohazard disposal container. This will leave ~250 μL of solution at the bottom of the tube. Resuspend cells by pipetting up and down until the cell pellet is no longer visible.
14. Add 1 mL of flow staining buffer and transfer sample to a 5 mL FACS tube. Vortex for 3–5 s on moderate speed to mix the cells. Take a 10–20 μL aliquot, mix 1:1 with trypan blue, and count using a standard hemocytometer.
15. Spin sample in 5 mL FACS tube at 350 × *g* for 5 min at 4 °C in a centrifuge. Cells should collect firmly at the bottom of the 5 mL FACS tube. Discard supernatant by pouring off into an appropriate biohazard disposal container. This will leave ~100 μL of solution at the bottom of the tube. Resuspend cells by vortex mixing for 5 s on moderate speed.

16. Perform Fc block and cell surface staining (*see* **Note 20**) as outlined in Subheading 3.2 ensuring cells are kept cold and handled with maximum sterility. Resuspend cells at a concentration of 10–20 million cells per mL of flow staining buffer, or adjust to the optimal concentration for your sorting machine. Include a viability dye (i.e. DAPI) at this stage.
17. Following the final wash and viability stain, filter samples through a FACS tube cell strainer cap (35 μm) by pipetting sample into the strainer cap on top of a new 5 mL FACS tube and centrifuging at 350 × *g* for 2 min at 4 °C.
18. Perform FACS and collect precursor populations of interest in a new 5 mL FACS tube with 1 mL of cold sterile RPMI (*see* **Note 21**). Record the number of cells that were sorted from the machine.
19. Centrifuge samples at 350 × *g*, discard supernatant, and resuspend cells as in **step 15** (*see* **Note 22**).
20. Resuspend progenitor cells at a concentration of 1500 cells per 100 μL of RPMI (*see* **Note 23**). Vortex gently for 2–3 s to ensure sample is mixed.
21. Using a 1 mL syringe with 0.1 mL gradations and a 30 G needle, aspirate 100 μL of sample per mouse to inject.
22. Anesthetize recipient mice (*see* **Note 24**) per approved institutional animal protocol. Once anesthetized, inject 100 μL of cells intravenously. We prefer the retroorbital route of delivery for ease and reproducibility of delivery, but tail vein is another common site for injection. Perform all procedures in accordance with approved institutional animal protocols.
23. Allow 3–12 weeks for reconstitution from the transferred progenitor cells, depending on the application. Analyze immunologic compartments of interest by preferred method (flow cytometry, immunofluorescence, etc.) to assess reconstitution of desired subsets of ILC and lineage potential of transferred progenitors (Fig. 5).

4 Notes

1. This cocktail is optimized for the bone marrow. Additional lineage markers should be considered for other tissues including CD3, CD4, CD8, αβTCR, γδTCR, NK1.1, and FcεRI. Of note, NK1.1 is not present on NK cells from BALB/c mice.
2. From each tibia and femur, we generally recover approximately 15 million live bone marrow cells from female BALB/c mice.

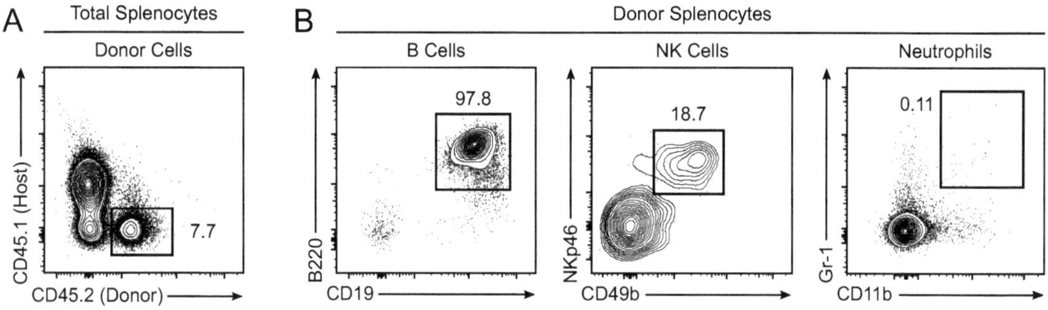

Fig. 5 Reconstitution of donor lineages after adoptive transfer of CLP. Donor CLP were isolated per the methods Subheading 3.3 from adult female CD45.2^{+} BALB/c mice and 1500 donor CLP were adoptively transferred via retroorbital injection into sublethally irradiated (5 Gy) adult female CD45.1^{+} BALB/c recipient mice. Mice were rested for 3 weeks to allow for partial leukocyte reconstitution. Mice were euthanized and splenocytes were evaluated for (A) donor CD45.2^{+} cells. (B) Subset analysis demonstrated reconstitution of donor CLP-derived adaptive lymphocytes (B cells, pregated for CD45.2^{+} CD3^{-}) and innate lymphocytes (NK cells, pregated CD45.2^{+} CD3^{-} CD19^{-}) but not myeloid cells (neutrophils, pregated CD45.2^{+}), consistent with known CLP lineage restriction. Values within flow plots indicate percentage of the parent population

3. In general, try to keep samples on ice (though some steps if performed quickly can be done on the benchtop as long as samples are immediately placed back on ice). For downstream applications that require live cells or maximal cell recovery, pay special attention to keeping all samples and solutions ice-cold.
4. Alternatively, if you are processing bone marrow from a large number of mice, you can consider centrifugation to expel the marrow. Briefly, puncture a hole in the bottom of a 0.5 mL microcentrifuge tube using an 18 G needle. Assemble the centrifugation construct by placing the bone and 250 μL cold RPMI inside the punctured 0.5 mL tube and placing this inside a 1.5 mL microcentrifuge tube. Centrifuge to expel marrow and vortex and/or aspirate repeatedly with a pipette to resuspend the cells.
5. Generally, complete expulsion of the marrow can be accomplished with one flush but if there appears to be marrow still present, consider additional flushes. You may have to rotate the syringe in the bone or inject through the opposite end of the bone to get all of the marrow.
6. If cells are still stuck to the bottom, try tapping the bottom of the tube against a hard surface to dislodge followed by additional vortex mixing. Overmixing may affect viability; only mix until cells are resuspended.
7. If analyzing intracellular antigens (i.e., transcription factor PLZF), replace flow staining buffer in this step with 1× PBS.

8. It is important to recognize that each control type has strengths and limitations reviewed in [29]. These limitations can be exacerbated by poor flow cytometry technique such as over-staining due to inadequate antibody titration. Moreover, these controls do not help define "high" versus "low" staining, which is important for certain antigens and populations. Thus, Isotype and FMO controls are best used in conjunction with the published literature and user experience/intuition to assemble gates.
9. Resuspending the cells is critical during preparation for flow cytometry. You may need to increase or decrease your vortex speeds, tap the tubes on the benchtop, or scrape the tubes on a rough surface (i.e., a grate) to loosen the cells.
10. For more than five million cells (i.e., *see* Subheading 3.3), multiply the volumes listed in the remainder of the protocol by increments of five million cells. For instance, if you have 6 million or 12 million cells, multiply by 2 or 3, respectively.
11. Cells will sink to the bottom during lengthy staining steps. If staining for >15 min, consider remixing the samples every 15 min by light vortexing.
12. Alternatively, many labs use fluorescent reporter mice for intracellular antigens (i.e. $PLZF^{GFPcre}$ [11] or $Id2^{RFP}$ [27] mice). In some cases, investigators have found surrogate surface markers that correlate with the intracellular antigen of interest, such as PD-1 as a surrogate for PLZF in ILCP [12].
13. ILC and ILC progenitors are rare cells in the bone marrow and require counting of large numbers of cells. We often collect 500,000–1,000,000 live cell events per sample.
14. Thoughtful selection of antibodies and fluorophores is crucial for optimal results, particularly given the size of the panels generally required to analyze ILC and ILC precursors. Working backward from your flow cytometer is good practice, as excitation wavelengths and filter combinations can differ meaningfully between machines altering the "brightness" of different fluorophores and the magnitude of compensation required. Generally, it is best to pair "dimmer" fluorophores with more highly expressed proteins and vice versa. Optimization of your initial cytometry panel will require an investment of time and resources but will provide demonstrably better and more consistent results.
15. It is important to use singlet gates (i.e., FSC-A vs. FSC-H) and a viability marker to exclude contamination of dead cells our doublets (Fig. 2).
16. Marrow can also be harvested from the hip if it is necessary to maximize per mouse yield.

17. The legs of up to five mice can be processed in the same 50 mL Falcon tube. Use an additional 50 mL Falcon tube for each set of up to five mice. In our hands, we recover 700–1000 CLP per WT BALB/c mouse.
18. If a significant number of mice are harvested, repeated aspirations and expulsions with a pipette may be necessary to dislodge the pellet.
19. Multiple LS Columns may be required depending on the sample size. Do not exceed 100 million cells per column as this will greatly increase the lineage positive flow through.
20. MACS is an important step to enrich for lineage negative cells and therefore significantly increase the efficiency of your FACS. Even with optimal MACS procedure, a percentage of lineage positive cells will flow through the column. Therefore, it is still important to surface stain at this step with lineage markers using fluorescently tagged antibodies for exclusion of contaminating lineage positive cells during FACS.
21. Keep samples chilled during the sort if your FACS machine has this capability.
22. Cell number will likely be too few to visually observe in the tube. Do not be alarmed if this is the case. Proceed to the next step.
23. Titration of cells to be transferred is important. Most published studies use 500–3000 cells per mouse, depending on the lab and the cell population being transferred.
24. Common recipient mice include mice genetically engineered to lack all lymphocytes including ILC such as $Rag2^{-/-}$ $Il2rg^{-/-}$ mice or CD45 allelic mismatch mice (i.e., CD45.1^{+} BALB/c or C57BL/6 mice when injecting CD45.2^{+} cells) that have been exposed to sublethal doses of radiation.

Acknowledgments

Optimization of the protocols described herein was supported by the National Institutes of Health (R01AI145265, R01AI124456, R01AI111820, R21AI145397, and U19AI095227) and the US Department of Veterans Affairs (I01BX004299).

References

1. Cherrier DE, Serafini N, Di Santo JP (2018) Innate lymphoid cell development: a T cell perspective. Immunity 48:1091–1103
2. Stier MT, Peebles RS (2017) Innate lymphoid cells and allergic disease. Ann Allergy Asthma Immunol 119:480–488
3. Ghaedi M, Steer CA, Martinez-Gonzalez I et al (2016) Common-lymphoid-progenitor-independent pathways of innate and T lymphocyte development. Cell Rep 15:471–480
4. Kondo M, Weissman IL, Akashi K (1997) Identification of clonogenic common

lymphoid progenitors in mouse bone marrow. Cell 91:661–672
5. Wong SH, Walker JA, Jolin HE et al (2012) Transcription factor RORα is critical for nuocyte development. Nat Immunol 13:229–236
6. Hoyler T, Klose CSN, Souabni A et al (2012) The transcription factor GATA-3 controls cell fate and maintenance of type 2 innate lymphoid cells. Immunity 37:634–648
7. Klose CSN, Flach M, Möhle L et al (2014) Differentiation of type 1 ILCs from a common progenitor to all helper-like innate lymphoid cell lineages. Cell 157:340–356
8. Satoh-Takayama N, Lesjean-Pottier S, Vieira P et al (2010) IL-7 and IL-15 independently program the differentiation of intestinal $CD3^{-}NKp46^{+}$ cell subsets from Id2-dependent precursors. J Exp Med 207:273–280
9. Moro K, Yamada T, Tanabe M et al (2010) Innate production of T(H)2 cytokines by adipose tissue-associated c-Kit(+)Sca-1(+) lymphoid cells. Nature 463:540–544
10. Yang Q, Li F, Harly C et al (2015) TCF-1 upregulation identifies early innate lymphoid progenitors in the bone marrow. Nat Immunol 16:1044–1050
11. Constantinides MG, McDonald BD, Verhoef PA et al (2014) A committed precursor to innate lymphoid cells. Nature 508:397
12. Yu Y, Tsang JCH, Wang C et al (2016) Single-cell RNA-seq identifies a PD-1hi ILC progenitor and defines its development pathway. Nature 539:102–106
13. Zook EC, Ramirez K, Guo X et al (2016) The ETS1 transcription factor is required for the development and cytokine-induced expansion of ILC2. J Exp Med 213:687–696
14. Stier MT, Zhang J, Goleniewska K et al (2018) IL-33 promotes the egress of group 2 innate lymphoid cells from the bone marrow. J Exp Med 215:263–281
15. de Kleer IM, Kool M, de Bruijn MJW et al (2016) Perinatal activation of the interleukin-33 pathway promotes type 2 immunity in the developing lung. Immunity 45:1285–1298
16. Saluzzo S, Gorki A-D, Rana BMJ et al (2017) First-breath-induced type 2 pathways shape the lung immune environment. Cell Rep 18:1893–1905
17. Steer CA, Martinez-Gonzalez I, Ghaedi M et al (2017) Group 2 innate lymphoid cell activation in the neonatal lung drives type 2 immunity and allergen sensitization. J Allergy Clin Immunol 140:593
18. Karta MR, Rosenthal PS, Beppu A et al (2018) β_2 integrins rather than β_1 integrins mediate Alternaria-induced ILC2 trafficking to the lung. J Allergy Clin Immunol 141(1):329–338.e12
19. Huang Y, Mao K, Chen X et al (2018) S1P-dependent interorgan trafficking of group 2 innate lymphoid cells supports host defense. Science 359:114–119
20. Kim MH, Taparowsky EJ, Kim CH (2015) Retinoic acid differentially regulates the migration of innate lymphoid cell subsets to the gut. Immunity 43:107–119
21. Gasteiger G, Fan X, Dikiy S et al (2015) Tissue residency of innate lymphoid cells in lymphoid and nonlymphoid organs. Science 350:981–985
22. O'Sullivan TE, Rapp M, Fan X et al (2016) Adipose-resident group 1 innate lymphoid cells promote obesity-associated insulin resistance. Immunity 45:428–441
23. Nussbaum JC, Van Dyken SJ, von Moltke J et al (2013) Type 2 innate lymphoid cells control eosinophil homeostasis. Nature 502:245–248
24. Schneider C, Lee J, Koga S et al (2019) Tissue-resident group 2 innate lymphoid cells differentiate by layered ontogeny and in situ perinatal priming. Immunity 50:1425
25. Bando JK, Liang H-E, Locksley RM (2015) Identification and distribution of developing innate lymphoid cells in the fetal mouse intestine. Nat Immunol 16:153–160
26. Koga S, Hozumi K, Hirano K et al (2018) Peripheral $PDGFR\alpha^{+}$ $gp38^{+}$ mesenchymal cells support the differentiation of fetal liver-derived ILC2. J Exp Med 215:1609–1626
27. Xu W, Cherrier DE, Chea S et al (2019) An Id2RFP-reporter mouse redefines innate lymphoid cell precursor potentials. Immunity 50:1054–1068.e3
28. Yu X, Wang Y, Deng M et al (2014) The basic leucine zipper transcription factor NFIL3 directs the development of a common innate lymphoid cell precursor. Elife 3. https://doi.org/10.7554/eLife.04406
29. Maecker HT, Trotter J (2006) Flow cytometry controls, instrument setup, and the determination of positivity. Cytometry A 69A:1037–1042

Chapter 3

Identification of Murine Innate Lymphoid Cell Subsets in Barrier Tissues and Their Draining Lymph Nodes Using Flow Cytometry

Emma E. Dutton and David R. Withers

Abstract

Innate lymphoid cells (ILCs) are present in most tissues within the body but have been most extensively studied within mucosal barriers such as the lung and intestine. Isolation of immune cells from such tissues requires enzymatic digestion, and the number and composition of the cells released are dependent upon robust protocols tailored to the tissue of study. Here, detailed methodologies to isolate ILCs from various barrier sites and their draining lymph nodes (LNs) are described. Flow cytometry staining and gating strategies for identification and quantification of ILCs are then provided. Combined, these provide an efficient means to study ILCs within the small intestine lamina propria, lung, ear skin, and LNs.

Key words ILCs, Flow cytometry, Ear skin

1 Introduction

The innate lymphoid cell (ILC) family brings together conventional (c)NK cells and lymphoid tissue inducer (LTi) cells, first described in the 1970s and 1990s, with other immune cells much more recently discovered following a flurry of studies approximately 10 years ago [1–3]. What all these cells have is their development from a lymphoid progenitor, independent of the Rag genes [4, 5]. The ILC family was formally described in 2013 with three subsets established on the basis of transcription factor (TF) expression and associated cytokine production, thus resembling innate counterparts of the adaptive Th1, Th2, and Th17 T cell populations [6, 7]. Group 1 ILCs combat intracellular infections through expression of the Th1 associated cytokines IFNγ and TNFα [8]. Their development is dependent upon the TF T-bet [9]. Group 2 ILCs can express the cytokines IL-4, IL-5, IL-9, IL-13 (which are associated with Th2 immunity) and contribute to promoting antihelminth responses but also allergic inflammation

Shoba Amarnath (ed.), *Innate Lymphoid Cells: Methods and Protocols*, Methods in Molecular Biology, vol. 2121,
https://doi.org/10.1007/978-1-0716-0338-3_3,

[1, 2, 10–12]. While all ILCs require GATA-3 expression for their development [13], ILC2 express the highest levels of this TF, but also depend upon RORα and Bcl11b [14–16]. Group 3 ILCs are best defined by their expression and developmental requirement for RORγt, the signature TF for Th17 immunity [17, 18]. While some ILC3 can make IL-17A and IL-17F, their signature cytokine is really IL-22, critically contributing to antibacterial immunity and epithelial barrier protection [19, 20]. The ILC3 subset is functionally diverse since it includes LTi cells that orchestrate secondary lymphoid tissue formation in the embryo through expression of $LT\alpha_1\beta_2$ and RANKL [21]. Postbirth, ILC3s resembling the LTi cells of the embryo regulate intestinal CD4 T cell responses dependent on MHCII expression [22, 23].

Since the initial characterization, further understanding of the developmental requirements has refined the description of these subsets somewhat. While the ILC1 subset contains cNK cells and ILC1, the former show distinct developmental requirements including dependency upon the transcription factor Eomes [9], while a common progenitor able to give rise to ILC1 (but not cNK), ILC2 and ILC3 has been identified [8]. Since cNK cells have cytotoxic activity but the other ILC family members do not, the term IL-7Rα$^+$ "helper" ILC has been used to better describe the features of different members of the family, again drawing parallels with T cell populations [8]. To further complicate ILC identification, it is now evident that these cells can show considerable plasticity in vitro [24, 25] and in vivo, most evident with the differentiation of ILC3 into ILC1 through the down regulation of RORγt and the upregulation of T-bet expression [26]. Definitive identification of such plasticity currently relies on murine models that enable fate mapping of TF expression [26].

Here we provide detailed methodology for the isolation of ILC populations from different tissues and identification of different ILC subsets using intracellular flow cytometry.

2 Materials

2.1 Specialist Equipment

1. Plastic funnel.
2. Nylon mesh.
3. Large petri dish.
4. Dissection forceps.
5. 50 mL Falcon tubes.
6. 37 °C Incubator (5% CO_2, Thermo Scientific).
7. 37 °C Incubator (5% CO_2, shaking incubator, Infors HT).
8. 70 μm cell strainer.

9. 100 μm cell strainer.
10. Pipette boy.
11. 10 mL glass pipettes.
12. 10 mL syringe.
13. Centrifuge.
14. Small petri dish.
15. Dissection microscope.
16. 2 mL syringe plunger.
17. Dissection scissors.
18. 1.5 mL eppendorf.
19. Paper tissues.
20. Plastic container.
21. Syringe needle 27 G (BD).
22. Thermomixer (Eppendorf).
23. 96-Well plate.
24. 50 μm CellTrics filters.
25. Fluorescence-activated cell sorting (FACS) tubes.

2.2 Media and Buffers

1. Roswell Park Memorial Institute 1640 Medium (RPMI).
2. Dulbecco's phosphate-buffered saline (DPBS).
3. Fetal bovine serum (FBS), heat inactivated.
4. Ethylenediaminetetraacetic acid solution (EDTA).
5. Hanks' balanced salt solution (HBSS).
6. Dulbecco's modified Eagle medium (DMEM).
7. Foxp3 TF staining buffer kit.

2.3 Enzymes

1. DNase1.
2. Collagenase/dispase.
3. Collagenase VIII.
4. Liberase thermolysin medium concentration (TM).

2.4 Antibiotics and Amino Acids

1. Penicillin–streptomycin solution (Invitrogen, Life Technologies).
2. L-Glutamine (Sigma-Aldrich).

2.5 Media and Buffer Preparation

1. Culture Media—RPMI, 0.98% penicillin–streptomycin solution, 0.98% L-Glutamine, 9.8% fetal bovine serum (heat inactivated).
2. Staining Buffer—DPBS [+] $MgCl_2$ (100 mg/L) and $CaCl_2$ (100 mg/L), 2.5 mM EDTA, 2% fetal bovine serum (heat inactivated).

3. Gey's solution—10 mL Solution A (35 g NH_4Cl, 1.85 g KCl, 1.5 g $Na_2HPO_4 \cdot 12H_2O$, 0.119 g KH_2PO_4, 5.0 g glucose, 25.0 g gelatin, 1.5 mL 1% phenol red), 2.5 mL of Solution B (4.2 g $MgCl_2 \cdot 6H_2O$, 1.4 g $MgSO_4 \cdot 7H_2O$, 3.4 g $CaCl_2$), 2.5 mL of Solution C (22.5 g $NaHCO_5$), and 35 mL of distilled water.
4. SI LP Digestion solution—15 mL of culture media, 15 mg of collagenase VIII.
5. Lung Digestion solution—3.5 mL of culture media, 10 μL of DNaseI, and 500 μL of Liberase TM (stock concentration of 424 μg/mL), this solution is for one lung.
6. Ear Digestion solution—3 mL of DMEM, 75 μL of DNaseI and 75 μL of Liberase TM (stock concentration of 25 mg/mL), this solution is for 1 ear.

2.6 Antibodies (See Note 1)

1. Viability dye, APC-eFluor780, 1:1000.
2. B220, clone RA3-6B2, FITC, 1:300.
3. CCR6, clone 29-2L17, BV605, 1:100.
4. CD3ε, clone 145-2C11, FITC, 1:100.
5. Intracellular CD3ε (iCD3ε), clone eBio500A2, AF700, 1:100.
6. CD4, clone RM4–5, BV510, 1:100.
7. CD5, clone 53–7.3, FITC, 1:100.
8. CD11b, clone M1/70, FITC, 1:300.
9. CD11c, clone N418, FITC, 1:300.
10. CD25, clone PC61, BV650, 1:200.
11. CD45.2, clone 104, BV785, 1:100.
12. GATA-3, clone TWAJ, PerCP710, 1:75.
13. IL-7Rα, clone A7R34, BV421, 1:50.
14. RORγt, clone AFKJS-9, PE, 1:100.
15. T-bet, clone eBio4B10, eF660, 1:50.

3 Methods

3.1 Small Intestine Lamina Propria Digestion Protocol

1. Dissect the small intestine (SI) from the mouse, cutting from below the stomach to above the caecum and place in a large petri dish containing HBSS 2% FCS.
2. Place the SI on paper tissue soaked in HBSS 2% FCS and carefully remove the fat and Peyer's patches (*see* **Note 2**).
3. Cut the SI longitudinally and wash out the contents into the large petri dish.

4. Transfer the SI into fresh HBSS 2% FCS, wash and then cut latitudinally into roughly 2 cm sections into a 50 mL Falcon tube containing 20 mL of HBSS.
5. Shake falcon tube vigorously for 20 s (*see* **Note 3**).
6. Line a funnel with nylon mesh and place over a plastic container.
7. Tip the SI into the nylon mesh lined funnel so that the media and SI are separated with the SI remaining within the mesh and the waste being discarded into the plastic container.
8. The SI then undergoes a series of incubations in various medias alongside washing steps:
   (a) The SI is removed from the mesh and placed in a 50 mL Falcon tube containing a specific media, shaken vigorously for 20 s and then incubated for a specific time at 37 °C, while being agitated in a shaking incubator (*see* **Note 4**).
   (b) The SI in then filtered again through the nylon mesh, placed in a 50 mL Falcon tube containing HBSS, shaken vigorously for 20 s before being filtered again.
9. **Step 8a-b** is conducted and repeated with the specific medias and incubation times stated below. Incubations with HBSS/EDTA strip the epithelial layer from the tissue to then enable digestion of the SI lamina propria (SI LP):
   (a) 20 mL HBSS 2 mM EDTA, 20 min.
   (b) 20 mL HBSS 2 mM EDTA, 20 min.
   (c) 15 mL prewarmed SI LP digestion solution (as described in Materials under Sect. 2.5, the Media and Buffer Preparation section), incubate for 10 min, check sample, shake vigorously for 20 s and leave to incubate for another 1–5 min depending on whether the tissue is fully digested or not (*see* **Note 5**). DO NOT REPEAT **STEP 9b** HERE, move straightaway to **step 10**.
10. Filter the digested SI through 100 μm cell strainer into a 50 mL Falcon tube and wash with 5 mL of staining buffer twice (*see* **Note 6**).
11. Repeat **step 10** through a 70 μm cell strainer.
12. Centrifuge samples in the 50 mL Falcon tube at 453 × *g* for 5 min at 4 °C and resuspend in 10 mL of staining buffer.
13. Repeat **step 11** (*see* **Note 7**).
14. Centrifuge samples in the 50 mL Falcon tube at 453 × *g* for 5 min at 4 °C and resuspend to create a total volume of 1 mL staining buffer.
15. For flow cytometry staining stain 1/5 of the cells isolated from the SI LP (200 μL of 1 mL suspension) (*see* **Note 8**).

3.2 Small Intestine Lamina Propria Gating Strategy to Identify ILCs (See Fig. 1)

Using the antibody panel above, ILCs can be identified as Live CD45$^+$IL-7Rα$^+$Lineage$^-$iCD3ε$^-$ cells (*see* **Note 9**). ILC subsets can then be identified using transcription factor staining with ILC1 identified as T-bet$^+$, ILC2 as GATA-3$^+$, and ILC3 as RORγt$^+$. Some ILC3 (the LTi-like subset) can express lower levels of CD45 so this needs to be taken into account when gating on CD45$^+$ cells.

3.3 Mesenteric Lymph Node Digestion Protocol

1. Isolate the mesenteric lymph nodes (mLN) from the mouse and place in 1 mL of RMPI.
2. Place the mLN in 3 mL of fresh RMPI in a small petri dish and using the dissection microscope and forceps clean the mLN, removing all of the fat and blood vessels (*see* **Note 10**).
3. Transfer the cleaned mLN into 3 mL of fresh RPMI in a new small petri dish (*see* **Note 11**).
4. Tease apart the mLN, releasing all of the cells into the RMPI media.

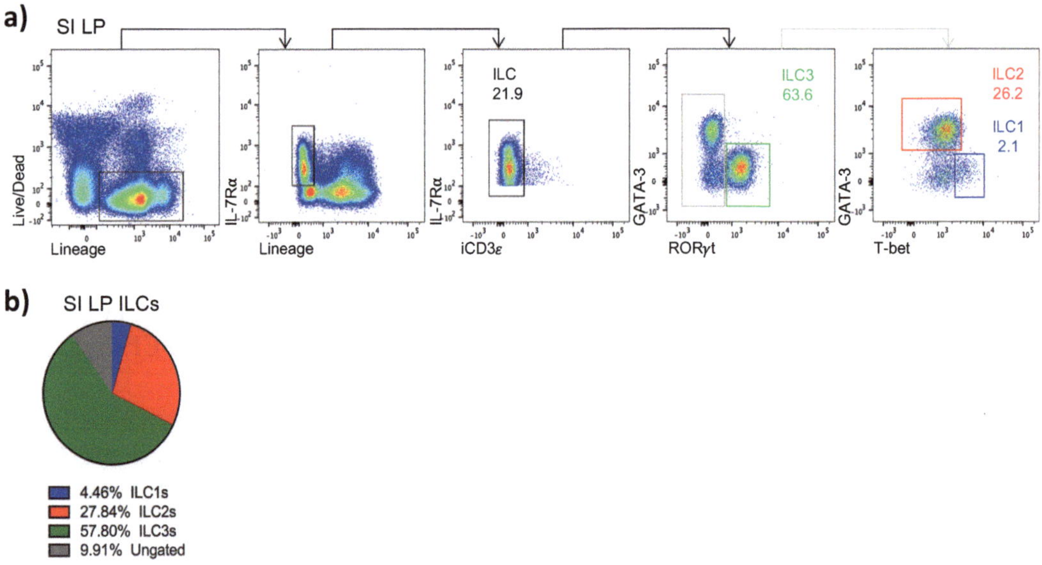

Fig. 1 Identification of ILC subsets within small intestine lamina propria. Cells were isolated from the SI LP from WT mice as described. When identifying ILCs the lineage markers include: B220, CD11c, CD11b, CD3ε, and CD5. ILCs were identified as IL-7Rα$^+$Lineage$^-$ and iCD3ε$^-$.. The transcription factors T-bet, GATA-3, and RORγt were then used to identify ILC1s, ILC2s, and ILC3s. (**a**) Representative flow cytometry plots showing the identification of ILCs (Live CD45$^+$ Lineage$^-$ IL-7Rα$^+$ iCD3ε$^-$), ILC1s (Live CD45$^+$ Lineage$^-$ IL-7Rα$^+$ iCD3ε$^-$ GATA-3$^-$ RORγt$^-$ T-bet$^+$), ILC2s (Live CD45$^+$ Lineage$^-$ IL-7Rα$^+$ iCD3ε$^-$ GATA-3$^+$ RORγt$^-$ T-bet$^-$), and ILC3s (Live CD45$^+$ Lineage$^-$ IL-7Rα$^+$ iCD3ε$^-$ GATA-3$^-$ RORγt$^+$). Values show proportion of given population. (**b**) Pie chart showing the mean percentage of ILC1s, ILC2s, ILC3s, and ungated ILCs (those that are RORγt$^-$, GATA-3$^-$, and T-bet$^-$) as a proportion of total ILCs within the SI LP

5. Add 7.5 μL of collagenase/dispase (final concentration, 0.25 mg/mL) and 7.5 μL of DNase (final concentration 0.025 mg/mL) and incubate at 37 °C for 20 min.
6. Add 60 μL of EDTA to each sample.
7. Add 5 mL of fresh RMPI to the mLN and then filter the sample through a 70 μm cell strainer into a 50 mL Falcon tube (*see* **Note 12**).
8. Use the plunger of a 2 mL syringe to crush the mLN through the filter (*see* **Note 13**).
9. Transfer 10 mL of fresh RMPI into the small petri dish, washing the remaining contents of the petri dish and transfer this through the filter.
10. Use the plunger again to crush the mLN on the filter.
11. Wash the petri dish again with 5 mL of fresh RMPI, then transfer this into the 50 mL Falcon tube through the filter.
12. Centrifuge samples in the 50 mL falcon tube at 453 × *g* for 5 min at 4 °C and resuspend to create a total volume of 400 μL staining buffer.
13. For flow cytometry staining, stain half of the mLN (200 μL of 400 μL suspension).

3.4 Mesenteric Lymph Node Gating Strategy to Identify ILCs (See Fig. 2)

Using the antibody panel above, ILCs can be identified as Live $CD45^+IL\text{-}7R\alpha^+Lineage^-iCD3\varepsilon^-$ cells; ILC subsets can then be identified using transcription factor staining with ILC1 identified as T-bet$^+$, ILC2 as GATA-3$^+$ and ILC3 as RORγt$^+$ (*see* **Note 14**).

3.5 Lung Digestion Protocol

1. When removing the lungs from the mouse, the tissue should be first perfused immediately after culling to clear the tissue of blood.
   (a) Pierce right atrium of the heart with a syringe needle and inject the left ventricle forcefully with 10 mL of DPBS (*see* **Note 15**).
   (b) Dissect out the lung and place in 1 mL of RPMI.
2. Place the lung in 1 mL of culture media in a small petri dish and clean the lung using the dissection microscope and forceps.
3. Tease apart the lung and transfer into a 50 mL Falcon tube (*see* **Note 16**).
4. Wash the petri dish with 1 mL of culture media and transfer into the Falcon tube.
5. Add 3 mL of the Lung Liberase TM/DNase solution (as described in Materials under Sect. 2.5, the Media and Buffer Preparation section) and incubate for 45 min at 37 °C while being agitated (use the shaker incubator).

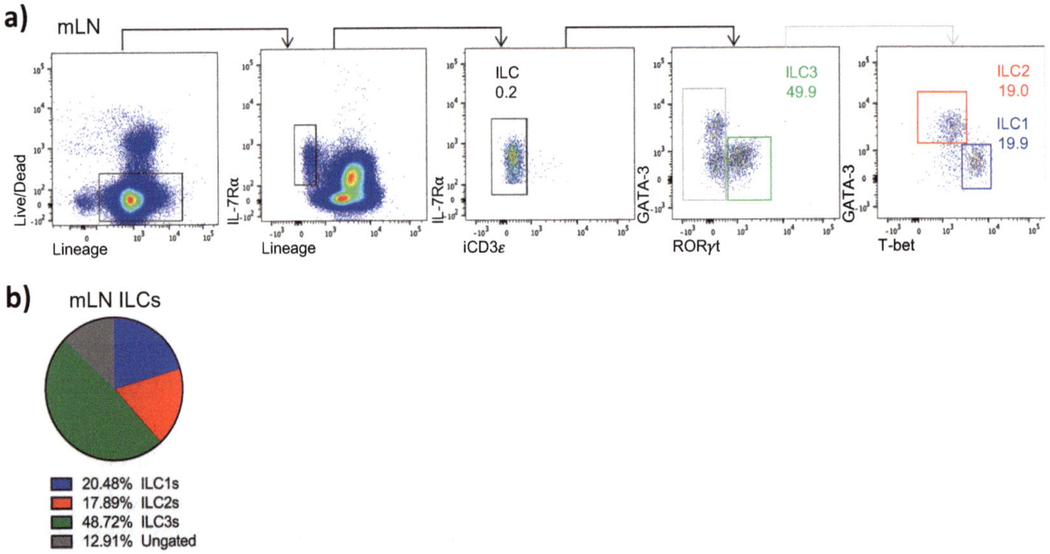

Fig. 2 Identification of ILC subsets within mesenteric lymph nodes. Cells were isolated from the mLN from WT mice as described. When identifying ILCs the lineage markers include: B220, CD11c, CD11b, CD3ε, and CD5. ILCs were identified as IL-7Rα$^+$Lineage$^-$ and iCD3ε$^-$. The transcription factors T-bet, GATA-3, and RORγt were then used to identify ILC1s, ILC2s, and ILC3s. (**a**) Representative flow cytometry plots showing the identification of ILCs (Live CD45$^+$ Lineage$^-$ IL-7Rα$^+$ iCD3ε$^-$), ILC1s (Live CD45$^+$ Lineage$^-$ IL-7Rα$^+$ iCD3ε$^-$ GATA-3$^-$ RORγt$^-$ T-bet$^+$), ILC2s (Live CD45$^+$ Lineage$^-$ IL-7Rα$^+$ iCD3ε$^-$ GATA-3$^+$ RORγt$^-$ T-bet$^-$), and ILC3s (Live CD45$^+$ Lineage$^-$ IL-7Rα$^+$ iCD3ε$^-$ GATA-3$^-$ RORγt$^+$). Values show proportion of given population. (**b**) Pie chart showing the mean percentage of ILC1s, ILC2s, ILC3s, and ungated ILCs (those that are RORγt$^-$, GATA-3$^-$, and T-bet$^-$) as a proportion of total ILCs within the mLN

6. Pour the sample into another 50 mL Falcon tube through a 70 μm cell strainer.
7. Crush the lung through the filter using a plunger from a 2 mL syringe.
8. Add 3 × 5 mL of culture media to the sample, washing the filter.
9. Centrifuge samples in the 50 mL Falcon tube at 453 × *g* for 5 min at 4 °C and resuspend in 3 mL of Gey's solution, then incubate for 5 min on ice.
14. Add 5 mL of culture media and then centrifuge samples at 453 × *g* for 5 min at 4 °C and resuspend in a total volume of 1 mL of staining buffer.
10. Filter the sample into a new 50 mL Falcon tube through a 70 μm cell strainer and wash with 5 mL of staining buffer.
11. Centrifuge samples in the 50 mL Falcon tube at 453 × *g* for 5 min at 4 °C and resuspend to create a total volume of 1200 μL staining buffer.
12. Stain 1/6 of the cells isolated from the lung (200 μL of 1200 μL suspension).

3.6 Lung Gating Strategy to Identify ILCs (See Fig. 3)

Using the antibody panel above, ILCs can be identified as Live CD45⁺IL-7Rα⁺Lineage⁻iCD3ε⁻ cells; ILC subsets can then be identified using transcription factor staining with ILC1 identified as T-bet⁺, ILC2 as GATA-3⁺ and ILC3 as RORγt⁺ (*see* **Note 17**).

3.7 Ear Skin Digestion Protocol

1. Cut the ear using dissection scissors along the hairline and place the ear in 1 mL of RPMI.
2. Place the ear into an empty 1.5 mL eppendorf tube and cut into small sections using the dissection scissors (*see* **Note 18**).
3. Add 1 mL of the Ear Liberase TM/DNase solution (as described in Materials under Sect. 2.5, the Media and Buffer Preparation section) and vortex the sample until resuspended in media.
4. Incubate the ear at 37 °C for 30 min while being agitated in the thermomixer.
5. Pour the contents of the eppendorf into a 50 mL Falcon tube through a 70 μm cell strainer, collecting the undigested tissue on the cell strainer.

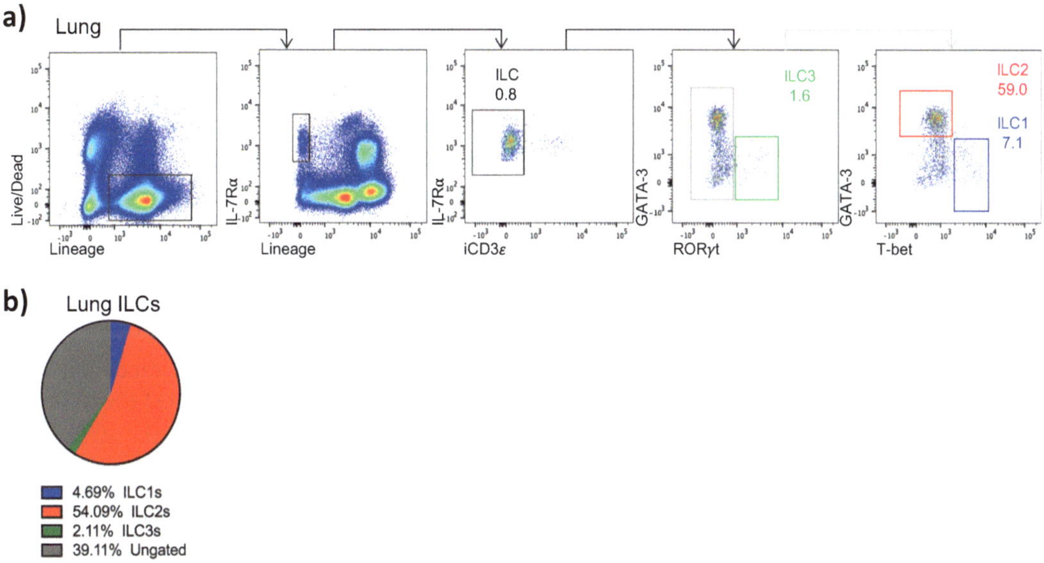

Fig. 3 Identification of ILC subsets within lung. Cells were isolated from the lung from WT mice as described. When identifying ILCs the lineage markers include: B220, CD11c, CD11b, CD3ε, and CD5. ILCs were identified as IL-7Rα⁺Lineage⁻ and iCD3ε⁻. The transcription factors T-bet, GATA-3, and RORγt were then used to identify ILC1s, ILC2s, and ILC3s. (**a**) Representative flow cytometry plots showing the identification of ILCs (Live CD45⁺ Lineage⁻ IL-7Rα⁺ iCD3ε⁻), ILC1s (Live CD45⁺ Lineage⁻ IL-7Rα⁺ iCD3ε⁻ GATA-3⁻ RORγt⁻ T-bet⁺), ILC2s (Live CD45⁺ Lineage⁻ IL-7Rα⁺ iCD3ε⁻ GATA-3⁺ RORγt⁻ T-bet⁻), and ILC3s (Live CD45⁺ Lineage⁻ IL-7Rα⁺ iCD3ε⁻ GATA-3⁻ RORγt⁺). Values show proportion of given population. (**b**) Pie chart showing the mean percentage of ILC1s, ILC2s, ILC3s, and ungated ILCs (those that are RORγt⁻, GATA-3⁻, and T-bet⁻) as a proportion of total ILCs within the lung

6. Wash the strainer with 5 mL of DMEM, collecting the wash through in the same 50 mL Falcon tube.
7. Remove the undigested tissue from the cell strainer and place back into the 1.5 mL eppendorf.
8. Repeat **steps 3–7**.
9. Repeat **steps 3–5**.
10. After the three digestions, crush the remaining undigested tissue (on the cell strainer) with the large end of a 2 mL syringe plunger.
11. Wash the cell strainer with 10 mL of DMEM and then centrifuge the 50 mL Falcon tube at 453 × *g* for 5 min at 4 °C and resuspend to create a total volume of 400 μL staining buffer.
12. For flow cytometry staining, stain at least 1/2 the cells isolated from the ear skin (200 μL of 400 μL suspension).

3.8 Ear Skin Gating Strategy to Identify ILCs (See Fig. 4)

Using the antibody panel above, ILCs can be identified as Live $CD45^{+}IL\text{-}7R\alpha^{+}Lineage^{-}iCD3\varepsilon^{-}$ cells; ILC subsets can then be identified using transcription factor staining with ILC1 identified as T-bet^{+}, ILC2 as GATA-3^{+}, and ILC3 as RORγt^{+} (*see* **Note 19**).

3.9 Flow Cytometry Staining Protocol

1. Transfer the samples that are resuspended in staining buffer, into a 96-well plate, 200 μL per well (*see* **Note 20**).
2. Centrifuge the plate at 453 × *g* for 2 min at 4 °C, remove the supernatant via flicking the plate upside-down over a sink.
3. Resuspend the samples in 200 μL of DPBS and repeat **step 2**.
4. Incubate samples for 20 min on ice in APC-eFluor780 viability dye diluted in 100 μL of DPBS (*see* **Note 21**).
5. Add 100 μL of DPBS and repeat **step 2**.
6. Resuspend in 200 μL of DPBS followed by **step 2**, repeat again.
7. Incubate samples for 30 min on ice with the extracellular antibodies, B220, CCR6, CD3ε, CD4, CD5, CD11b, CD11c, CD25, CD45.2, and IL-7Rα diluted in 100 μL of staining buffer.
8. Repeat **steps 5** and **6** with staining buffer instead of DPBS.
9. Fix the samples, using the Foxp3 TF staining buffer kit, by incubating them in 100 μL of Foxp3 fixation/permeabilization buffer (1:3 concentrate–diluent ratio) for 30 min on ice (*see* **Note 22**).
10. Repeat **steps 5** and **6** with Permeabilization buffer (1:10 10× Permeabilization buffer–dH_2O ratio) instead of DPBS.
11. Incubate samples for 45 min at room temperature with the intracellular antibodies, iCD3ε, GATA-3, T-bet, and RORγt (*see* **Note 23**).

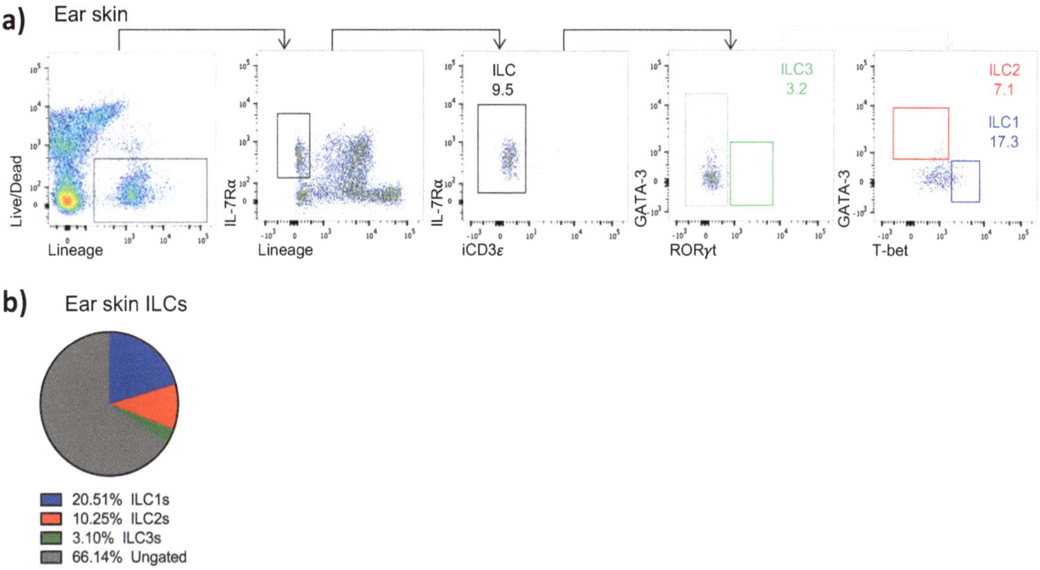

Fig. 4 Identification of ILC subsets within ear skin. Cells were isolated from the ear from WT mice as described. When identifying ILCs the lineage markers include: B220, CD11c, CD11b, CD3ε, and CD5. ILCs were identified as IL-7Rα$^+$Lineage$^-$ and iCD3ε$^-$. The transcription factors T-bet, GATA-3, and RORγt were then used to identify ILC1s, ILC2s, and ILC3s. (**a**) Representative flow cytometry plots showing the identification of ILCs (Live CD45$^+$ Lineage$^-$ IL-7Rα$^+$ iCD3ε$^-$), ILC1s (Live CD45$^+$ Lineage$^-$ IL-7Rα$^+$ iCD3ε$^-$ GATA-3$^-$ RORγt$^-$ T-bet$^+$), ILC2s (Live CD45$^+$ Lineage$^-$ IL-7Rα$^+$ iCD3ε$^-$ GATA-3$^+$ RORγt$^-$ T-bet$^-$), and ILC3s (Live CD45$^+$ Lineage$^-$ IL-7Rα$^+$ iCD3ε$^-$ GATA-3$^-$ RORγt$^+$). Values show proportion of given population. (**b**) Pie chart showing the mean percentage of ILC1s, ILC2s, ILC3s, and ungated ILCs (those that are RORγt$^-$, GATA-3$^-$, and T-bet$^-$) as a proportion of total ILCs within the ear

12. Repeat **steps 5** and **6** with Permeabilization buffer instead of DPBS.
13. Resuspend samples in 200 μL of staining buffer and repeat **step 2**.
14. Resuspend samples in 200 μL of staining buffer and transfer into FACS tubes through 50 μm CellTrics filters.
15. Wash samples with 100 μL of staining buffer and transfer final volume through CellTrics filters into FACS tubes.

4 Notes

Materials

1. A variety of different lineage markers have been used to help to identify ILCs, with discrimination from T cells paramount. Within this panel IL-7Rα is used to positively identify ILCs, alongside the "lineage" markers B220, CD11b, CD11c, CD3, and CD5. An extended "lineage" channel can be used (*see*

below), although we find that this makes very little difference to the ILC populations that can be identified. We have titrated the Abs in the "lineage" cocktail to ensure that a distinct population of IL-7Rα^{+}Lineage^{-} cells can be identified. The use of the intracellular CD3ε Ab further helps clearly distinguish ILC from all T cell populations.

(a) B220, CD11b, CD11c, CD3, CD5, CD19, Ter119, Gr1, F4/80, FcεRI, and CD49b.

Methods

2. It is important that you do not let the SI dry out at any stage. To prevent this, you can cut the SI laterally into two sections and clean one section at a time. Frequently wash the SI in HBSS 2% FCS while cleaning to prevent drying out and cell death.
3. **Steps 2–6** must be completed immediately after the mouse is culled to prevent excess cell death.
4. To remove the SI from the mesh. Place the mesh flat on a small pile of tissue paper then simply use forceps to remove the SI from the mesh and place the tissue into a 50 mL Falcon tube.
5. At this stage it is important to make sure you do not "over-digest" the SI LP. After incubating the tissue for 10 min and vigorously shaking (first part of **step 9c**), examine the sample to determine how digested the tissue is. You are aiming for there to be no large chunks of SI LP within the suspension, excluding small bits of fat. If the digestion is not complete after 10 min, then incubate for a further period between 1 and 5 min (no longer than 5 min) depending on how digested the tissue is. Once the tissue has digested then stop the procedure and move onto **step 10**.
6. During this stage the 100 μm filter may become blocked. It is advised to take this step slowly, as to not over load the strainer. If the strainer remains blocked, you can use a pipette to gently agitate the contents of the strainer. Pipetting up and down can help remaining liquid pass through the strainer.
7. This is an extremely important step in our experience as without this second round of filtering, we find that the cell yield drops dramatically due to clumping of the cells.
8. The proportions of sample for flow cytometry staining given within this Methods section for all tissues are recommended. The amounts suggested here give robust ILC populations in our hands.
9. Within the SI LP staining, one should aim for 65–80% Live CD45^{+} cells (although this will obviously depend somewhat on how the initial gate of FSC vs SSC is drawn). If the proportion

of Live CD45$^+$ cells is much less than 65%, this likely reflects suboptimal digestion of the tissue.

10. Remove the fat carefully trying not to disturb/pierce the mLN.
11. This is an important step as fat released during the cleaning of the mLN if not removed can cause cells to clump together and reduce the cell yield.
12. Upon adding the 5 mL of fresh RMPI to the sample, swirl the small petri dish and then carefully pour the sample through the filter. Any large segments of mLN that remain within the petri dish move to the filter using the dissection forceps.
13. Gently run the top end of the plunger over the filter, be careful not to split the filter.
14. ILC3s mainly comprise LTi-like cells within the mLN and other LNs. These can be distinguished by their expression of CCR6 and lack of expression of NKp46.
15. The lungs should inflate and turn white.
16. Pour the teased apart lung in 1 mL of culture media into the Falcon tube. If there are any segments of lung remaining in the petri dish, transfer them into the Falcon tube using dissection forceps.
17. Within the lung there is a large proportion of ILCs lacking expression of the three defining transcription factors (GATA-3, RORγt, and T-bet) and thus cannot be categorized as ILC1s (T-bet$^+$), ILC2s (GATA-3$^+$), or ILC3s (RORγt$^+$). We think this population at least partly reflects an artifact of the digestion process whereby ILC appear to lose expression of signature transcription factors. For more information on this population please refer to [27].
18. Put the dissection scissors in the eppendorf and cut up the ear at the bottom of the eppendorf into smaller sections. Make sure no ear is stuck to the scissors after cutting.
19. Similar to the lung there is a large proportion of ILCs that lack expression of the transcription factors RORγt, GATA-3, or T-bet [27].
20. Use the volumes and proportions of tissues suggested at the final step of each digestion protocol.
21. *See* suggested antibody clones, fluorochromes, and dilutions in Subheading 2.6.
22. Use the Foxp3 TF staining buffer kit following the manufacturer's instructions.
23. Samples are stained for iCD3ε to negatively gate out any immature T cells when identifying ILCs.

References

1. Moro K, Yamada T, Tanabe M et al (2010) Innate production of T(H)2 cytokines by adipose tissue-associated c-Kit(+)Sca-1(+) lymphoid cells. Nature 463:540–544
2. Neill DR, Wong SH, Bellosi A et al (2010) Nuocytes represent a new innate effector leukocyte that mediates type-2 immunity. Nature 464:1367–1370
3. Sawa S, Cherrier M, Lochner M et al (2010) Lineage relationship analysis of RORgammat+ innate lymphoid cells. Science 330:665–669
4. Artis D, Spits H (2015) The biology of innate lymphoid cells. Nature 517:293–301
5. Serafini N, Vosshenrich CA, Di Santo JP (2015) Transcriptional regulation of innate lymphoid cell fate. Nat Rev Immunol 15:415–428
6. Spits H, Artis D, Colonna M et al (2013) Innate lymphoid cells—a proposal for uniform nomenclature. Nat Rev Immunol 13:145–149
7. Walker JA, Barlow JL, Mckenzie AN (2013) Innate lymphoid cells—how did we miss them? Nat Rev Immunol 13:75–87
8. Klose CS, Flach M, Mohle L et al (2014) Differentiation of type 1 ILCs from a common progenitor to all helper-like innate lymphoid cell lineages. Cell 157:340–356
9. Gordon SM, Chaix J, Rupp LJ et al (2012) The transcription factors T-bet and Eomes control key checkpoints of natural killer cell maturation. Immunity 36:55–67
10. Halim TY, Krauss RH, Sun AC et al (2012) Lung natural helper cells are a critical source of Th2 cell-type cytokines in protease allergen-induced airway inflammation. Immunity 36:451–463
11. Price AE, Liang HE, Sullivan BM et al (2010) Systemically dispersed innate IL-13-expressing cells in type 2 immunity. Proc Natl Acad Sci U S A 107:11489–11494
12. Wilhelm C, Hirota K, Stieglitz B et al (2011) An IL-9 fate reporter demonstrates the induction of an innate IL-9 response in lung inflammation. Nat Immunol 12:1071–1077
13. Yagi R, Zhong C, Northrup DL et al (2014) The transcription factor GATA3 is critical for the development of all IL-7Ralpha-expressing innate lymphoid cells. Immunity 40:378–388
14. Halim TY, Maclaren A, Romanish MT et al (2012) Retinoic-acid-receptor-related orphan nuclear receptor alpha is required for natural helper cell development and allergic inflammation. Immunity 37:463–474
15. Hoyler T, Klose CS, Souabni A et al (2012) The transcription factor GATA-3 controls cell fate and maintenance of type 2 innate lymphoid cells. Immunity 37:634–648
16. Walker JA, Oliphant CJ, Englezakis A et al (2015) Bcl11b is essential for group 2 innate lymphoid cell development. J Exp Med 212:875–882
17. Cherrier M, Sawa S, Eberl G (2012) Notch, Id2, and RORgammat sequentially orchestrate the fetal development of lymphoid tissue inducer cells. J Exp Med 209:729–740
18. Eberl G, Marmon S, Sunshine MJ et al (2004) An essential function for the nuclear receptor RORgamma(t) in the generation of fetal lymphoid tissue inducer cells. Nat Immunol 5:64–73
19. Sonnenberg GF, Monticelli LA, Alenghat T et al (2012) Innate lymphoid cells promote anatomical containment of lymphoid-resident commensal bacteria. Science 336:1321–1325
20. Takatori H, Kanno Y, Watford WT et al (2009) Lymphoid tissue inducer-like cells are an innate source of IL-17 and IL-22. J Exp Med 206:35–41
21. Mebius RE (2003) Organogenesis of lymphoid tissues. Nat Rev Immunol 3:292–303
22. Hepworth MR, Fung TC, Masur SH et al (2015) Immune tolerance. Group 3 innate lymphoid cells mediate intestinal selection of commensal bacteria-specific CD4(+) T cells. Science 348:1031–1035
23. Hepworth MR, Monticelli LA, Fung TC et al (2013) Innate lymphoid cells regulate CD4+ T-cell responses to intestinal commensal bacteria. Nature 498:113–117
24. Bernink JH, Krabbendam L, Germar K et al (2015) Interleukin-12 and -23 control plasticity of CD127(+) group 1 and group 3 innate lymphoid cells in the intestinal lamina propria. Immunity 43:146–160
25. Bernink JH, Peters CP, Munneke M et al (2013) Human type 1 innate lymphoid cells accumulate in inflamed mucosal tissues. Nat Immunol 14:221–229
26. Klose CS, Kiss EA, Schwierzeck V et al (2013) A T-bet gradient controls the fate and function of CCR6-RORgammat+ innate lymphoid cells. Nature 494:261–265
27. Dutton EE, Camelo A, Sleeman M et al (2017) Characterisation of innate lymphoid cell populations at different sites in mice with defective T cell immunity. Wellcome Open Res 2:117

Chapter 4

Identification and Functional Characterization of Murine Group 3 Innate Lymphoid Cell (ILC3) Subsets in the Intestinal Tract and Associated Lymphoid Tissues

Felipe Melo-Gonzalez and Matthew R. Hepworth

Abstract

Group 3 innate lymphoid cells (ILC3s) are critical mediators of innate immune responses at mucosal barriers in both health and disease. ILC3s rapidly respond to environmental cues to reinforce barrier function and foster a mutualistic microbiota. ILC3s are defined by the expression of the master transcription factor RORγt, but can be further subdivided by the surface expression of the chemokine receptor CCR6 or the natural killer cell-associated receptor NKp46, as well as through the coexpression of the transcription factor T-bet. Importantly, while these subsets exhibit overlapping functions such as the secretion of the cytokines IL-17A and IL-22, they also differ significantly transcriptionally, functionally and by their localization within tissues. Thus, it is critical that studies investigating ILC3 biology consider the heterogeneity and tissue specificities of these subsets. Here, we describe common tools used to dissect and characterize ILC3s subset phenotypes and functions by flow cytometry and strategies for cell sorting of these cells in both the gastrointestinal tract and associated lymph nodes. Together, these approaches provide a tool kit for researchers aiming to dissect ILC3 subset responses at homeostasis, during infection, or in the context of inflammation.

Key words ILC3, Mucosal barrier, mLN, Peyer's patches, Flow cytometry, IL-22

1 Introduction

The innate lymphoid cell (ILC) family has been the focus of intense research over the last decade. ILC family members are innate, transcriptionally primed, and polarized cells that respond rapidly to environmental cues and danger signals and play key roles in infection and inflammation at barrier surfaces [1]. Current nomenclature classifies ILC into three different groups based on the expression of different transcription factors, which mirror $CD4^{+}$ T helper (Th) cell subsets [1]. Thus, group 1 ILC (ILC1), ILC2 and ILC3 are defined by the expression of the transcription factors T-bet, GATA3 and Retinoid-related orphan receptor γt (RORγt), respectively. ILC3s play important roles in early life in the

Shoba Amarnath (ed.), *Innate Lymphoid Cells: Methods and Protocols*, Methods in Molecular Biology, vol. 2121, https://doi.org/10.1007/978-1-0716-0338-3_4,

formation of lymphoid tissues, while in adult life ILC3s are crucial for the maintenance of tissue health and homeostasis at mucosal barrier sites—in particular the gastrointestinal tract. Here, ILC3 modulate responses toward both commensal and pathogenic bacteria through production of cytokines such as IL-22 [2, 3]. Additionally, ILC3s can modulate adaptive immune responses through the production of soluble factors or via direct cellular crosstalk and antigen presentation [4–12].

ILCs can be globally defined by their lack of classical lineage markers, surface expression of common gamma chain cytokine receptors (e.g., IL-7Rα, IL-2R). ILC3 can be identified by their expression of RORγt, while distinct subsets of ILC3s have been described that differ functionally and transcriptionally and which are differentially distributed across tissues—where they inhabit distinct microenvironments [3, 13]. In the adult mouse, there are at least two major subsets of ILC3s, canonically defined by differential expression of the NK-cell associated receptors—most notably NKp46 (termed NCR^+ ILC3)—and the expression of the chemokine receptor CCR6 (termed lymphoid tissue inducer (LTi)-like ILC3) [3]. Moreover, NCR+ ILC3 are further distinguished from their LTi-like ILC3 counterparts through coexpression of the transcription factor T-bet—and have been reported to arise from an NCR^- $CCR6^-$ "double negative" tissue progenitor in the small intestine [14, 15].

These two subsets have overlapping functions, including the secretion of the cytokines IL-17A, GM-CSF, and IL-22 in response to myeloid cell-derived signals (e.g., IL-1β and IL-23) and play important roles in modulating antimicrobial responses to both commensal and pathogenic bacteria [16–20]. However, ILC3 subsets are also increasingly appreciated to have divergent functional capacities, including the ability to modulate adaptive immune responses in lymphoid tissues (comprehensively reviewed in [3, 12]), and when dysregulated may instead contribute to disease pathology through release of proinflammatory cytokines.

Here we describe common tools and approaches to dissect and characterize ILC3 subset frequencies and functions by flow cytometry, as well as strategies for sort-purification of these cells from the gastrointestinal tract and associated lymph nodes. The aim of this chapter is to provide an overview of the methodologies available to characterize the distribution of ILCs in the intestinal tract and associated lymphoid tissues, with a focus on the functional and phenotypic heterogeneity of ILC3 subsets. Together these methods provide a basic tool kit for researchers making their initial forays into ILC3 biology.

2 Materials

2.1 Mice

1. The protocols described herein are designed for use with wild type or transgenic mice on a C57BL/6 background or with RORγt-eGFP reporter mice (mice used in this study were originally a kind gift from Gerard Eberl, Institut Pasteur, Paris). Mice on other genetic backgrounds may differ in the distribution and phenotype of ILC subsets.

2.2 Equipment

1. Large petri dish.
2. Dissection forceps.
3. Dissection scissors.
4. 50 ml Falcon tubes.
5. 5 ml Bijou containers.
6. 37 °C/5% CO_2 incubator.
7. 37 °C/5% CO_2 incubator shaker.
8. 70 μm cell strainer.
9. Centrifuge.
10. 2 ml syringe plunger.
11. 96-Well plates.
12. Flow cytometry tubes.
13. Flow cytometer.
14. Cell sorter.

2.3 Buffers and Cell Media

1. *Complete RPMI media*: RPMI-1640 supplemented with 1% L-glutamine, 1% Pen-Strep and 10% fetal bovine serum (FBS).
2. *Stripping buffer*: 1 mM EDTA, 1 mM dithiothreitol (DTT) and 5% FBS.
3. *Flow cytometry (FACS) buffer*: PBS with 2% FBS and 1 mM EDTA.
4. *Gut digestion buffer*: 1 mg/ml collagenase/dispase (Roche) and 20 μg/ml Deoxyribonuclease I type IV from bovine pancreas (DNase I) (*see* also details below and associated Notes).
5. Liberase TL (Roche), reconstituted in RPMI at a concentration of 0.33 mg/ml.
6. *Cell stimulation media*: 50 ng/ml phorbol 12-myristate 13-acetate (PMA), 750 ng/ml ionomycin, 10 μg/ml Brefeldin A diluted in complete RPMI. Stimulation cocktail can also be modified to include ILC3 activating cytokines. The representative data sets contained in this article utilized 20 ng/ml recombinant mouse IL-1β and 20 ng/ml recombinant mouse IL-23, which together enhance ILC3 production of IL-22 ex vivo.

Table 1
Example flow cytometry reagents used in this protocol (*see* also Note 1)

Antibody	Clone	Suggested conjugate	Supplier	Working concentration
Total innate lymphoid cell identification				
LIVE/DEAD Fixable aqua dead cell stain	N/A	N/A	Life Technologies	As per manufacturer's instructions
CD45	30-F11	Brilliant violet 650	BioLegend	1 μg/ml
CD3	145-2C11	PerCP-Cy5.5	ThermoFisher	1 μg/ml
CD5	53–7.3	PerCP-Cy5.5	ThermoFisher	1 μg/ml
NK1.1	PK136	PerCP-Cy5.5	ThermoFisher	1 μg/ml
B220	RA3-6B2	APC-eFluor 780	ThermoFisher	1 μg/ml
CD11b	M1/70	APC-eFluor 780	ThermoFisher	1 μg/ml
CD11c	N418	APC-eFluor 780	ThermoFisher	1 μg/ml
CD90.2	30-H12	Alexa Fluor 700	BioLegend	2.5 μg/ml
CD127	A7R34	FITC	ThermoFisher	2.5 μg/ml
ILC subset identification (surface)				
CCR6	29-2L17	PE-Cy7	BioLegend	1 μg/ml
NKp46	29A1.4	PE-eFluor 610	ThermoFisher	1 μg/ml
Biotin anti-ST2	RMST2-33	–	ThermoFisher	2.5 μg/ml
Fluorochrome-conjugated streptavidin	N/A	APC	ThermoFisher	1 μg/ml
KLRG1	2F1	FITC	ThermoFisher	2.5 μg/ml
MHCII	M5/ 114.15.2	eFluor450	ThermoFisher	1 μg/ml
CD4	GK1.5	BUV395	BD	1 μg/ml
Intracellular staining buffer				
eBioscience FoxP3/Transcription factor staining buffer set	–	–	ThermoFisher	As per manufacturer's instructions
ILC subset identification (transcription factors)				
RORγt	2BD	PE	ThermoFisher	2 μg/ml

(continued)

Table 1
(continued)

Antibody	Clone	Suggested conjugate	Supplier	Working concentration
T-bet	eBio4B10	eFluor660	ThermoFisher	2 μg/ml
ILC3 intracellular cytokine staining				
IL-17A	eBio17B7	Alexa Fluor 488	ThermoFisher	5 μg/ml
IL-22	Poly5164	Alexa Fluor 647	BioLegend	2 μg/ml

2.4 Flow Cytometry Reagents

A full list of fluorescent-labeled antibodies and flow cytometry reagents are listed in Table 1 (*see* **Note 1**).

3 Methods

3.1 Isolation of ILC3s from Mouse Lymph Nodes

1. Carefully isolate mouse mesenteric lymph nodes (*see* **Note 2**), removing all the fat, into 2 ml of *complete RPMI* and put on ice until processing.
2. Generate a single cell suspension by gently dispersing the tissue through a 70 μm cell strainer, using the plunger of a 2 ml syringe, and wash with *complete RPMI*. Alternatively, lymph nodes can be briefly enzymatically digested to increase cell yields (*see* **Note 3**).
3. Centrifuge cells at 450 × *g* for 5 min at 4 °C and resuspend the cells in *complete RPMI* for culture or *FACs buffer* for direct flow cytometry analysis. Cells can be kept at 4 °C or on ice before flow cytometric staining.

3.2 Isolation of Single Cell Suspensions from Mouse Peyer's Patches

1. Carefully isolate Peyer's patches from the small intestine in a tube containing 500 μl of Liberase TL (0.33 mg/ml) and DNase (final conc. of 400 μg/ml) and store on ice until further processing. Typically, the small intestine has around 6–10 Peyer's patches per mouse.
2. Mince Peyer's patches using scissors and digest tissue for 20 min at 37 °C, shaking occasionally.
3. Generate a single cell suspension by gently dispersing the tissue through a 70 μm cell strainer, using the plunger of a 2 ml syringe, and wash with *complete RPMI*.
4. Centrifuge cells at 450 × *g* for 5 min at 4 °C and resuspend the cells in *complete RPMI* for culture or *FACS buffer* for direct flow cytometry analysis. Cells can be kept at 4 °C before flow cytometric staining.

3.3 Isolation of Single Cell Suspensions from Intestinal Tissues

1. Isolate small and large intestines (*see* **Note 4**), carefully removing associated fat (*see* **Note 5**) and Peyer's patches (as above), and open tissue longitudinally. Wash tissue thoroughly in a Petri dish containing PBS to remove feces and loose intestinal contents. Collect the tissue in 15 ml PBS on ice until further processing (*see* **Note 6**).
2. Vigorously vortex intestinal tissue for 10 s (full speed), remove PBS supernatant by passing tissue over a sieve, collecting the tissue and returning to the tube. Repeat wash and vortex step twice with 15 ml fresh PBS to fully remove luminal contents and surface mucus.
3. Add 15 ml of a prewarmed *stripping buffer* for 10 min at 37 °C in a rotating shaker or mixing water bath. This step should remove remaining mucus while further stripping epithelial cells and intraepithelial lymphocytes.
4. Vortex vigorously for 10 s, remove supernatant by passing tissue over a sieve and collect the tissue and return to the tube. Repeat incubation with *stripping buffer* for a further 20 min.
5. Vortex as above and discard supernatant. Wash tissue with a large volume of PBS to remove residual stripping buffer and discard supernatant by passing over a sieve.
6. Add 10 ml of *digestion buffer* and incubate for 45 min at 37 °C in a rotating shaker (*see* **Note** 7).
7. Collect the buffer containing liberated lamina propria lymphocytes (LPL) by filtering the supernatant through a 70 μm cell strainer into a new tube. Remaining tissue collected in the filter at this point can be discarded. Wash cells by adding fresh *complete RPMI*. Centrifuge cells at 450 × *g* for 5 min and resuspend the cell pellet in *complete RPMI* or *FACs buffer* and proceed to stimulation cultures or flow cytometry staining.

3.4 In Vitro ILC3 Stimulation

1. To stimulate cytokine production in ILC3s, incubate up to five million total cells in complete media containing 50 ng/ml PMA, 750 ng/ml ionomycin, 10 μg/ml Brefeldin A (*see* **Note 8**) with or without IL-1β and IL-23 (20 ng/ml each), or stimulus of choice (*see* **Note 9**), for 4 h at 37 °C in a 5% CO_2 incubator.
2. After the incubation, centrifuge cells at 450 × *g* for 5 min and proceed with flow cytometric staining.

3.5 Flow Cytometric Staining

1. For flow cytometry, incubate cells in an antibody cocktail containing ILC markers and lineage markers (*see* Subheading 2, Table 1 and **Note 1**). Stain samples for 30 min at 4 °C in *FACS buffer*. Additional markers for ILC1s, ILC2 and ILC3 subsets can be included (*see* **Note 10**). Example flow cytometry data for mLN ILCs and gating strategies based on these panels are

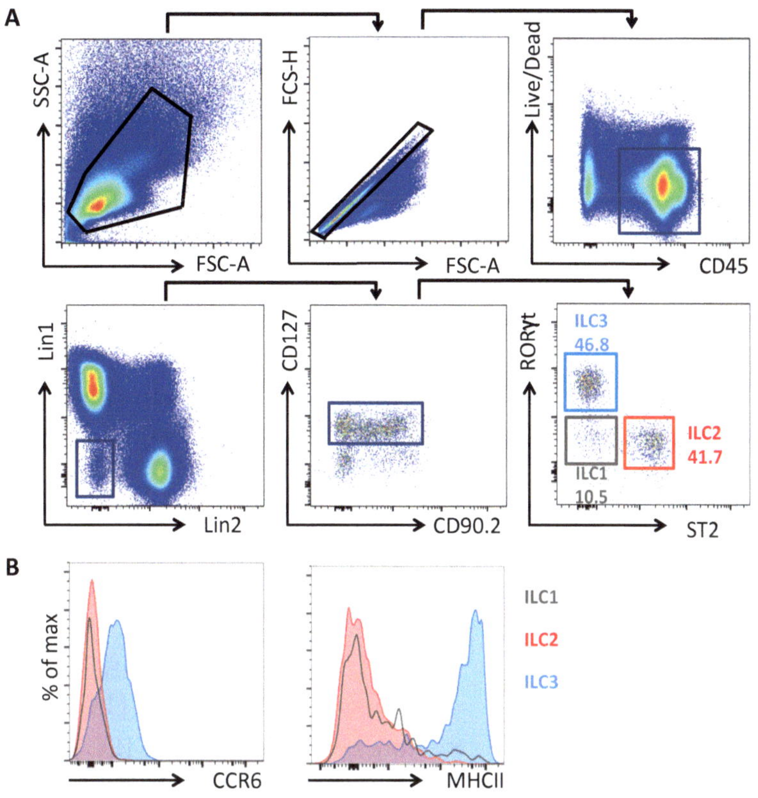

Fig. 1 Representative gating strategy for identification of ILC subsets. Representative flow cytometry plots of mesenteric lymph node (mLN) cells. (**a**) Total ILCs gated as live CD45^{+}, lineage negative (Lin 1 = CD3 CD5 NK1.1 and Lin 2 = B220 CD11b CD11c) cells and CD127^{+} CD90$^{-/+}$ cells (*see* **Note 12**). ILC in the mLN can be further subsetted and classified as RORγt^{+} LTi-like ILC3 (blue gate), ST2^{+} ILC2 (red gate), and RORγt^{-} ST2^{-} ILC1/ex-ILC3 (gray gate) (*see* **Note 13**). (**b**) CCR6 and MHCII further define mLN-resident LTi-like ILC3 in the mLN. ILC1 (gray line), ILC2 (red fill), and ILC3 (blue fill)

shown in Fig. 1. In the mLN ILC3s are primarily CCR6+ LTi-like ILC3 and additionally express high levels of MHC class II (Fig. 1), as previously described [4, 5, 10].

2. Wash cells with FACS buffer and centrifuge for 5 min at 450 × *g*.
3. If performing intracellular staining (e.g., for transcription factors or cytokines) add Fix/Perm buffer from Foxp3 intracellular kit and incubate for 30 min at 4 °C.
4. Wash with Perm/Wash buffer from Foxp3 intracellular kit.
5. Add intracellular stain (each antibody diluted in Perm/Wash at the concentration indicated in Table 1) for 30 min at 4 °C.
6. Wash cells and resuspend in 200 μl *FACS buffer*.
7. Acquire samples in a flow cytometer (*see* **Note 11**).

3.6 Gating Strategy

1. Gate cells according to their size and granularity using forward scatter (FSC) and side scatter (SSC), excluding debris (*see* Fig. 1).
2. Gate single cells using the area and height properties of the FSC (FCS-A and FCS-H) to exclude doublets.
3. Gate on live, $CD45^+$ hematopoietic cells.
4. Gate on lineage negative cells, lacking the lymphoid markers (CD3, CD5, NK1.1, lineage 1) and B cell/myeloid markers (B220, CD11b, CD11c, lineage 2) (*see* **Note 12**).
5. Gate total ILCs as $CD127^+$ $CD90.2^+$ cells (*see* **Note 13**).
6. If using intracellular transcription factor staining, proceed by gating total ILC3s based on RORγt expression and excluding $ST2^+$ ILC2 and RORγt$^-$ $ST2^-$ ILC (ILC1/ex-ILC3) (*see* Fig. 2, **Notes 10** and **12**).
7. Gate major ILC3 subsets based on the expression of CCR6 and NKp46. In the small intestinal three distinctive subsets of ILC3s can be identified: $CCR6^+$ LTi-like ILC3, $NKp46^+$ ILC3, and double negative (DN) ILC3 (*see* Fig. 2a and **Note 14**). Alternatively, ILC3 subsets can be discriminated based on the coexpression of T-bet [14, 15] (*see* Fig. 2b and **Note 14**).
8. Note that distribution of ILC groups, as well as ILC3 subsets, differs across regions of the gastrointestinal tract and within associated lymphoid tissues (*see* Fig. 2c, d and **Note 15**).

3.7 Cell Sorting

1. Similar gating strategies can be used for cell sorting (*see* **Note 16**). In order to accurately identify ILC3s, while maintaining viability for downstream analyses, we recommend the use of a RORγt-reporter mouse [21] (Fig. 3) in combination with surface markers to subset ILC3. Nonetheless, sorting for non-reporter mice is also possible by using additional surface markers for ILC3 subsets and ensuring a thorough exclusion of other potential contaminant ILC subsets (ILC1 and ILC2). For example, ILC3 subsets can be sorted based on the expression of CCR6 and NKp46 if additional markers are used to exclude ILC1 and classical NK cells.

4 Notes

1. We have validated the indicated concentrations and conjugations for the antibodies, listed in Table 1, but each user should determine the optimal concentration for their samples and flow cytometer. Additionally, antibody clones from other commercial vendors or alternative conjugates are acceptable but should be tested and optimized.

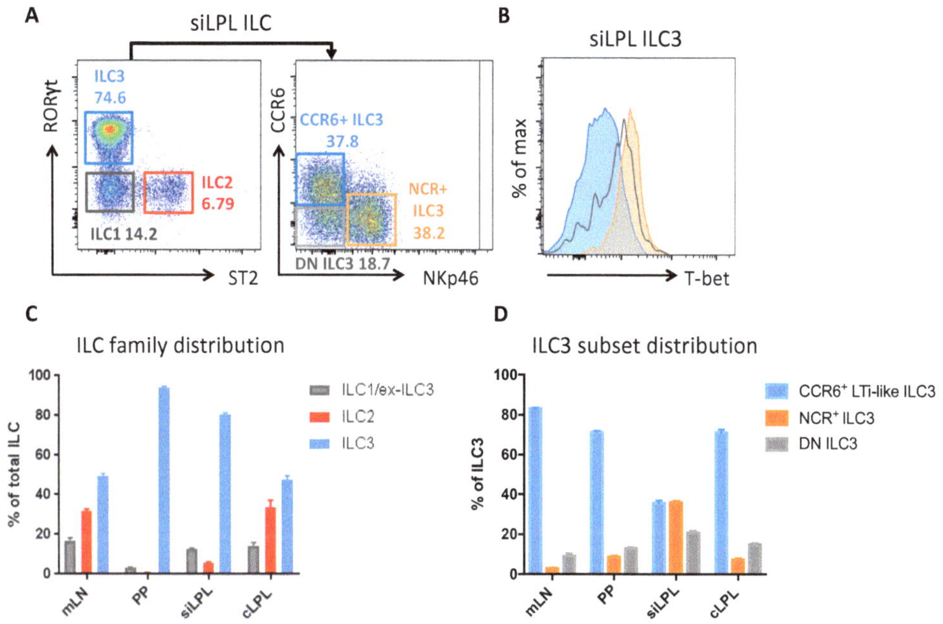

Fig. 2 Distribution of ILC family members and subsets within the gastrointestinal tract and associated lymphoid tissues. (**a**) Representative plots of small intestinal lamina propria (siLPL) ILC, gated as RORγt$^+$ ILC3 (blue gate), ST2$^+$ ILC2 (red gate), and RORγt$^-$ ST2$^-$ ILC1/ex-ILC3 (dark gray gate). The full heterogeneity of ILC3 subsets is particularly evident in the siLPL, and subsets can be identified as CCR6$^+$ LTi-like ILC3 (blue gate), NKp46$^+$ NCR$^+$ ILC3 (orange gate) and CCR6$^-$ NKp46$^-$ double negative (DN) ILC3 (light gray gate). (**b**) Expression of the transcription factor T-bet further distinguishes CCR6$^+$ and CCR6$^-$ ILC3; LTi-like ILC3 (blue fill), DN ILC3 (gray line) and NCR$^+$ ILC3 (orange fill). (**c**) Distribution of ILC groups in the gastrointestinal tract (small intestinal and colonic/cecal lamina propria, siLPL, and cLPL) and lymphoid associated tissues (mLN and PP) (*see* **Note 14**). (**d**) Distribution of ILC3 subsets in the gastrointestinal tract (siLPL and cLPL) and lymphoid associated tissues (mLN and PP) (*see* **Note 15**)

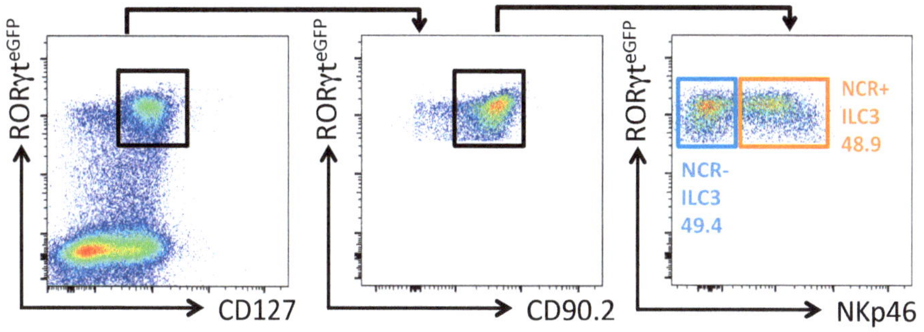

Fig. 3 Sorting strategy for isolation of small intestinal lamina propria ILC3 using RORγt-eGFP reporter mice. ILC3, identified as RORγt^{eGFP+} CD127$^+$, can be further gated based on the expression of NKp46, CCR6 or other markers for cell sorting (*see* **Note 16**). Plots pregated as in Fig. 1

2. This protocol refers to mouse mesenteric lymph nodes but can be used for other peripheral lymph nodes.
3. If higher cell yield is needed from the mLN (i.e., for cell sorting), tissue can be digested using Liberase TL (0.33 mg/ml) and DNase (400 μg/ml) for 20 min at 37 °C.

4. *Optional*: We typically include cecum in our large intestine lamina propria preparations and ILC subsets in the cecum are similar to the ones observed in the colon. However, this should be omitted if pure colon lamina propria cell population is required.
5. It is crucial to remove all the fat from the gut, to maximize cell viability and yield.
6. Intestinal preparations are time sensitive. Tissue can be left on ice in cold PBS during the dissection but should be processed as quickly as possible.
7. Collagenase/dispase digest is relatively gentle and only liberates cells from the lamina propria; after 45 min most of the tissue will remain intact with this digestion protocol. Using collagenase/dispase, cell yield are 5–10 million cells from the small intestine and 1–2 million cells from the colon. This cell yield is typically enough to perform at least two staining panels. For ILC cell sorting (from collagenase/dispase digest), cell yield is usually low (around 1000–10,000 ILC3 per mouse/tissue) so we recommend pooling at least three mice per tissue. An alternative digest protocol to improve cell yield from the small intestine (i.e., for cell sorting) is to instead utilize Collagenase from *Clostridium histolyticum* type VIII (collagenase VIII, Sigma). However, it should be noted this enzyme will fully digest the tissue, resulting in a mixed preparation containing both lamina propria cells and cells from the muscularis layer. Please note that digestion enzymes can interfere with certain surface markers and as such it should be confirmed that any additional antibodies not specified in this protocol are compatible with the digestion protocol.
8. Alternatively, commercial cell stimulation cocktails are available which contain a premixed combination of all the reagents needed for the stimulation (PMA, ionomycin, brefeldin A, and monensin).
9. Detection of cytokine from naive ILC3s under homeostatic conditions can be difficult without the addition of stimulatory cytokines (IL-1β and IL-23) (Fig. 4). During inflammation and infection, ILC3-associated cytokines may be detected without the addition of activating cytokine due to in vivo activation. Additionally, other cytokines such as IL-2 and IL-7 (20 ng/ml each) can be added in culture to improve cell survival during stimulation, while additional stimuli such as TL1A and neuropeptides have also been reported to activate ILC3 to produce effector cytokines. In our experimental conditions, we usually observe production of IL-22 by adult ILC3s, but with limited

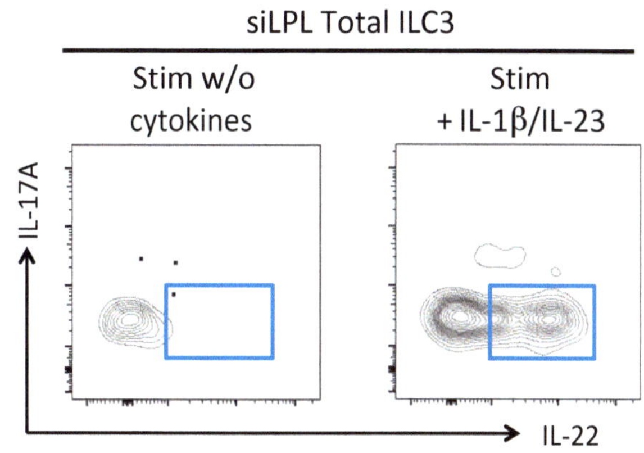

Fig. 4 Ex vivo stimulation of small intestinal ILC3 for the detection of intracellular cytokines. Representative plots of siLPL cells gated as in Figs. 1 and 2 and stimulated in the presence or absence of IL-1β and IL-23, in combination with PMA/ionomycin and Brefeldin A. Data shows intracellular detection of IL-22 and IL-17A in total ILC3 after 4 h

IL-17A. However, IL-17A expression may be enhanced in the context of intestinal infection and inflammation, while additional ILC3-derived cytokines such as GM-CSF and IL-2 may be detected via similar stimulation assays.

10. Additional markers can be utilized to identify ILC1 including CD49a, NK1.1, NKp46 and T-bet (as well as Eomes to distinguish from classical NK cells). Similarly, additional markers for ILC2 such as KLRG1, ST-2, ICOS, and GATA3 can be used in a similar manner. In our experience, directly conjugated ST2 antibodies display a very weak fluorescence, so we recommend using a biotinylated ST2 followed by a staining with fluorochrome conjugated streptavidin to improve the fluorescence intensity. Additional markers to distinguish LTi-like ILC3 from other ILC3 subsets include CD4, MHCII, CD117 (c-kit), and neuropilin-1.
11. It is best practice to filter samples in a 40 μm cell strainer before acquiring to prevent blocking the flow cytometer. We recommend acquisition of at least two million events in the flow cytometer to ensure an appropriate analysis and resolution of ILC subsets. For some tissues such as large intestine it may be necessary to acquire all available cells to ensure appropriate analysis.
12. While this lineage cocktail works well in combination with positively expressed ILC markers (i.e., CD127, CD90) for isolating total ILC across multiple tissues, users can consider using alternative or additional markers (e.g., TCRβ, TCRγδ, FceRI) for completion. In addition, NK1.1 can be omitted as

inclusion in the lineage cocktail removes a considerable fraction of ILC1 from the total ILC gate.

13. mLN ILC are $CD127^+$ $CD90^{-/+}$, whereas intestinal ILC and PP ILC are $CD127^+$ $CD90.2^+$ ILC.
14. NCR^+ ILC3 are predominantly found in the small intestine, while $CCR6^+$ LTi-like ILC3 are the predominant ILC3 subset in the mLN, PP, and cLPL. NCR^+ ILC3 can be further distinguished based on their high levels of T-bet compared to $CCR6^+$ LTi, which are T-bet^-. DN ILC3 express intermediate levels of T-bet, which reflects their transitional state (*see* also Fig. 2).
15. Distribution of ILC subsets may vary depending on the mouse background, strain and environmental conditions (e.g., microbiota composition and diet). ILC3s are the predominant ILC group in the small intestine and PP and a more diverse composition of ILC family members is observed in the mLN and colon.
16. If sorted cells will be used for RNA extraction, we recommend sorting straight into a cell lysis buffer (e.g., RLT buffer included in the RNeasy kit (Qiagen)) to maximize RNA yield. As cell yields are usually low, the micro RNeasy kit (Qiagen) is recommended for RNA extraction.

Acknowledgments

We thank members of the Hepworth lab for critical input and David Withers and Shoba Amarnath for suggestions. The Hepworth Laboratory is supported by a Royal Society and Wellcome Trust Sir Henry Dale Fellowship (Grant Number 105644/Z/14/Z) and a Lister Institute of Preventative Medicine Prize.

References

1. Spits H, Artis D, Colonna M et al (2013) Innate lymphoid cells—a proposal for uniform nomenclature. Nat Rev Immunol 13:145–149
2. Sonnenberg GF, Fouser LA, Artis D (2011) Border patrol: regulation of immunity, inflammation and tissue homeostasis at barrier surfaces by IL-22. Nat Immunol 12:383–390
3. Melo-Gonzalez F, Hepworth MR (2017) Functional and phenotypic heterogeneity of group 3 innate lymphoid cells. Immunology 150:265–275
4. Hepworth MR, Monticelli LA, Fung TC et al (2013) Innate lymphoid cells regulate CD4+ T-cell responses to intestinal commensal bacteria. Nature 498:113–117
5. Hepworth MR, Fung TC, Masur SH et al (2015) Immune tolerance. Group 3 innate lymphoid cells mediate intestinal selection of commensal bacteria-specific CD4(+) T cells. Science 348:1031–1035
6. Mackley EC, Houston S, Marriott CL et al (2015) CCR7-dependent trafficking of ROR-gamma(+) ILCs creates a unique microenvironment within mucosal draining lymph nodes. Nat Commun 6:5862
7. Von Burg N, Chappaz S, Baerenwaldt A et al (2014) Activated group 3 innate lymphoid cells promote T-cell-mediated immune responses. Proc Natl Acad Sci U S A 111: 12835–12840

8. Withers DR, Gaspal FM, Mackley EC et al (2012) Cutting edge: lymphoid tissue inducer cells maintain memory CD4 T cells within secondary lymphoid tissue. J Immunol 189:2094–2098
9. Castellanos JG, Woo V, Viladomiu M et al (2018) Microbiota-induced TNF-like ligand 1A drives group 3 innate lymphoid cell-mediated barrier protection and intestinal T cell activation during colitis. Immunity 49:1077–1089.e1075
10. Melo-Gonzalez F, Kammoun H, Evren E et al (2019) Antigen-presenting ILC3 regulate T cell-dependent IgA responses to colonic mucosal bacteria. J Exp Med 216:728–742
11. Zhou L, Chu C, Teng F et al (2019) Innate lymphoid cells support regulatory T cells in the intestine through interleukin-2. Nature 568:405–409
12. Sonnenberg GF, Hepworth MR (2019) Functional interactions between innate lymphoid cells and adaptive immunity. Nat Rev Immunol 19:599
13. Vivier E, Artis D, Colonna M et al (2018) Innate lymphoid cells: 10 years on. Cell 174:1054–1066
14. Klose CS, Kiss EA, Schwierzeck V et al (2013) A T-bet gradient controls the fate and function of CCR6-RORgammat+ innate lymphoid cells. Nature 494:261–265
15. Vonarbourg C, Mortha A, Bui VL et al (2010) Regulated expression of nuclear receptor RORgammat confers distinct functional fates to NK cell receptor-expressing RORgammat (+) innate lymphocytes. Immunity 33:736–751
16. Sonnenberg GF, Monticelli LA, Alenghat T et al (2012) Innate lymphoid cells promote anatomical containment of lymphoid-resident commensal bacteria. Science 336:1321–1325
17. Giacomin PR, Moy RH, Noti M et al (2015) Epithelial-intrinsic IKKalpha expression regulates group 3 innate lymphoid cell responses and antibacterial immunity. J Exp Med 212:1513–1528
18. Qiu J, Heller JJ, Guo X et al (2012) The aryl hydrocarbon receptor regulates gut immunity through modulation of innate lymphoid cells. Immunity 36:92–104
19. Rankin LC, Groom JR, Chopin M et al (2013) The transcription factor T-bet is essential for the development of NKp46+ innate lymphocytes via the Notch pathway. Nat Immunol 14:389–395
20. Satoh-Takayama N, Vosshenrich CA, Lesjean-Pottier S et al (2008) Microbial flora drives interleukin 22 production in intestinal NKp46 + cells that provide innate mucosal immune defense. Immunity 29:958–970
21. Lochner M, Peduto L, Cherrier M et al (2008) In vivo equilibrium of proinflammatory IL-17+ and regulatory IL-10+ Foxp3+ RORgamma t+ T cells. J Exp Med 205:1381–1393

Chapter 5

Identification of Murine and Human Innate Lymphoid Cells in Frozen Tissue Sections Using Immunofluorescence

Emma E. Dutton and David R. Withers

Abstract

ILCs interact with multiple cell types within their local environment to integrate a wealth of different signals into coordinated responses that regulate tissue homeostasis as well as immune responses upon challenge. While the development and function of ILCs has been extensively studied, principally using flow cytometry, there is limited understanding of the precise composition of cellular niches within which ILCs reside. While this might be optimally studied using dynamic live imaging approaches, immunofluorescence staining of tissue sections can provide fundamental basic information regarding the nature of these microenvironments. Here, a methodology enabling the identification of murine and human ILC populations in frozen tissue sections using immunofluorescence is described.

Key words ILCs, Immunofluorescence, Fetal mLN ILC

1 Introduction

The characterization of ILC populations has been facilitated by complex flow cytometry where many different parameters can be simultaneously assessed to enable precise detection of ILCs among all the other immune cell populations. Obviously such approaches provide no information of the location of these cells within tissues and the microenvironments in which ILC reside remain poorly understood. This really reflects the challenges of identifying ILCs using far fewer parameters than those available for flow cytometry. Since ILC most closely resemble effector T cell populations, identification of the signature transcription factors expressed by ILC (e.g., RORγt, GATA-3) in cells lacking expression of T cell markers (e.g., CD3) has provided some basic information regarding the location of some ILC populations [1–4]. Here we describe approaches to identify murine and human ILC3 (and also ILC2) populations in frozen tissue sections.

Shoba Amarnath (ed.), *Innate Lymphoid Cells: Methods and Protocols*, Methods in Molecular Biology, vol. 2121,
https://doi.org/10.1007/978-1-0716-0338-3_5,

2 Materials

2.1 Specialist Equipment

1. Cryostat.
2. Slide tray.
3. Tissue-Tek O.C.T. compound.
4. Multispot (×4) Microscope slides.
5. Superfrost Plus Adhesion Microscope Slides.
6. Glass staining dish and metal slide rack.
7. Filter papers.

2.2 Buffers

1. Staining Buffer: DPBS [+] $MgCl_2$ (100 mg/L) and $CaCl_2$ (100 mg/L), 1% BSA.

2.3 Additional Reagents

1. Acetone Laboratory reagent grade.
2. Dulbecco's Phosphate Buffered Saline (DPBS).
3. Bovine serum albumin (BSA, Merck).
4. Avidin/Biotin Blocking Kit (SP-2001, Vector Laboratories).
5. Prolong Gold antifade reagent (P36930, Thermo Fisher Scientific).
6. Mouse serum (M5905, Merck).
7. Rat serum (R9759, Merck).
8. Horse serum (H0146, Merck).
9. DAPI, diluted to 20 μg/ml in PBS (D1306, Thermo Fisher Scientific).
10. Streptavidin-Alexa Fluor (AF) 555 (S21381, Thermo Fisher Scientific).

2.4 Antibodies

1. Anti-human/mouse RORγt, clone AFKJS-9, purified, 1:25.
2. Anti-human/mouse GATA-3, clone TWAJ, purified, 1:25.
3. Anti-mouse CD127 (IL-7Rα), clone A7R34, efluor660 conjugated, 1:25.
4. Anti-mouse CD3ε, clone eBio500A2, biotin conjugated, 1:100.
5. Anti-rat IgG (H + L), donkey polyclonal, FITC conjugated, 1:150.
6. Anti-fluorescein/Oregon green, rabbit polyclonal, AF488 conjugated, 1:200.
7. Anti-rabbit IgG, donkey polyclonal, AF488 conjugated, 1:200.
8. Anti-human CD3ε, clone UCHT1, biotin conjugated, 1:25.
9. Anti-human CD127 (IL-7Rα), clone A019D5, AF647 conjugated, 1:25.

3 Methods

3.1 Detection of RORγt, CD3, and CD127 by Immunofluorescence Protocol for Murine Frozen Tissue Samples (See Fig. 1)

1. Prepare frozen tissue sections on cryostat (6–8 μm thick), mount on 4-spot slides, air-dry (30 min), fix in cold acetone (+4 °C for 20 min), then air-dry and store at −20 °C in sealed bags (*see* **Note 1**).
2. On day of staining, remove slides from −20 °C and air-dry for approximately 30 min.
3. Wash the slides with PBS for approximately 5 min to rehydrate.
4. Carefully dry the slides around the edges (using paper towel) and between each spot (using folded filter paper) (*see* **Note 2**).
5. OPTIONAL—this is tissue dependent. Block endogenous biotin in the tissue using the avidin/biotin blocking kit as per kit instructions. Add one drop of "Avidin solution" to each spot and incubate for 15 min. Tip off the excess block onto a paper towel and wash slides for 2 min in PBS. Then add one drop of "Biotin solution" to each spot and incubate for 15 min. Tip off the excess block onto a paper towel and wash slides for 2 min in PBS (*see* **Note 3**).
6. Apply horse serum blocking solution (10% horse serum in PBS/1% BSA) to the section. Typically we use 70 μl solution per section to be stained. Incubate for approximately 15 min in a sealed humidified slide rack (*see* **Note 4**).
7. Aspirate excess blocking solution from each spot. Apply diluted primary Abs—anti-human/mouse RORγt and anti-mouse CD3ε in a volume of 70 μl per section and incubate for 1 h at RT (*see* **Note 5**).

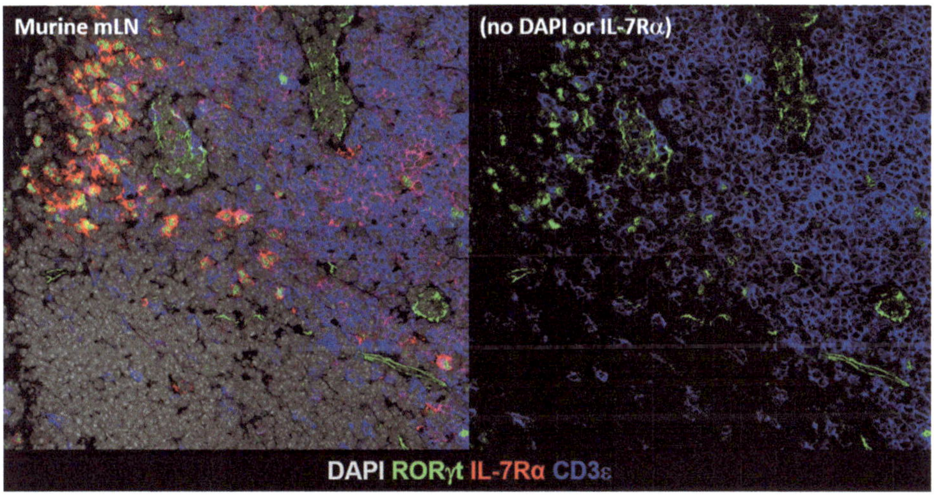

Fig. 1 Murine ILC3 in the mLN reside in the interfollicular space. Immunofluorescence staining to detect ILC3 (RORγt$^+$IL-7Rα$^+$CD3ε$^-$) in sections of murine mLN

8. While incubating with Primary Abs, cross-adsorb the secondary Abs required to detect the anti-human/moue RORγt-Purified Abs, the donkey anti-rat IgG FITC, in 10% mouse Serum (MS). Use an initial total volume of 100 μl (therefore 10 μl MS in PBS/1% BSA), for 30 min before increasing the volume to the desired concentration with PBS/1% BSA (*see* **Note 6**).
9. Wash the slides (tipping off excess Abs before onto paper towel), changing the PBS once and washing for a total of approximately 5 min. Then dry slides as before and carefully add 70 μl of the appropriately diluted secondary Abs (donkey anti-Rat IgG FITC) per spot. Incubate for at least 30 min as before.
10. While incubating with Secondary Ab, cross-adsorb tertiary reagent, rabbit anti-FITC AF488, following the strategy used to cross-adsorb secondary Abs. After 30 min incubation, dilute to the required concentration with PBS/1% BSA and add the remaining primary Abs: anti-mouse IL-7Rα-eFluor660 (*see* **Note** 7).
11. Wash slides then add 70 μl rat serum block (rat 10% serum in PBS/1% BSA) to each spot. Incubate for 10–15 min (*see* **Note 8**).
12. Aspirate off block and then add 70 μl of diluted tertiary Abs (plus residual primary Abs) to appropriate spots and incubate for 30 min as before.
13. While incubating with tertiary Abs, cross-adsorb quaternary reagents (donkey anti-rabbit AF488 and streptavidin-AF555) in 10% MS as described already, then dilute to the required concentration with PBS/1% BSA.
14. Wash slides and add diluted quaternary Abs to all spots 70 μl per spot, incubate for 30 min.
15. Wash selected coverslips in ethanol and allow to air-dry. This helps remove dust particles which are very fluorescent and can introduce artifacts into the analysis.
16. Wash slides and then stain for DAPI;
    - (a) Remove slide 1 from PBS and submerge in DAPI solution stain for 30 s.
    - (b) Move slide 1 into PBS for 30 s and repeat PBS wash ×2.
    - (c) Dry slide 1 using paper towel and filter paper.
    - (d) Repeat this with all slides.
17. Add two drops of Prolong Gold antifade reagent (stored at room temperature) per spot and then gently lower the dried coverslip onto the slide, avoiding bubbles (*see* **Note 9**).

18. Allow to dry and then seal the coverslip with clear nail varnish around the edges (*see* **Note 10**).
19. Leave overnight in a dry container in darkness at room temperature.

3.2 Detection of RORγt, CD3, and CD127 by Immunofluorescence Protocol for Human Frozen Tissue Samples (See Fig. 2)

1. Prepare frozen tissue sections on cryostat (6–8 μm thick), mount on 4-spot slides, air-dry (30 min), fix in cold acetone (+4 °C for 20 min), then air-dry and store at −20 °C in sealed bags (*see* **Note 1**).
2. On day of staining, remove slides from −20 °C and air-dry for approximately 30 min.
3. Wash the slides with PBS for approximately 5 min to rehydrate.
4. Carefully dry the slides around the edges (using paper towel) and between each spot (using folded filter paper) (*see* **Note 2**).
5. OPTIONAL—this is tissue dependent. Block endogenous biotin in the tissue using the avidin/biotin blocking kit as per kit instructions. Add one drop of "Avidin solution" to each spot and incubate for 15 min. Tip off the excess block onto a paper towel and wash slides for 2 min in PBS. Then add one drop of "Biotin solution" to each spot and incubate for 15 min. Tip off the excess block onto a paper towel and wash slides for 2 min in PBS.
6. Apply horse serum blocking solution (10% horse serum in PBS/1% BSA) to the section. Typically we use 70 μl solution per section to be stained. Incubate for approximately 15 min in a sealed humidified slide rack (*see* **Note 4**).

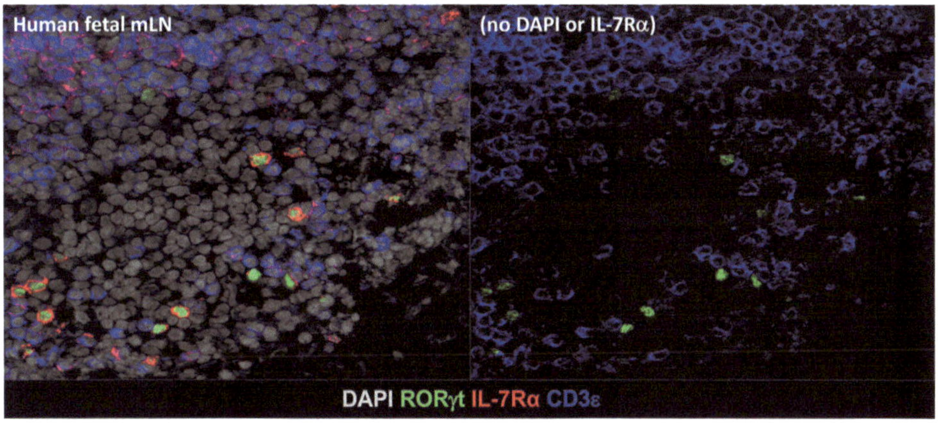

Fig. 2 Detection of human ILC3 in fetal mLN. Immunofluorescence staining to detect ILC3 (RORγt$^+$IL-7Rα$^+$CD3ε$^-$) in sections of human fetal mLN

7. Aspirate excess blocking solution from each spot. Apply diluted primary Abs—anti-human/mouse RORγt, anti-human IL-7Rα AF647 and anti-human CD3ε biotin in a volume of 70 μl per section and incubate for 1 h at RT (*see* **Note 5**).
8. While incubating with primary Abs, cross-adsorb the secondary Abs required to detect the anti-human/mouse-RORγt purified Abs, the donkey anti-rat IgG FITC, in 10% Human Serum (HS). Use an initial total volume of 100 μl (therefore 10 μl HS in PBS/1% BSA), for 30 min before increasing the volume to the desired concentration required with PBS/1% BSA (*see* **Note 6**).
9. Wash the slides (tipping off excess Abs before onto paper towel), changing the PBS once and washing for a total of approximately 5 min. Then dry slides as before and carefully add 70 μl of the appropriately diluted Secondary Abs (donkey anti-rat IgG FITC) per spot. Incubate for at least 30 min as before.
10. While incubating with secondary Ab, cross-adsorb tertiary reagent, rabbit anti-FITC AF488, following the strategy used to cross-adsorb secondary Abs. After 30 min incubation, dilute to the required concentration with PBS/1% BSA.
11. Aspirate off block and then add 70 μl of diluted tertiary Abs to appropriate spots and incubate for 30 min as before.
12. While incubating with tertiary Abs, cross-adsorb quaternary reagents (donkey anti-rabbit AF488 and streptavidin-AF555) in 10% HS as described already, then dilute to the required concentration with PBS/1% BSA.
13. Wash slides and add diluted quaternary Abs to all spots 70 μl per spot, incubate for 30 min.
14. Wash selected coverslips in ethanol and allow to air-dry. This helps remove dust particles which can be highly fluorescent.
15. Wash slides and then stain for DAPI;
    (a) Remove slide 1 from PBS and submerge in DAPI solution stain for 30 s.
    (b) Move slide 1 into PBS for 30 s and repeat PBS wash ×2.
    (c) Dry slide 1 using paper towel and filter paper.
    (d) Repeat this with all slides.
16. Add two drops of Prolong Gold antifade reagent (stored at room temperature) per spot and then gently lower the dried coverslip onto the slide, avoiding bubbles (*see* **Note 9**).
17. Allow to dry and then seal the coverslip with clear nail varnish around the edges (*see* **Note 10**).
18. Leave overnight in a dry container in darkness at room temperature.

4 Notes

Methods

1. Multisport slides are very useful for enabling multiple different stains (up to four stains have been tested) on a given slide and work well for spleen and lymph node sections. For larger tissue sections or tissues that seem to adhere less well to the four-spot slides, we use Superfrost Plus Adhesion Microscope Slides.
2. When using the multisport slides it is essential to thoroughly dry the area between spots (folded filter paper is great at this). If done correctly there is no need to use wax pens to separate the multiple sections on the slide, surface tension will do this.
3. The requirement for using the avidin–biotin blocking steps is very tissue dependent. In murine lymph node or spleen we find no benefit for it, but for lung, intestine, skin, or tumor, it is very important and without it, we get substantial background issues.
4. While we routinely use horse serum for blocking, we have also used goat serum and not observed any problems.
5. The protocol here describing detection of RORγt expression also works well to detect GATA-3 expression, in both human and murine samples. It is simply a case of swapping the primary Abs—both are purified rat IgG Abs and detected in the same way. For positive control sections of tissue, we find lymph nodes or spleen from "T-cell deficient" genetically modified mice (e.g., TCRα$^{-/-}$, ZAP70$^{-/-}$) work best where numbers of ILC2 are particularly elevated and background staining is minimal. Both RORγt and GATA-3 show punctate nuclear staining. Nonnuclear staining is sometimes observed but should be ignored (we think it is an artifact). Any nonrat IgG primary Abs can also be added at this stage (e.g., anti-mouse CD3ε biotin is a hamster IgG). For human tissue, primary Abs are mouse IgG and so cross reactivity with anti-rat IgG is not an issue.
6. Final volumes for secondary (and tertiary/quaternary) Abs depend on number of sections to be stained. We routinely use approximately 70 μl of staining solution, but obviously this can be adjusted if the tissue section is very small or very large.
7. Since detection of the purified anti-RORγt Abs requires a secondary Ab that is anti-rat IgG, use of rat IgG primary Abs to CD127 (or other surface molecules) can only be added to the sections after addition of the secondary Ab (donkey anti-rat IgG FITC, Subheading 3.1, **step 9**) and subsequent blocking with 10% rat serum (Subheading 3.1, **step 11**) to prevent cross

binding. Failure to do this will give artificial colocalization of the CD127 and RORγt signals in the murine protocol.

8. The 10% Rat serum block after incubation with the secondary Abs is essential for detection of other rat IgG primary Abs (e.g., Anti-CD127). This block prevents the secondary Abs binding all rat IgG primary Abs.
9. We find that very few air bubbles are formed if the coverslip is gently lowered such that it touches the drops of antifade reagent before releasing.
10. While the antifade reagent is meant to set solid, this does not always happen in our hands, maybe due to dilution in residual PBS on the section. We routinely seal around the coverslip with nail varnish to avoid leakage.

Acknowledgments

We acknowledge Dr. Emma Mackley for her efforts refining detection of ILC in murine tissue sections. We thank Dr. Tom Cupedo for provision of frozen human fetal mesenteric lymph node.

References

1. Halim TYF, Rana BMJ, Walker JA et al (2018) Tissue-restricted adaptive type 2 immunity is orchestrated by expression of the costimulatory molecule OX40L on group 2 innate lymphoid cells. Immunity 48:1195–1207.e1196
2. Longman RS, Diehl GE, Victorio DA et al (2014) CX(3)CR1(+) mononuclear phagocytes support colitis-associated innate lymphoid cell production of IL-22. J Exp Med 211:1571–1583
3. Mackley EC, Houston S, Marriott CL et al (2015) CCR7-dependent trafficking of ROR-gamma(+) ILCs creates a unique microenvironment within mucosal draining lymph nodes. Nat Commun 6:5862
4. Withers DR, Gaspal FM, Mackley EC et al (2012) Cutting edge: lymphoid tissue inducer cells maintain memory CD4 T cells within secondary lymphoid tissue. J Immunol 189:2094–2098

Chapter 6

Assessing Phosphorylation of STAT Transcription Factors in Mouse Innate Lymphoid Cells

Gianluca Scarno, Giuseppe Pietropaolo, Chiara Di Censo, Giovanna Peruzzi, and Giuseppe Sciumè

Abstract

Innate lymphoid cells (ILCs) ensure protection against pathogens by quickly reacting to the alterations of the cytokine milieu taking place upon infection. More than 50 cytokines and growth factors activate the Janus kinases (JAKs), leading to phosphorylation of members of the signal transducer and activator of transcription (STAT) family. Activation of STATs induces specific transcriptional programs which are associated with distinct cellular outcomes. Thus, an efficient measurement of rapid STAT phosphorylation enables not only to dissect the spectrum of cytokine sensitivity among ILC subsets but also to pinpoint specific transcriptional programs and cellular functions initiated after activation. Using this method, we have previously dissected the downstream events of Interleukin (IL)-23 and IL-12 signaling in ILCs, shedding light on the differential usage of STATs among ILC subsets. Here, we provide an optimized and detailed protocol describing how to analyze phosphorylation of STAT transcription factors in murine NK and ILC subsets isolated from different tissues.

Key words Innate lymphoid cells, Natural killer cells, Cytokines, STATs, Phosphorylation, Intracellular staining

1 Introduction

1.1 Innate Lymphoid Cells

Innate lymphoid cells (ILCs) are strategically localized at environmental interfaces, regulating key homeostatic processes and providing rapid protection upon pathogen entry. ILCs are endowed with a wide range of effector functions, and based on their similarity with adaptive T cells, are divided in four distinct lineages, which comprise natural killer (NK) cells, ILC1, ILC2, and ILC3 [1, 2].

NK cells represent the innate counterpart of cytotoxic T lymphocytes and are characterized by the expression of the lineage defining transcription factors (LDTFs) EOMES and T-BET. These cells are also potent producers of the type 1 signature cytokine, Interferon (IFN)-γ, and are commonly included within group

Shoba Amarnath (ed.), *Innate Lymphoid Cells: Methods and Protocols*, Methods in Molecular Biology, vol. 2121,
https://doi.org/10.1007/978-1-0716-0338-3_6,

1 ILCs, along with ILC1. The latter are specialized helper ILCs, having distinct phenotypes according to the tissue where they reside. Beyond the production of type 1 cytokines, NK cells and ILC1 have several other traits in common, including the expression of surface markers, as natural cytotoxicity receptors (NCRs) and transcription factors (TFs). In this regard, both NK cells and ILC1 express T-BET, while EOMES is expressed at high levels by NK cells [3–6].

Helper ILCs producing type 2 cytokines (Interleukin (IL)-5, IL-9, IL-13) and type 3 cytokines (IL-17 and IL-22) have been also characterized and named ILC2 and ILC3, respectively. ILC2 can be discriminated by the high levels of the LDTF GATA-3, the receptors for IL-25 and IL-33, and the killer cell lectin-like receptor subfamily G member 1 (KLRG1, expressed at lower levels also on NK cells). Finally, the ILC3 group is characterized by cells depending on the LDTF RORγt and includes two prototypical subsets: lymphoid-tissue inducer (LTi)-like cells and NCR^+ ILC3. The latter subset does not express IL-17 and requires T-BET for development [7–11].

1.2 JAK/STAT Pathway at a Glance

The preprimed features of ILC responses are determined at epigenetic and transcriptional levels, which superimpose both phenotypic and functional diversification of the ILC subsets. Of consequence, programmed high basal expression levels of LDTFs, cytokine receptors, and signal dependent transcription factors (SDTFs), altogether underlie the ability of ILCs to quickly respond to changes of the microenvironment [12–14].

Cytokines activating the Janus kinases (JAKs) and members of the signal transducer and activator of transcription (STAT) family are important players during lymphoid development, homeostasis, and activation. At molecular level, the interaction between cytokines and their cognate receptors induces JAKs to phosphorylate each other, as well as the receptor subunits and STATs. The juxtaposition of JAKs and STATs drives formation of dimers (both homo- or heterodimers) which dissociate from the membrane complex and translocate to the nucleus to regulate gene expression [15–17].

In mammals, there are seven genes encoding for STATs, *STAT1–4*, two genes encoding for STAT5, namely, *STAT5A* and *STAT5B*, and *STAT6*; and each STAT is specialized in the control of different aspects of the immune response. Acting downstream of IL-15 and IL-7, STAT5 has a central role during lymphoid development, including the whole ILC compartment, and in regulation of lymphoid homeostasis. STAT4 is mainly induced by IL-12 and represents a key player in driving the type 1 response. This SDTF is expressed at high levels in NK cells, ILC1, and NCR^+ ILC3. The type 2 response is regulated by the IL-4–STAT6 axis in T cells, while their role in ILC2 activation seems to be secondary if compared to the effect that IL-25 and IL-33 exert on these cells. However, a role for STAT6 in promoting ILC2 proliferation and effector functions has recently emerged. Finally, Th17/22 differentiation and ILC3 activation are regulated by STAT3, activated downstream of IL-23; while STAT1 and STAT2 are commonly associated to the signaling pathway of IFNs (*reviewed in* [18–20]).

Although it is true that cytokines can have a bias in mainly inducing activation of one selected STAT, it is also true that this system is more complex and sophisticated, and each cytokine can activate distinct spectra of STATs, with different degrees of phosphorylation. This feature increases exponentially the diversity of the transcriptional programs that can be elicited downstream of distinct cytokines.

1.3 Method Description/Rationale

The method described below is based on previous protocols [21–23] and has been optimized here for the analysis of phosphorylated STATs (pSTATs) in distinct subsets of NK cells and ILCs isolated from the spleen (sp), liver (li) (*see* Fig. 1), and small intestinal lamina propria (siLP) (*see* Fig. 2). The increasing variety of ILCs present in the tissues requires more complex combination of antibodies, recognizing both surface receptors and LDTFs, necessary to dissect the distinct subsets. However, the detrimental effects of methanol fixation on several epitopes and fluorochromes, as well as the limiting numbers of siLP ILCs require a very accurate optimization of both the procedures and the flow cytometry panels used.

By using this method, we have previously identified the selective ability of IL-23 to induce both STAT3 and STAT4 phosphorylation in NCR$^+$ ILC3, demonstrating that different ILC3 subsets can have distinct expression profiles of STATs [24]. Here, we show pSTAT3, pSTAT4, and pSTAT5 levels in different subsets of NK cells and ILCs exposed to a short stimulation with distinct cytokines including IL-7, IL-12, IL-15, and IL-23. We established flow cytometry panels comprising anti-NKp46 and anti-NK1.1 antibodies to gate NCR$^+$ ILCs and evaluated the expression of different surface markers and LDTFs, including CD49a, CD49b, CD11b, KLRG-1, CD4, Eomes, T-bet, Gata-3, and Rorγt. The complete list of tested antibodies is described in Subheading 2, while the detailed gating strategy is outlined in figure captions.

2 Materials

2.1 Mice

C57Bl/6j female mice (6–12 weeks old) were used in the experiments described below. All animal procedures were performed according to the NIH guidelines for the use and care of live animals, the Directive 2010/63/UE and were approved by the Italian Ministry of Health.

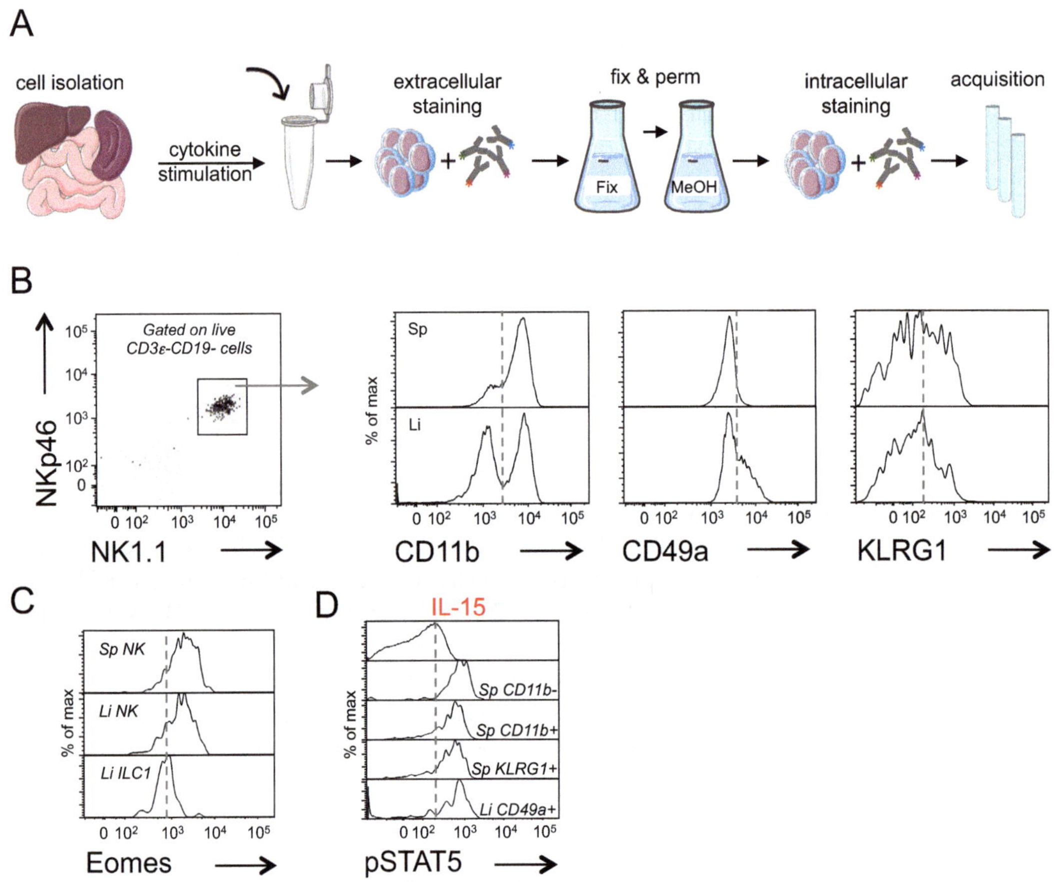

Fig. 1 Evaluation of pSTATs in distinct Sp NK cell subsets and Li ILC1. (**a**) Schematic representation of the protocol used for pSTAT staining. Briefly, cells were isolated from sp., li, and siLP, and stimulated 30 min in complete medium with IL-15 (100 ng/ml for sp., li, and siLP) or IL-7, IL-12, and IL-23 (100 ng/ml for siLP only). Next, cells were stained for extracellular markers, fixed, permeabilized, and stained for intracellular proteins as described in Subheading 3.2. (**b**) Lymphocytes were electronically gated based on SSC-A and FSC-A parameters, doublets were excluded based on FSC-A and FSC-H, and live (Fixable Viability Stain 780 negative) $CD3\varepsilon^{-}CD19^{-}$ (both BV605) cells were plotted for NKp46 (eFluor450) and NK1.1 (BV650) expression. $NKp46^{+}NK1.1^{+}$ cells from sp. and li were dissected for CD11b (APC-R700), CD49a (BB700), and KLRG-1 (BV786) expression. (**c**) NK cells were defined in sp. and li as $NKp46^{+}NK1.1^{+}CD49a^{-}$ cells, while ILC1 as $NKp46^{+}NK1.1^{+}CD49a^{+}$ cells. Levels of Eomes (PE) in NK cells and ILC1 are shown in histogram plot. (**d**) NK cells from sp. gated as $NKp46^{+}NK1.1^{+}Eomes^{+}CD49a^{-}$ cells were divided in three subsets based on CD11b and KLRG1 expression ($CD11b^{-}KLRG1^{-}$; $CD11b^{+}KLRG1^{-}$; $CD11b^{+}KLRG1^{+}$). Levels of pSTAT5(pY694) (AF647) after stimulation with IL-15 in sp. NK subsets and li ILC1 are represented by histogram plot, relative to control

2.2 Buffers and Solutions

15 and 50 ml polypropylene tubes; 1.5 ml microcentrifuge tubes; cell strainers; HBSS (w/o calcium and magnesium); DPBS (w/o calcium and magnesium); PBS 10×; EDTA; HEPES Buffer; Liberase TL; DNase I; Percoll; Lympholyte®-M Cell Separation Media; red blood cell lysing buffer; RPMI-1640 medium; fetal

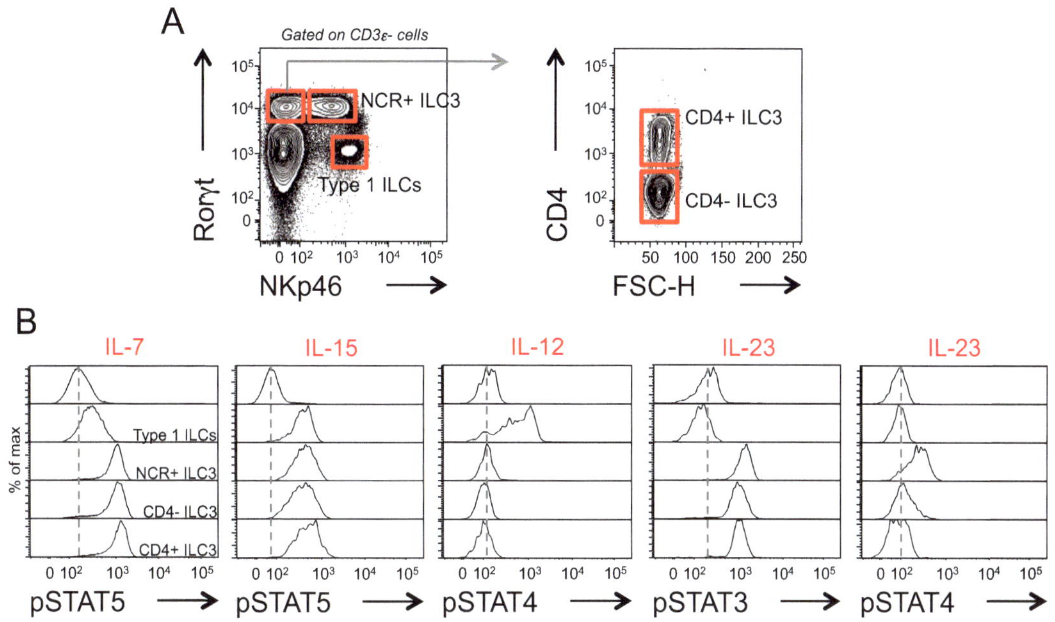

Fig. 2 pSTAT detection in siLP ILC3 subsets compared with type 1 ILCs. (**a**) CD3ε$^-$ (FITC) siLP lymphocytes, isolated as described in Subheading 3.1, were dissected for NKp46 (APC), Rorγt (PE), and CD4 (V500) expression, defining four different subsets. Type 1 ILCs were defined as NKp46$^+$Rorγt$^-$ cells, while NCR$^+$ ILC3 were gated as NKp46$^+$Rorγt$^+$ cells. NKp46$^-$Rorγt$^+$ ILC3 were further divided into two distinct subsets depending on the expression of CD4. (**b**) siLP lymphocytes were stained for pSTAT3(pY705) (PE), pSTAT4 (pY693) (AF488), and pSTAT5(pY694) (AF488). Histogram plot shows the differential STAT phosphorylation upon selective cytokine stimulation in distinct ILC subsets, relative to control

bovine serum (FBS); bovine serum albumin (BSA); β-mercaptoethanol; glutamine; penicillin–streptomycin.

2.3 Cytokines and Media

Recombinant murine IL-7; IL-12 p70; IL-15; IL-23; RPMI-1640 medium supplemented with 10% FBS, glutamine 2 mM, 100 IU/ml penicillin, 0.1 mg/ml streptomycin, HEPES buffer 20 mM (pH 7.2–7.5), and β-mercaptoethanol 2 mM; Staining Buffer (PBS with 0.5% bovine serum albumin and EDTA 2 mM); Brilliant Staining Buffer; BD Cytofix/Cytoperm™; BD Perm/Wash™; methanol.

2.4 Antibodies and Dyes

Live-Dead (BD Horizon™ Fixable Viability Stain 780). For evaluation of surface markers: 2.4G2 antibody (anti-Mouse CD16/CD32, clone 2.4G2), Normal Mouse Serum, anti-CD3ε (clone 145-2C11), CD19 (clone 1D3), CD4 (clone RM4–5), NKp46 (CD335) (clone 29A1.4), NK1.1 (clone PK136,), CD49a (clone Ha31/8), CD49b (clone DX5), CD11b (clone M1/70), KLRG-1 (clone 2F1). For evaluation of LDTF: Rorγt (clone B2D), Eomes (clone Dan11mag), T-bet (clone eBio4B10), Gata-3 (clone L50-823). For evaluation of pSTAT levels: pSTAT3(pY705)

Table 1
List of antibodies used for immunofluorescence stainings

Antibody	Clone	Dye
CD3ε	145-2C11	BV605
CD19	1D3	BV605
CD4	RM4-5	V500
NKp46 (CD335)	29A1.4	eFluor450
NK1.1	PK136	BV650
CD49a	Ha31/8	BB700
CD49b	DX5	APC
CD11b	M1/70	APC-R700
KLRG1	2F1	BV786
Rorγt	B2D	PE
Rorγt	B2D	PerCP-eFluor710
Eomes	Dan11mag	PE
T-bet	eBio4B10	APC
Gata-3	L50-823	AF-488
pSTAT3 (pY705)	4/P-STAT3	PE
pSTAT4 (pY693)	38/p-Stat4	AF-488
pSTAT5 (pY694)	47/Stat5(pY694)	AF-488
pSTAT5 (pY694)	47/Stat5(pY694)	AF647
IgG2a, κ Isotype Control	MOPC-173	PE
IgG2b, κ Isotype Control	27–35	AF-488
IgG1 κ Isotype Control	MOPC-21	AF-488
IgG1 κ Isotype Control	MOPC-21	AF-647

(clone 4/P-STAT3), pSTAT4(pY693) (clone 38/p-Stat4), pSTAT5(pY694) (clone 47/Stat5(pY694)), Mouse IgG2a, κ Isotype Control (clone MOPC-173), Mouse IgG2b, κ Isotype Control (clone 27–35), Mouse IgG1 κ Isotype Control (clone MOPC-21), Mouse IgG1 κ Isotype Control (clone MOPC-21). *See* Table 1 for dyes.

2.5 Flow Cytometric Analysis

The flow cytometry experiments were performed using Becton Dickinson (BD) instruments LSRFortessa or FACS CANTOII according to the antibody panel setup. The experiments aimed to characterize pSTAT5 levels in NK cell/ILC1 subsets (Fig. 1) were performed using flow cytometer BD LSRFortessa that is equipped with five different lasers and can detect up to 18 distinct colors, simultaneously. The lasers are 488, 561, 639, 405 nm, and UV 355 nm. The 561 nm laser, Yellow/Green, allows better resolution of PE and PE-like fluorophores being excited by a laser closer to the peak of max emission fluorescence. With this instrument the simultaneous multiparametric panel was performed as follow: FITC/AF488, BB700/PerCP-eFluor710 (488 nm laser); PE (561 nm laser); APC/AF647, APC-R-700, Live-Dead FVS 780 (639 nm laser); eFluor 450, BV605, BV650, BV786 (405 nm laser). The experiments aimed to dissect pSTAT3, pSTAT4 and pSTAT5 levels in intestinal ILCs (Fig. 2) were performed using the flow cytometer FACS CANTOII, which is equipped with three different lasers: 488, 633, and 405 nm and can detect up to eight colors simultaneously. In both cases, cells were acquired using FACSDiva software (BD Biosciences, San Jose, CA, USA). In general, cells were first gated using Forward vs. Side Scatter parameter (FSC-A vs. SSC-A) strategy, followed by doublets exclusion using FSC-A vs. FSC-H. Live (FVS780 negative cells) and CD3ε/CD19 negative cells were then analyzed for subsequent subsets (*see* Figure caption for details). Postacquisition analysis was performed using FlowJo software.

3 Methods

3.1 ILC Preparation

In this section, we briefly describe three conventional methods for isolation of leukocytes from the small intestinal lamina propria (siLP), spleen (sp), and liver (li).

3.1.1 ILC Isolation from siLP

Euthanize mice and, after incision of skin and peritoneum, remove the small intestine by cutting in proximity of the duodenum and caecum (*see* **Note 1**). Since isolation of siLP leukocytes is a long procedure (about 2.5 h) which includes several steps, work fast and use cold solutions to preserve cell viability. After removal, place the small intestine in a petri dish with cold DPBS, clear the organ from residual stools, and remove mesenteric fat and Peyer's patches (PPs, *see* **Note 2**). Next, longitudinally open the small intestine by inserting scissors throughout the lumen and wash in HBSS for three times (*see* **Note 3**). Transfer the small intestine in a 50 ml tube containing HBSS solution with EDTA 5 mM and Hepes buffer 10 mM and place the tube horizontally in a shaker at 37 °C, 100 rpm. After 30 min discard the solution to proceed with isolation of siLP cells (*see* **Note 4**). Repeat washing with HBSS for three times, as described in **Note 3**. For tissue dissociation, use a cold solution containing RPMI-1640 medium (w/o FBS), Liberase TL (0.25 mg/ml) and DNase I (0.5 mg/ml). Use a volume of 5 ml for each small intestine processed. While keeping the small intestine in cold Liberase TL/DNase I containing medium, mince rapidly the tissue in small pieces with scissors and then transfer the solution in a new 50 ml tube (*see* **Note 5**). Next, place the tube horizontally in a shaker at 37 °C, 100 rpm. After 15 min, add a volume of 20 ml of complete RPMI-1640 medium to block the enzymatic reaction and filter the cell suspension using a 100 μm cell strainer. Centrifuge for 5 min at 500 × *g*. Finally, to separate leukocytes, resuspend cells in 40% Percoll and centrifuge for 15 min, 500 × *g*, room temperature (RT) (*see* **Note 6**); then wash with DPBS.

3.1.2 ILC Isolation from Liver

Euthanize mice, perfuse the liver with 20 ml of DPBS solution through the portal vein; remove the organ from the peritoneal cavity and place it into a 50 ml centrifuge tube with DPBS. Then, remove the gallbladder and smash the organ using a 100 μm cell strainer. Wash twice in DPBS (500 × *g*, RT for 5 min). Resuspend in 8 ml of DPBS (RT) and separate leukocytes using Lympholyte Cell Separation Media (*see* **Note 7**).

3.1.3 ILC Isolation from Spleen

Euthanize the mouse and expose the left side, open the peritoneal cavity and remove the spleen. Place the organ in a 15 ml centrifuge tube, with DPBS. Then, press the spleen through a 70 μm cell strainer. Centrifuge for 5 min at 500 × *g* (4 °C). For red cell lysis, resuspend in 1 ml of red blood cell lysing buffer for 1 min, then wash once with DPBS.

3.2 ILC Stimulation for pSTAT5 Activation

Briefly, cells were isolated from sp., li, and siLP, and stimulated for 30 min in complete medium with IL-15 (100 ng/ml for sp., li and siLP) or IL-7, IL-12, and IL-23 (100 ng/ml for siLP only). Cells were then harvested and used for pSTAT analysis.

3.3 Detection of pSTATs in ILCs

The procedure described below is used to determine the phosphorylation levels of distinct STATs in subsets of NK cells and ILCs, using flow cytometry. This method enables a quick analysis of ILC responses to cytokines following acute stimulation. Herein, we provide the detailed protocol used which is based on the combination of a formaldehyde-based buffer (Fixation/Permeabilization reagent from BD Cytofix/Cytoperm) and pure methanol to fix and permeabilize labeled cells.

1. After isolation, wash and resuspend cells to 0.5–1 × 10^7 cells/ml in cold staining buffer (*see* **Note 8**).
2. Transfer 50 μl of cell suspension into a 1.5 ml microcentrifuge tube (*see* **Note 9**).
3. Stain for Fixable Viability Dye for 10 min; wash once with cold DPBS.
4. Add 2.4G2 antibody, 1 μg/10^6 cells and incubate for 10 min at 4 °C (*see* **Note 10**).

5. Add the antibody cocktail, mix well and incubate at 4 °C for 15 min.
6. Wash once with cold staining buffer (*see* **Note 11**).
7. Vortex the tube and fix the cells using 100 μl of standard/commercial BD Cytofix/Cytoperm buffer (*see* **Notes 12** and **13**), mix well and incubate at 4 °C for 15 min.
8. Wash once with cold staining buffer.
9. Permeabilize cells by adding pure cold methanol (300 μl), pipette extensively to avoid cell aggregates; incubate for 15 min, RT (*see* **Notes 14** and **15**).
10. Wash twice with staining buffer.
11. Add purified mouse serum to reduce background by blocking nonspecific antibody binding.
12. Stain for pSTATs and TFs by adding the appropriate fluorochrome-conjugated antibodies resuspended in Staining Buffer and mix well (*see* **Note 16**).
13. Incubate for 30 min, RT.
14. Wash with staining Buffer.
15. Resuspend cells in PBS and transfer the sample into a 5 ml FACS tube.
16. Acquire samples using a flow cytometer.

4 Notes

1. In the proposed protocol, we separate the small from the large intestine. To remove the small intestine, cut from the proximal tract of duodenum to the ileocecal junction. To isolate the large intestine, grab the caecum, remove the fat and excise the rectum until the anus. Keep in mind that distribution of ILC subsets varies among the different intestinal tracts.
2. Peyer's patches are located on the surface of the small intestine and can be independently collected and processed for leukocyte extraction, as previously described [25]. To reach a high lymphocyte yield, the whole mesenteric fat has to be removed. In this regard, it is possible to isolate lymphocytes from fat tissue as described in ref. 26.
3. Wash by placing the intestinal layer in a 50 ml tube containing 25 ml of HBSS. Shake extensively for 30 s and repeat three times with fresh HBSS. If performed correctly, the HBSS left after the last wash will be completely clear of intestinal residues.
4. This solution, in combination with the shearing forces, allows to break epithelial junctions and remove residual mucus. After incubation, this buffer will be highly enriched for

intraepithelial cells, which can be analyzed after filtration using a cell strainer (100 μm) and an 80–40% Percoll gradient.

5. If several samples need to be processed, it is preferred to add the Liberase TL only after all intestines are chopped.
6. To remove red cells, an 80–40% Percoll gradient is usually used. However, red cells represent only a minor fraction of this preparation, in spite of the higher yield obtained by using a 40% Percoll gradient only.
7. Add the suspension containing liver cells very gently. Centrifuge with minimum acceleration and no deceleration/brake, at 500 × *g*, 20 min, RT.
8. Define the optimal starting number of cells for the staining according to the relative frequency of the ILC subset that has to be analyzed for pSTAT levels. The lower the frequency of the subset the higher will be the number of cells to start with. Suggested cell numbers ranges from 2×10^6 cells/test for spleen and liver to 5×10^6 cells/test for small intestine.
9. Microcentrifuge tubes are preferred to FACS tubes, since they allow an optimal removal/disposal of the different buffers after each wash, as well as more precise volumes of the reagents used throughout the next steps. However, in case of a high number of samples, 96 wells plates can be used instead, to speed up the staining procedure.
10. 2.4G2 is routinely used in flow cytometric analysis to neutralize nonspecific binding of the antibody cocktail used for the extracellular staining to mouse FcγIII (CD16), FcγII (CD32) receptors and FcγI receptor (CD64), which are widely expressed on immune cells.
11. To enhance membrane staining efficiency, DPBS with FBS or DPBS containing 0.5% BSA, EDTA 2 mM are preferred instead of DPBS alone. Supplementation of sodium azide is recommended for a longer preservation of the buffer. To avoid spill over among BV- and BUV-conjugated antibodies, commercially available BD Brilliant Staining Buffer is recommended for the preparation of the antibody cocktail.
12. Cells can be fixed using 2% PFA (paraformaldehyde) instead of BD Cytofix/Cytoperm buffer.
13. Whether performing flow cytometry analysis to simultaneously evaluate cytokine expression and pSTATs levels, follow the next steps: after fixation use a saponin-containing buffer, as the BD Perm Wash reagent, for permeabilization. Then, stain cells with selected anti-cytokine antibodies resuspended in BD Perm/Wash for 30 min, wash and proceed with **Note 9**.
14. Methanol fixation is the most critical step of the whole protocol. 300 μl of prechilled methanol (−20 °C) must be shot onto

the cells, paying attention to discard most of the washing buffer, as methanol dilution heavily drops the pSTAT staining efficiency and brightness. Since methanol has a huge impact on several fluorochromes, we performed experiments aimed to scale down methanol concentration. However, we observed a reduction of pSTAT fluorescence intensity at 90% methanol already, and loss of fluorescence intensity at 80%. Thus, we use the concentration of 100% in our experiments. List of tested fluorochromes which are stable after methanol permeabilization: FITC, BB700, AF647, Pacific Orange, Pacific Blue, V450, V500, BV421/eFluor450, BV510, BV605, BV650, BV786, APC, APC-R700. List of tested fluorochromes sensitive to methanol: PE, PerCP-Cy5.5, PE-Cy7, APC-Cy7.

15. It is possible to reduce the time for fixation and permeabilization up to 5 min for each step, without any major loss of detection in pSTAT intensity. This aspect is particularly useful when cells express Green Fluorescent Protein (GFP) and are to be analyzed for pSTATs.
16. Detection of the LDTFs Eomes, T-bet, Gata-3, and Rorγt expression is still effective after methanol fixation, although the anti-T-bet used works more efficiently with BD Cytofix/Cytoperm and BD permeabilization buffer.

Acknowledgments

G.S. is supported by the Italian Association for Cancer Research (AIRC), MFAG 2018 (Project Code: 21311); and Institut Pasteur (France), Transversal Research Program; PTR-113-17.

References

1. Spits H, Artis D, Colonna M et al (2013) Innate lymphoid cells—a proposal for uniform nomenclature. Nat Rev Immunol 13:145–149. https://doi.org/10.1038/nri3365
2. Vivier E, Artis D, Colonna M et al (2018) Innate lymphoid cells: 10 years on. Cell 174:1054–1066. https://doi.org/10.1016/j.cell.2018.07.017
3. Sun JC, Lanier LL (2011) NK cell development, homeostasis and function: parallels with CD8$^+$ T cells. Nat Rev Immunol 11:645–657. https://doi.org/10.1038/nri3044
4. Cortez VS, Colonna M (2016) Diversity and function of group 1 innate lymphoid cells. Immunol Lett 179:19–24. https://doi.org/10.1016/j.imlet.2016.07.005
5. Vosshenrich CAJ, Di Santo JP (2013) Developmental programming of natural killer and innate lymphoid cells. Curr Opin Immunol 25:130–138. https://doi.org/10.1016/j.coi.2013.02.002
6. Vivier E, Raulet DH, Moretta A et al (2011) Innate or adaptive immunity? The example of natural killer cells. Science 331:44–49. https://doi.org/10.1126/science.1198687
7. McKenzie ANJ (2014) Type-2 innate lymphoid cells in asthma and allergy. Ann Am Thorac Soc 11(Suppl 5):S263–S270. https://doi.org/10.1513/AnnalsATS.201403-097AW
8. Kabata H, Moro K, Koyasu S (2018) The group 2 innate lymphoid cell (ILC2) regulatory network and its underlying mechanisms. Immunol Rev 286:37–52. https://doi.org/10.1111/imr.12706

9. Cherrier M, Eberl G (2012) The development of LTi cells. Curr Opin Immunol 24:178–183. https://doi.org/10.1016/j.coi.2012.02.003
10. Withers DR, Hepworth MR (2017) Group 3 innate lymphoid cells: communications hubs of the intestinal immune system. Front Immunol 8:1298. https://doi.org/10.3389/fimmu.2017.01298
11. von Moltke J, Locksley RM (2014) I-L-C-2 it: type 2 immunity and group 2 innate lymphoid cells in homeostasis. Curr Opin Immunol 31:58–65. https://doi.org/10.1016/j.coi.2014.09.009
12. Sciumè G, Shih H-Y, Mikami Y et al (2017) Epigenomic views of innate lymphoid cells. Front Immunol 8:1579. https://doi.org/10.3389/fimmu.2017.01579
13. Klose CSN, Diefenbach A (2014) Transcription factors controlling innate lymphoid cell fate decisions. Curr Top Microbiol Immunol 381:215–255. https://doi.org/10.1007/82_2014_381
14. De Obaldia ME, Bhandoola A (2015) Transcriptional regulation of innate and adaptive lymphocyte lineages. Annu Rev Immunol 33:607–642. https://doi.org/10.1146/annurev-immunol-032414-112032
15. Levy DE, Darnell JE (2002) Stats: transcriptional control and biological impact. Nat Rev Mol Cell Biol 3:651–662. https://doi.org/10.1038/nrm909
16. Villarino AV, Kanno Y, O'Shea JJ (2017) Mechanisms and consequences of Jak-STAT signaling in the immune system. Nat Immunol 18:374–384. https://doi.org/10.1038/ni.3691
17. Kovanen PE, Leonard WJ (2004) Cytokines and immunodeficiency diseases: critical roles of the gamma(c)-dependent cytokines interleukins 2, 4, 7, 9, 15, and 21, and their signaling pathways. Immunol Rev 202:67–83. https://doi.org/10.1111/j.0105-2896.2004.00203.x
18. Stabile H, Scarno G, Fionda C et al (2018) JAK/STAT signaling in regulation of innate lymphoid cells: the gods before the guardians. Immunol Rev 286:148–159. https://doi.org/10.1111/imr.12705
19. Huntington ND (2014) The unconventional expression of IL-15 and its role in NK cell homeostasis. Immunol Cell Biol 92:210–213. https://doi.org/10.1038/icb.2014.1
20. Gotthardt D, Sexl V (2016) STATs in NK-cells: the good, the bad, and the ugly. Front Immunol 7:694. https://doi.org/10.3389/fimmu.2016.00694
21. Krutzik PO, Nolan GP (2003) Intracellular phospho-protein staining techniques for flow cytometry: monitoring single cell signaling events. Cytometry A 55:61–70. https://doi.org/10.1002/cyto.a.10072
22. Miyagi T, Lee S-H, Biron CA (2010) Intracellular staining for analysis of the expression and phosphorylation of signal transducers and activators of transcription (STATs) in NK cells. Methods Mol Biol 612:159–175. https://doi.org/10.1007/978-1-60761-362-6_11
23. Villarino AV, Sciumè G, Davis FP et al (2017) Subset- and tissue-defined STAT5 thresholds control homeostasis and function of innate lymphoid cells. J Exp Med 214:2999–3014. https://doi.org/10.1084/jem.20150907
24. Mikami Y, Scarno G, Zitti B et al (2018) NCR^+ ILC3 maintain larger STAT4 reservoir via T-BET to regulate type 1 features upon IL-23 stimulation in mice. Eur J Immunol 48:1174. https://doi.org/10.1002/eji.201847480
25. Takahashi H, Kanno T, Nakayamada S et al (2012) TGF-β and retinoic acid induce the microRNA miR-10a, which targets Bcl-6 and constrains the plasticity of helper T cells. Nat Immunol 13:587–595. https://doi.org/10.1038/ni.2286
26. Moro K, Ealey KN, Kabata H et al (2015) Isolation and analysis of group 2 innate lymphoid cells in mice. Nat Protoc 10:792–806. https://doi.org/10.1038/nprot.2015.047

Chapter 7

Determining Coreceptor Expression and Function in Murine ILC2 Through Flow Cytometry Characterization and Coculture Techniques

Christian Schwartz and Padraic G. Fallon

Abstract

Group 2 innate lymphoid cells are important innate effectors and regulators of adaptive immunity in response to parasitic infections and allergic inflammation. Their low frequency in vivo during steady state condition may complicate research on the cells. During type 2 biased immune responses they are activated, increase in frequency and release cytokines as well as regulate T cell functions through direct interactions including MHC class II–T cell receptor interactions. Importantly, coreceptors significantly influence the ILC2-T cell cross talk and shape the adaptive immune response.

Here, we provide an experimental framework to study the function of coreceptors expressed on tissue ILC2. In brief, we describe flow cytometric analysis of the coreceptor of interest, the isolation and culture of mouse pulmonary ILC2 and splenic T cells, as well as approaches to manipulate their coculture. Finally, downstream readout options are outlined.

Key words ILC2, Coculture, T helper cells, MHC2, Type 2 immune response, Stimulation, Cytokines, Transcription factors, Innate immunity, Regulation

1 Introduction

Group 2 innate lymphoid cells (ILC2) are critical regulators of adaptive immune responses—in particular during type 2 immune response generally associated with parasitic infections and allergic hypersensitivity. ILC2 instruct CD4$^+$ T helper cell responses to alter adaptive immunity by numerous ILC2-derived signals. In a landmark publication ILC2 have been shown to express MHC class II molecules with the capacity to both acquire and process antigens [1]. While the capacity of ILC2 to present antigens is lower than that of conventional dendritic cells, the efficiency of ILC2 to induce antigen-specific responses and proliferation is comparable to plasmacytoid DCs and naïve B cells. Additionally, as ILC2 are localized within inflamed tissue, the cells are therefore optimally posed to fine-tune immune responses in a tissue-specific manner. On antigen

Shoba Amarnath (ed.), *Innate Lymphoid Cells: Methods and Protocols*, Methods in Molecular Biology, vol. 2121, https://doi.org/10.1007/978-1-0716-0338-3_7,

presenting cells (APC) the expression of coreceptors such as CD80, CD86, or ICOS-L can activate or inhibit T cell differentiation or proliferation through the interaction with the cognate receptors CD28, CTLA-4 or ICOS on T cells. ILC2 have been shown to have the functional expression of CD80, CD86, ICOS, OX40L, PD-L1, and PD-1 with different effects on the emerging T cell response [1–5]. While monoclonal antibody (mAb) blockade of CD80 or CD86 (on ILC2) only had modest effects on the capacity to induce antigen-specific T cell proliferation, combined administration of anti-CD80- and anti-CD86-mAb impaired T cell proliferation [1]. While disruption of CD80 and CD86 in knockout animals had no effect on pulmonary ILC2 homeostasis [6], deletion of PD-1 led to an unchecked expansion of ILC2 [5]. ICOS–ICOSL interactions have been shown to promote airway hyperreactivity and affect homeostasis of ILC2 by ILC2–ILC2 interaction [4]. Although induced regulatory T cells can also suppress pulmonary ILC2 through ICOS–ICOSL interactions [7], there was no impact on ILC2 homeostasis observed in ICOSL −/− animals [6]. These conflicting results highlight the complexity of coreceptors: ILC2 can regulate T cells (CD80, CD86, PD-L1, OX40L); they can regulate themselves (ICOS, ICOSL); or be regulated by T cells (ICOSL). Therefore, detailed analysis of coreceptors is warranted to understand their role during immune homeostasis and inflammation.

Here, we describe a protocol to investigate the expression and function of coreceptors on lung-derived ILC2, which—using appropriate protocols—can be adapted for other tissues, such as skin or intestine. We will describe the isolation of ILC2 from the mouse lung, flow cytometric analysis, in vitro stimulation, coculture with T cells, and downstream analysis of T cells (proliferation, cytokine production).

2 Materials

2.1 Reagents

70% ethanol.

Phosphate-buffered saline (PBS).

Complete (c)RPMI: RPMI 1640, 10% fetal bovine serum (FBS), 2 mM L-glutamine, 100 IU/ml penicillin, 100 μg/ml streptomycin.

Collagenase D.

Ficoll (or other separation medium with density 1.077 g/cm^3).

PharmLyse.

FACS buffer: PBS, 2% FBS, 0.1% (w/v) NaN_3.

mAb for flow cytometry (*see* Tables 1 and 2).

Table 1
Multicolor flow cytometry mAb panel for analysis of ILC2

	Clone	μg/ml final
Anti-CD3	17A2	1
Anti-TCRβ	H57-597	1
Anti-CD5	53-7.3	1
Anti-CD19	ebio1D3	1
Anti-γδ TCR	ebioGL3	1
Anti-CD11b	M1/70	1
Anti-CD11c	N418	1
Anti-FcεRIα	MAR-1	1
Anti-Ter-119	TER-119	1
Anti-Gr-1	RB6-8C5	1
Anti-F4/80	BM8	1
Anti-NK1.1	PK136	1
Anti-T1/ST2	DJ8	2
Anti-KLRG1	2F1	1
Anti-CD45	30-F11	1
Anti-CD90.2	53-2.1	0.5
Anti-MHC2	M5/114.15.2	1
Anti-coreceptor 1		tbd
Anti-coreceptor 2		tbd

CD16/CD32 mAb: Fc block (clone 93).

LIVE/DEAD™ Fixable Aqua Dead Cell Stain Kit.

Optional: Naive T cell isolation kit for magnetic separation (Miltenyi Biotec) or similar.

2.2 Equipment

Surgical dissection kit (scissors and forceps).

Orbital shaker.

Incubator.

Flow cytometer.

FACS sorter.

Optional: AutoMACS or manual magnetic separation system.

Table 2
Multicolor flow cytometry mAb panel for T cell analysis

Epitope	Clone	μg/ml final
Anti-CD3	17A2	1
Anti-CD4	RM4-5	1
Anti-CD25	PC61	1
Anti-CD44	IM7	1
Anti-CD45	30-F11	1
Anti-CD62L	MEL-14	1
Intracellular staining		
Anti-GATA3	TWAJ	5 μl/sample
Anti-FoxP3	FJK-16s	1
Anti-RorγT	Q31-378	1

2.3 Mice

C57BL/6.

Balb/c.

Optional: Mice deficient in coreceptor of interest or conditional knockout.

2.4 Cell Isolation and Preparation

26 G needle.

10 ml plastic syringe.

5 ml plastic syringe.

50 ml polypropylene tubes.

70 μm cell strainer.

2.5 In Vitro Culture

96-Well U-bottom plate.

Cytokines (IL-2, IL-7, IL-33).

Anti-CD3 mAb (clone 145-2C11).

Anti-CD28 mAb (clone 37.51).

CellTrace™ Violet and CFSE Cell Proliferation Kit.

2.6 Downstream Analysis

Optional (Depends on Coreceptor)
qPCR and/or Western blot equipment.

mAb against coreceptor of interest.

ELISA kits for cytokines of interest.

Phorbol myristate acetate (PMA), ionomycin, Brefeldin A.

FoxP3-staining buffer kit.

Fluorochrome-conjugated mAb for staining transcription factors (*see* Table 2).

3 Methods

3.1 Isolation from Mouse Lungs

1. Preparation.

   Prepare a syringe fitted with a 26 G needle filled with cold 10 ml PBS (per mouse). Keep enough PBS, on ice, ready for all mice. Spray dissecting instruments with 70% ethanol.

2. Dissection and handling.

   Kill mouse according to the local legislature (CO_2 inhalation, cervical dislocation, anesthetic overdose) (*see* **Note 1**). Spray animal with 70% ethanol. Using scissors, remove the ventral skin beginning at the midline, cut open the peritoneum and open the rib cage by cutting laterally through the ribs and the diaphragm attaching to the ribs (*see* **Notes 2** and **3**). Lift the sternum, cut and completely remove the front of the ribcage.

   Cut the aorta below the lungs close to the diaphragm to allow exsanguination of the mouse. Gently insert needle in to the right ventricle and slowly (!) inject 10 ml of cold PBS to flush the pulmonary circulation and remove nontissue cells. The lung should inflate slightly and appear almost white (*see* **Note 4**).

   Use forceps (spray with 70% ethanol) to hold individual lobes by their bronchi and remove lobes by cutting the lobar bronchi with a pair of scissors (*see* **Note 5**). Place all lobes in a 50 ml tubes with 5 ml cRPMI and place on ice (*see* **Note 6**).

3. Prepare single cell suspensions.

   Under a sterile workbench, use sharp scissors to mince the lung (still in the 50 ml tube). Add 1 mg/ml collagenase D (Roche) and incubate shaking (150–200 rpm) at 37 °C for 30 minutes (*see* **Notes** 7 and **8**) on an orbital shaker.

   Pour the digested lung tissue into a 70 μm cell strainer placed on top of a fresh 50 ml tube. Using the plunger of a 5 ml syringe gently pass the tissue through the cell strainer. Flush the cell strainer with 10–20 ml of cRPMI.

4. Ficoll cleanup and RBC lysis.

   Centrifuge 50 ml tube at 300 × *g* for 5 min. Discard supernatant.

   Resuspend pellet in 5 ml 30% Ficoll.

   Centrifuge at 300 × *g* for 5 min. Discard supernatant.

   Resuspend pellet in BD PharmLyse buffer to lyse red blood cells (*see* **Note 9**). Incubate for 5 min at room temperature then add 5× volume of cRPMI to stop lysis.

   Centrifuge at 300 × *g* for 5 min. Discard supernatant.

   Resuspend cells in cRPMI. Count live cells using Trypan Blue (mix 20 μl of the cell suspension with 180 μl Trypan Blue) and a Neubauer counting chamber or an automated cell counter. Proceed with flow cytometry analysis or cell sorting.

3.2 Flow Cytometry Staining for Coreceptor Expression

1. Use 3×10^6 cells per sample in PBS (*see* **Note 10**). Include the following groups of control samples for all mAb used is best practice: unstained, single stained, fluorescence-minus-one (FMO) best practice. If cells are scares ensure at least control samples to test conjugated-mAb specific to dim antigens (i.e., IL7Ra, ST2) and also the antigen of interest (i.e., coreceptors) are analyzed. If possible, include a fully stained biological negative control (i.e., coreceptor knockout mouse).
2. Add LIVE/DEAD Aqua stain to each sample. Mix by pipetting up and down. Incubate for 20 min at 4 °C in the dark (*see* **Note 11**). Add 3–5× volume PBS to wash away excess stain. Centrifuge at 300 × *g* for 5 min. Discard supernatant.
3. Resuspend cells in 50 μl FACS buffer. Add Fc-Block to each sample. Incubate for 10 min at 4 °C in the dark.
4. Prepare mAb staining cocktail (Table 1, *see* **Note 12**).
5. Add 50 μl of staining cocktail to each sample. Mix/vortex. Incubate at 4 °C for 20 min in the dark. Add 3–5× volume to wash away excess stain. Centrifuge at 300 × *g* for 5 min. Discard supernatant.
6. Resuspend cells in FACS buffer and immediately proceed to analysis. Representative flow cytometry images of gating of PD-L1 expression on ILC2 (Fig. 1).

3.3 Flow Cytometry Staining for Isolation

Optional Step: Magnetic Bead Enrichment of Lineage Negative Cells (*See* **Note 13**)

1. Resuspend cells in sterile FACS buffer. Per 3×10^6 cells add 2 μl Fc-Block. Incubate for 10 min at 4 °C. Prepare mAb staining cocktail (Table 1).

   Per 3×10^6 cells add 50 μl of staining cocktail. Incubate for 30 min at 4 °C (or on ice) in the dark. Wash excess mAb with 3–5× volume of FACS buffer. Centrifuge at 300 × *g* for 5 min. Discard supernatant.

   Resuspend in FACS buffer, filter through a 40 μm cell strainer and proceed to sorting.
2. After the sort, centrifuge and resuspend in cRPMI. Count cells.
3. Now cells can be used for further analysis:
   (a) Coculture with other cells, such as T cells (*see* Subheading 3.4).
   (b) Verify coreceptor expression by Western Blot or quantitative RT-PCR. Both techniques require appropriate positive and negative controls (refer to appropriate protocols, such as [8, 9]).
   (c) Stimulate ILC2: Seed 5×10^3 cells in 100 μl per well in a 96-well U-bottom plate. Add 100 μl cRPMI supplemented with 20 ng/ml IL-2 and 20 ng/ml IL-7 (final

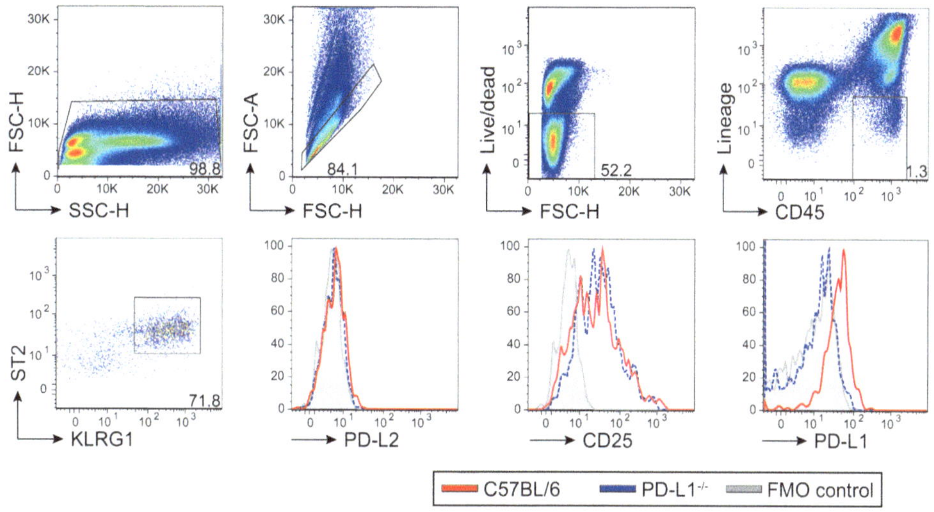

Fig. 1 Coreceptor staining on ILC2. Flow cytometry staining of pulmonary ILC2 identified as single live lineage$^-$ CD45$^+$ST2$^+$KLRG1$^+$ cells. C57BL/6 (red) and PD-L1$^{-/-}$ (blue) ILC2 were analyzed for expression of PD-L2, CD25, and PD-L1. Fluorescence-minus-one (FMO) was used to determine unspecific background fluorescence

concentration 10 ng/ml for each) and add cytokines/ stimulating agents to the cultures. Verify coreceptor expression by flow cytometry, Western blot, or qPCR.

(d) Expand ILC2: Seed 5×10^3 cells in 100 μl per well in a 96-well U-bottom plate. Add 100 μl cRPMI supplemented with 20 ng/ml IL-2, 20 ng/ml IL-7, and 20 ng/ml IL-33 (final concentration 10 ng/ml for each) to expand ILC2 (*see* **Note 14**). Replace 100 μl of media with freshly made cytokine-containing media every 3–4 days.

3.4 In Vitro Coculture and Readout

1. Preparation:

   The day before stimulation of T cells coat the required number of wells in a 96-well plate (U-bottom) with 50 μl sterile PBS containing 5 μg/ml anti-CD3 (clone 145-2C11) mAb (*see* **Note 15**). Cover plate and use Parafilm or cling wrap to prevent evaporation. Incubate overnight at 4 °C.

2. Isolation of interacting cells:

   In order to set up cocultures of ILC2 and other cell populations, the interacting cell population need to be isolated from mice or generated in vitro. This can be achieved through sort purification of CD4 T cells or B cells via FACS (for high purity; suitable for subsets) or MACS (lower purity; suitable for naive and bulk). Macrophages and dendritic cells can be derived from bone marrow cells [10, 11].

3. Optimization of coculture conditions:

Depending on the cell type used the optimal coculture conditions need to be tested and optimized (*see* **Note 16**). Here, we will outline the protocol used for the analysis of cocultures consisting of ILC2 and naïve CD4 T cells.

4. Isolation of naïve $CD4^+$ T cells:
   (a) Kill mouse, spray with 70% ethanol. Make a small incision on the left ventral flank of the abdomen, remove fur and skin and open the peritoneal cavity. Careful hold the spleen with forceps and cut away connective tissue and remove the spleen (*see* **Note 17**). Store the spleen in cRPMI on ice.
   (b) Disrupt splenic tissue in a 70 μm cell strainer on top of a 50 ml tube using the plunger of a 5 ml syringe. Flush cell strainer with ≥10 ml of cRPMI. Centrifuge at 300 × *g* for 5 min. Discard supernatant. Loosen the cell pellet by flicking tube, vortexing or running the tube along a tube stand.
   (c) Resuspend pellet in BD Pharmlyse buffer to lyse red blood cells (*see* **Note 9**). Incubate for 5 min at room temperature. Centrifuge at 300 × *g* for 5 min. Discard supernatant and loosen cell pellet. Count cells.
   (d) For fluorescence-activated cell sorting (FACS) (positive selection): Resuspend cells in sterile FACS buffer and add 0.25 μg Fc-Block per 10^6 cells. Incubate for 10 min at 4 °C. Add 50 μl of FACS buffer containing anti-CD3, anti-CD4, anti-CD25, anti-CD44, anti-CD45, and anti-CD62L conjugated-mAb (*see* Table 2). Vortex. Incubate for 20 min at 4 °C in the dark. Fill the tube with FACS buffer. Centrifuge at 300 × *g* for 5 min. Discard supernatant. Resuspend cells in FACS buffer and proceed to FACS sorting. Naïve CD4 T cells are identified as $CD45^+CD3^+CD4^+CD25^+CD44^-CD62L^-$.
   (e) For MACS purification using an AutoMACS sampler (negative selection): Use the "Naïve $CD4^+$ T Cell Isolation Kit" (Miltenyi Biotec). Resuspend cells in MACS buffer (40 μl per 10^7 cells). Add 10 μl (per 10^7 cells) of biotinylated mAb cocktail. Mix and incubate for 5 min at 4 °C. Add 20 μl MACS buffer, 20 μl anti-biotin beads, and 10 μl CD44 microbeads per 10^7 cells. Incubate for 10 min at 4 °C. Add 1–2 ml MACS buffer per 10^8 cells, centrifuge at 300 × *g* for 5 min, and aspirate supernatant. Resuspend in 500 μl MACS buffer.

       Proceed to AutoMACS. Use program "depletes." Naïve T cells will be in the negative fraction, position B.

       Magnetic purification can be adapted to various conventional magnet-based systems with the appropriate naïve T cell isolation kits (e.g., StemCell Technologies, BioLegend).

(f) Optional: Label cells with cell trace dyes. To use cell proliferation as a readout assay for the coculture both cell populations need to be labelled with a cell trace dye before coculture. Different fluorescent dyes are available to track proliferation. Here, we describe the use of "Cell-Trace Violet Cell Proliferation Kit" (Invitrogen), in our hands this kit yields robust results. Prepare CellTrace stock solution at 5 mM in DMSO.

Remove supernatants from cell pellet. Resuspend cells in PBS (up to 10^6 per ml). Add CellTrace Violet (1:1000 in PBS; 5 μM working concentration) to the cells and incubate for exactly(!) 20 min at 37 °C while protected from light. Then add 5× the original volume of cRPMI (or media containing at least 1% protein) and incubate for 5 min to remove excess dye. Centrifuge for 5 min at 300 × *g*. Discard supernatant. Resuspend in prewarmed media (*see* **Note 18**).

(g) For nonpolarizing conditions incubate the cocultures on anti-CD3 mAb (clone 145-2C11)-coated plates and add soluble anti-CD28 mAb (1 μg/ml; *see* **Note 15**). Adjust ILC2 to 1–2 × 10^5 cells per ml. Adjust T cells to 0.4–1 × 10^6 cells per ml. Add 50 μl of each cell population per well according to experimental condition. For ILC2 only and T cell only control wells add 50 μl of culture media to the respective cell populations.

5. In order to manipulate the interaction between ILC2 and, in this case, CD4 T cells, two options may be available:

(a) Using knockout models of the coreceptor of interest:

- Isolate ILC2 according to this protocol from mice deficient in the coreceptor of interest (Fig. 1). For the proper control, isolate T cells from mice deficient in the respective interacting protein (*see* **Note 19**). Incubate wild-type or coreceptor knockout ILC2 together with wild-type or coreceptor-ligand knockout T cells. This will yield eight experimental groups, as both ILC2 only and T cell only control groups should be included for both wild-type and knockout animals.

- Generate transient knockout models by CRISPR/Cas9, shRNA, siRNA, retroviral/lentiviral transduction. Refer to the literature on how to apply these methods.

(b) Using blocking mAb:

- In order to block the interaction, blocking mAb can be added to the culture targeting one side or the other of the receptor–ligand system (*see* **Note 20**). mAb need to be added directly to the culture at a concentration 1–10 μg/ml to effectively block interaction.

6. Analysis of interaction.
   (a) Different readout assays may be employed to get insights into the coculture system. Typically, 3–4 days of culture are sufficient.
      - Centrifuge the plate for 3 min at 300 × *g* and carefully transfer the supernatant into a fresh plate. These supernatants can be stored at −80 °C for downstream analysis using an enzyme-linked immunosorbent assay. Typical parameters to analyze are cytokines associated with T cell polarization (IL-4, IFNγ, IL-17A, etc.). However, this may result in a mixture of cytokines derived from T cells ***and*** ILC2 (for cytokines like IL-5, IL-9, IL-13). Therefore, it may be necessary to separate T cells and ILC2 (FACS sort) and restimulate isolated T cells on a freshly coated plate for 24 h.
      - Analyze cell proliferation by flow cytometry. If the cells have been stained with a proliferation dye this can now be analyzed.
      - Analyze cytokine production on a per cell basis using intracellular cytokine staining: To this end, restimulate cells for 4–5 h with 40 ng/ml phorbol 12-myristate 13 acetate and 1 μg/ml ionomycin and for the last 2 h in the presence of Brefeldin A (*see* **Note 21**). Wash cells for flow cytometry.
      - Analyze transcription factor expression in T cells: The most common approach to determine the expression of transcription factors is by intranuclear staining for flow cytometry (*see* Table 2). This will allow for analysis of expression on a per cell basis instead of mean expression levels using qPCR or Western blotting.

4 Notes

1. Cervical dislocation may lead to bleeding in the chest cavity contaminating the lungs with blood.
2. Be careful not to cut the lung while removing the front of the ribcage.
3. Be aware that sharp scissors will become dull when cutting bones.
4. If foam builds around the mouse's mouth the integrity of the lung has been destroyed (due to resistance and too much pressure) and subsequent analysis may be impaired.
5. Make sure that the tracheal and mediastinal lymph nodes are NOT removed together with the lung lobes.

6. For FACS sorting, you may pool the lungs of 3–5 mice to reduce downstream handling time; adjust collagenase accordingly.
7. The gentleMACS (Miltenyi Biotec) can also be used. Please refer to [12].
8. Check whether collagenase activity interferes with surface expression of the molecule of interest.
9. Optionally, ammonium chloride–potassium (ACK: H_2O, 150 mM NH_4Cl, 10 mM $KHCO_3$, 0.1 mM Na_2EDTA, pH 7.2–7.4) buffer can be used.
10. Staining can be performed in 5 ml plastic tubes or 96-well V-bottom plates when handling many samples.
11. The amine groups in the LIVE/DEAD stain may bind to proteins present in FBS, therefore refrain from using FACS buffer.
12. mAb concentrations should be titrated per clone and fluorophore conjugate according to the laboratories' SOP for flow cytometry staining. Final concentrations in Tables 1 and 2 are given as a good starting point for mAb titration.
13. In order to reduce sorting time, lineage negative cells can be enriched by negative selection via magnetic beads. To this end, the lineage cocktail should be conjugated to a fluorochrome for which magnetic beads exist, such as anti-PE- or anti-APC beads. We would suggest to only label lineage-positive cells (20 min, 4 °C, dark), wash (MACS buffer), incubate with anti-fluorochrome beads and retain the lineage-positive fraction using a magnet. As different companies offer different systems, please adhere to the manufacturer's instructions (Miltenyi Biotec columns, BioLegend MojoSort, STEMCELL Technologies EasySep).
14. OP9-DL1 cells can be used as feeder cells to improve expansion [13].
15. Concentrations of stimulating anti-CD3 mAb can range from 1 to 10 μg/ml, and from 0.5 to 5 μg/ml for soluble anti-CD28 mAb. CD4 T cells without stimulation through the TCR will die rapidly.
16. As many factors can affect coculture quality, the following aspects need to be considered: Are both cell types viable in the culture media used? Will cytokines added for one cell type impact on the other (i.e., IFNγ for M1 macrophages will inhibit ILC2)? Do the cells need to be preactivated/rested/pretreated? How can the cells be distinguished in downstream analyses?

17. To facilitate downstream analysis, congenic mouse strains can be used (e.g., CD90.1, CD45.1).
18. To have a baseline control for staining intensity it is good practice to analyze cells by flow cytometry now. Ensure to allow at least 10 min for acetate hydrolysis of the dye in the stained cells.
19. Be aware that isolation of cells from any knockout mouse strain may introduce bias into the experiment as the cells developed in a knockout environment and may have adopted compensatory mechanisms masking possible functions. Consider using conditional knockout animals with a cell-restricted knockout of the coreceptor of interest.
20. Blocking mAb for in vitro studies must be cell culture grade reagents.
21. For some cytokines it may be necessary to use monensin instead of brefeldin A. Refer to the literature or online resources provided by mAb manufacturers.

References

1. Oliphant CJ, Hwang YY, Walker JA et al (2014) MHCII-mediated dialog between group 2 innate lymphoid cells and CD4(+) T cells potentiates type 2 immunity and promotes parasitic helminth expulsion. Immunity 41:283–295
2. Halim TYF, Rana BMJ, Walker JA et al (2018) Tissue-restricted adaptive type 2 immunity is orchestrated by expression of the costimulatory molecule OX40L on group 2 innate lymphoid cells. Immunity 48:1195–1207.e1196
3. Schwartz C, Khan AR, Floudas A et al (2017) ILC2s regulate adaptive Th2 cell functions via PD-L1 checkpoint control. J Exp Med 214:2507–2521
4. Maazi H, Patel N, Sankaranarayanan I et al (2015) ICOS:ICOS-ligand interaction is required for type 2 innate lymphoid cell function, homeostasis, and induction of airway hyperreactivity. Immunity 42:538–551
5. Taylor S, Huang Y, Mallett G et al (2017) PD-1 regulates KLRG1(+) group 2 innate lymphoid cells. J Exp Med 214:1663–1678
6. Dutton EE, Camelo A, Sleeman M et al (2017) Characterisation of innate lymphoid cell populations at different sites in mice with defective T cell immunity. Wellcome Open Res 2:117
7. Rigas D, Lewis G, Aron JL et al (2017) Type 2 innate lymphoid cell suppression by regulatory T cells attenuates airway hyperreactivity and requires inducible T-cell costimulator-inducible T-cell costimulator ligand interaction. J Allergy Clin Immunol 139:1468–1477.e1462
8. Kurien BT, Scofield RH (eds) (2015) Western blotting. Methods Mol Biol 1312
9. Biassoni R, Raso A (eds) (2014) Quantitative Real-Time PCR. Methods Mol Biol 1160
10. Assouvie A, Daley-Bauer LP, Rousselet G (2018) Growing murine bone marrow-derived macrophages. Methods Mol Biol 1784:29–33
11. Segura E, Onai N (eds) (2016) Dendritic cell protocols. Methods Mol Biol 1423
12. Nussbaum JC, Ortiz-Carpena JF (2018) Isolation and identification of group 2 innate lymphoid cells in settings of type 2 inflammation. Methods Mol Biol 1799:93–107
13. Seehus C, Kaye J (2016) In vitro differentiation of murine innate lymphoid cells from common lymphoid progenitor cells. Bio Protoc 6:e1770

Chapter 8

Protocol for Determining the Effect of Neuroendocrine Hormones on Murine ILC Function

Linda Quatrini and Sophie Ugolini

Abstract

Neuroendocrine hormones are recognized as important mediators of inflammation that participate in the regulation of the magnitude and length of the immune response. It was demonstrated that endogenous glucocorticoids control the function of innate lymphoid cells (ILCs), and this regulatory mechanism is both cell type- and tissue-specific and is required for host protection during infections. We describe here how to analyze in vitro the effects of corticosterone on murine ILCs, using flow cytometry. The protocols described allow for the identification of the specific combination of stimuli with which glucocorticoids cooperate to regulate the function of ILCs. These methods are instrumental to understanding the molecular mechanisms downstream of glucocorticoid receptor activation and can explain the tissue specificity of ILC response to glucocorticoids.

Key words NK cells, ILCs, Glucocorticoids, Flow cytometry, IFN-γ, Cytotoxic granule exocytosis

1 Introduction

Glucocorticoids are neuroendocrine hormones released into systemic circulation upon activation of the hypothalamic–pituitary–adrenal (HPA) axis. They exert their biological function through binding to the glucocorticoid receptor, a cytosolic receptor ubiquitously expressed that regulates the transcription of many target genes through its ability to bind DNA either directly and indirectly, through interaction with other transcription factors [1]. Thus, the effect of these hormones on a given cell target is strictly dependent on the other signals the cell receives at the same time. Upon inflammation, the HPA axis is activated and the action of GCs on the immune cells is responsible for the control of the inflammatory response [2].

Innate lymphoid cells (ILCs) are a first line of defense against pathogens and are sensitive to GC regulation because they express GR [3]. This regulation is dependent on the tissue microenvironment in vivo [4]. It is possible to mimic the tissue

Shoba Amarnath (ed.), *Innate Lymphoid Cells: Methods and Protocols*, Methods in Molecular Biology, vol. 2121,
https://doi.org/10.1007/978-1-0716-0338-3_8, © Springer Science+Business Media, LLC, part of Springer Nature 2020

microenvironment and reproduce the stimuli ILCs receive in vivo by engaging, in vitro, different combinations of receptors, including cytokine receptors. To understand the effect of the combination of these signals with the GCs in ILCs, receptor engagement is achieved in the presence or absence of corticosterone (the principal endogenous glucocorticoid released in rodents). The possible modulatory effect of GCs is verified by comparing the effector functions activated by these stimuli in these two conditions. As a control of the specificity of the effect of the compound, in the control condition the vehicle in which corticosterone is dissolved (ethanol) is added.

In this chapter, we describe a method to study the effect of corticosterone, but it can be used to test other neurohormones, such as epinephrine and norepinephrine. It is important, in any case, to titrate the compound to find a concentration that is not toxic and is compatible with cell viability. Depending on the subset of ILCs to be studied, a specific kind of stimuli is used, and a specific consequent effector function is analyzed. In this chapter, we focus on murine NK cells, therefore we describe a protocol to mimic NK cell recognition of a target cell by triggering an activating receptor (NK1.1, *see* **Note 1**), and to stimulate cells with the cytokines that are capable of inducing NK cell activation (IL-12 and IL-18). To engage NK1.1 surface receptor on NK cells, the flat bottom of a 96-well plate is coated with anti-NK1.1 antibody, using a carbonate buffer that allows protein adsorption to the plastic. NK stimulation with IL-12 and IL-18 is achieved by adding cytokines to the cell culture media. After NK1.1 antibody-mediated activation and stimulation with cytokines, the main NK cell effector functions are analyzed to verify whether corticosterone has a regulatory effect. These effector functions are the production of IFN-γ, and the exocytosis of lytic granules [5].

Flow cytometry is the primary measuring tool used to analyze NK cell effector functions upon in vitro activation. In a flow cytometer, fluid containing a cell suspension is forced through a nozzle to make a fine stream containing single cells. This single cell population flows past a set of laser light sources and fluorescence and light scattered from cells is recorded. A number of lasers are used with different wavelength, capable to excite many fluorochromes, and each of them has a particular emission spectrum, allowing to discriminate their signals through specific filters [6]. By staining cells with antibodies conjugated with different fluorochromes, it is possible to detect specific signals associated to the bound antigens, making this instrument suitable for detection of many parameters on a single cell at the same time. Prior to cytometric analysis, the sample is stained with a mix of antibodies designed to identify a specific cell population and detect the expression of antigens induced by the stimulation. In this protocol we describe the procedure of staining stimulated murine splenocytes samples with antibodies that allow the discrimination of NK cells from the other lymphocytes, and the analysis of their effector functions. To measure cytokine production, the sample is fixed and permeabilized and stained with an anti-IFN-γ antibody, to detect the intracellular expression of this cytokine. Exocytosis of lytic granules is measured by quantifying the expression of a marker of cytotoxic granules that is expressed at NK cell surface upon degranulation (CD107a). To do this, an anti-CD107a antibody conjugated with a fluorochrome (fluorescein isothiocyanate, FITC) is added to the cell suspension media from the beginning of cell stimulation, to be able to detect cell degranulation over time. To enhance CD107a and IFN-γ detectability with flow cytometry analysis, the addition of the protein transport inhibitors brefeldin A and monensin to the media is required. Brefeldin A blocks the protein transport from the Golgi apparatus to the endoplasmic reticulum resulting in the accumulation of cytokines in the cell. Monensin neutralizes the acidic pH of endosomes and thereby prevents degradation of fluorochrome conjugates of endocytosed anti-CD107a antibodies [7–11].

As a positive control of cell activation, splenocytes are activated with strong stimuli that act independently of surface receptor engagement: phorbol myristate acetate (PMA, a potent protein kinase C activator) and ionomycin (that raises intracellular Ca^{2+} concentration). As a negative control, unstimulated splenocytes are used.

2 Materials

2.1 Cell Stimulation

- 70 μm cell strainers to obtain single cell suspensions from spleen.
- Red Blood Cell Lysis (RBCL) Buffer.
- Complete medium: RPMI medium with 10% FCS, 100 μg/ml penicillin/streptomycin, 2 mM L-glutamine, 1 mM sodium pyruvate and 0.01 M Hepes.
- Corticosterone (reconstituted in ethanol at a stock concentration of 1 mM).
- 96 well plate U bottom for stimulation with cytokines and PMA/ionomycin, and 96 well plate flat bottom (Immulon 2HB) for stimulation with anti-NK1.1 antibody.
- Carbonate buffer (Na_2CO_3 15 mM and $NaHCO_3$ in H_2O, pH 9.6).
- Cytokines: IL-12 and IL-18.
- PMA and ionomycin.
- Anti-CD107a FITC conjugated antibody (clone 1D4B) to add during stimulation.

- GolgiPlug™ and GolgiStop™.
- Anti NK1.1 antibody (PK136) and IgG2a isotype-matched control antibody (eBM2a).

2.2 Preparation of Samples for Flow Cytometry Analysis

- FACS tubes: Round-Bottom Polypropylene Tubes, 5 mL.
- FACS buffer: PBS 5 mM EDTA.
- LIVE/DEAD™ Fixable Blue Dead Cell Stain Kit, for UV excitation (*see* **Note 2**).
- eBioscience Intracellular Fixation and Permeabilization Buffer Set.
- Antibodies: Purified Rat Anti-Mouse CD16/CD32 (2.4G2, blocks non-antigen-specific binding of immunoglobulins to the FcγIII and FcγII, and possibly FcγI, Receptors); anti-CD3 PE-CF594 (145-2C11) and anti-CD19 PE CF594 (1D3); anti-NKp46 BV421 (29A1.4) and anti-IFN-γ APC (XMG1.2).
- Flow cytometer: BD LSRII.

3 Methods

Carry out all procedures until FACS staining in sterile conditions to prevent cell contamination during stimulation. In each well the final volume will be 150 μl, deriving from the addition of 50 μl of three solutions (splenocytes + cytokines + corticosterone). The principle of the setup of the plate prior incubation is to prepare each of these three solutions at a concentration of 3× so that the final concentration is 1× each.

3.1 *In Vitro* Stimulation of Murine Splenocytes

1. Preparation of splenocyte cell suspension (*see* **Note 3**).
   (a) Disrupt the spleen, forcing it through a 70 μm cell strainer using the plunger of a syringe. Rinse the cell strainer with PBS 5 mM EDTA, wash cells by centrifugation at 600 × *g* for 5 min at RT and resuspend the pellet in 1 ml RBCL buffer (prewarmed at RT). Incubate 1 min at RT.
   (b) Wash cells with warm complete medium twice by centrifugation at 600 × *g* for 5 min at RT.
   (c) Resuspend cells at a concentration of 20×10^6 cells/ml in complete media, so that in each well 10^6 cells are plated in 50 μl.
2. For NK1.1 receptor stimulation, incubate a flat bottom 96 well plate (Immulon 2HB) overnight at 4 °C in carbonate buffer with anti NK1.1 antibody or isotype control antibody (25 μg/ml) to allow protein adsorption the day before the experiment (*see* **Note 4**). Once the splenocytes are ready, wash the coated plate three times with PBS and put 50 μl complete media/well.

3. For cytokine stimulation, prepare a mix with a concentration 3× of IL-12 and IL-18 in complete media, so that the final concentration is 25 ng/ml for IL-12 and 20 ng/ml for IL-18 (*see* **Note 5**). As for cytokines, a mix containing PMA/ionomycin at a concentration 3× must be prepared so that the final concentration is 200 ng/ml for PMA and 1 μg/ml for ionomycin.
4. Prepare a solution of corticosterone 3× and vehicle 3× in complete media so that the final dilution is 1:2000 (500 nM for corticosterone). These two solutions must also contain the anti-CD107a-FITC antibody (final dilution 1 μl/well) and GolgiPlug and GolgiStop 3× (final dilution 1:1000 for GolgiPlug and 1:500 for GolgiStop).
5. Plate 50 μl of the cells prepared in 1 and 50 μl of the solutions prepared in 4 (Corticosterone/vehicle + anti-CD107a antibody + GolgiPlug + GolgiStop) in each well of the two plates. Add 50 μl of the solutions prepared in 3 (cytokine mix or PMA/ionomycin) or of complete medium alone (as unstimulated control) in each well of the round bottom 96 well plate. The final volume in each well must be 150 μl.
6. Incubate the plates at 37 °C for 4 h (*see* **Note 6**). At the end of the stimulation, wash cells with cold FACS buffer by centrifugation at 600 × *g* for 5 min at 4 °C and transfer them in tubes for antibody staining and FACS analysis.

3.2 FACS Analysis of Activated ILCs

1. Prepare a mix of antibodies in PBS EDTA (*see* **Note 7**):
   (a) Fc blocking antibody (2.4G2) (1:200).
   (b) Fixable blue dead cell stain kit (1:1000) (*see* **Note 2**).
   (c) Anti-CD3 PE-Texas red (1:500) (*see* **Note 8**).
   (d) Anti-CD19 PE-Texas red (1:500) (*see* **Note 8**).
   (e) Anti-NKp46 BV421 (1:50).
2. Add to the cell pellet the antibody mix (50 μl/well) and incubate the tubes at 4 °C for 30 min in the dark.
3. Wash the cells with FACS buffer by centrifugation at 600 × *g* for 5 min at 4 °C.
4. Resuspend the cell pellet in cold Fixation Buffer and incubate 30 min at RT in the dark or at 4 °C ON.
5. Wash cells 2× with cold Permeabilization Buffer by centrifugation at 1000 × *g* for 5 min at 4 °C.
6. After the final wash, add to the pellet the anti-IFN-γ antibody APC diluted 1:100 in Permeabilization Buffer and incubate tubes at 4 °C for 1 h in the dark.
7. Wash cells 2× with cold FACS buffer by centrifugation at 1000 × *g* for 5 min at 4 °C.

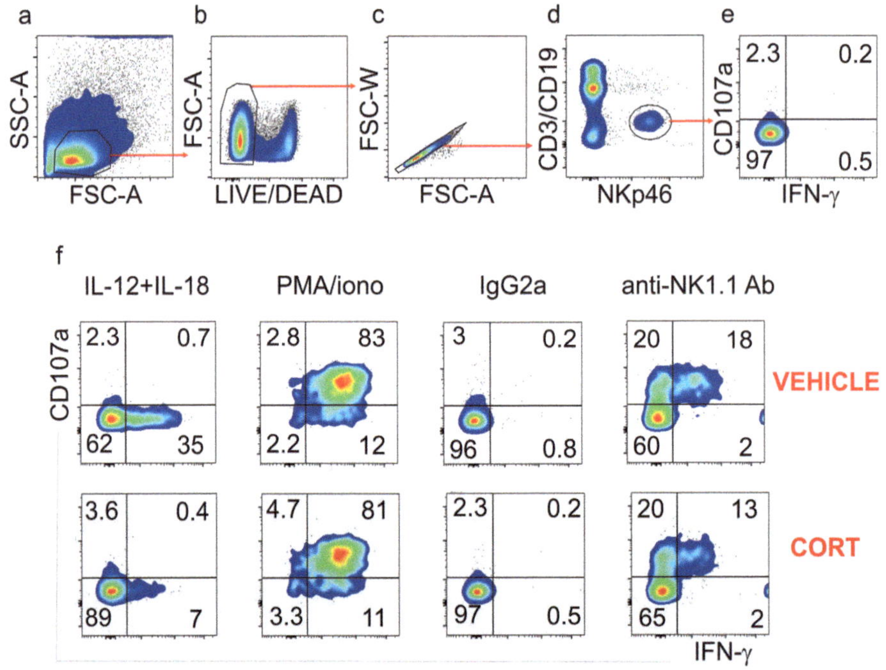

Fig. 1 Gating strategy to analyze NK cell effector functions among splenocytes. On the basis of the physical parameters of the cells in the sample (FSC-A vs SSC-A plot), a first gate is done on the lymphocyte population (**a**) (*see* **Note 10**). Dead cells are then excluded in the LIVE/DEAD vs FSC-A plot (**b**) (*see* **Note 2**), and then single cells are gated in the FSC-A vs FSC-W plot (**c**) (*see* **Note 3**). NK cells are gated as NKp46^{+}CD3^{-}CD19^{-} cells (**d**), and on the NK cell population CD107a expression and intracellular IFN-γ are measured by plotting IFN-γ vs CD107a (**e**). The gates divide the plot in four quadrants and the numbers indicate the % of NK cells in each quadrant. The gates are drawn on the negative control sample, that is, unstimulated stained splenocytes, so that they define the limit between positive and negative signal. In panel (**f**) the same plots as in (**e**) is shown for stimulated samples in both control (vehicle) and corticosterone (cort) conditions. They show that cytokine (IL-12 + IL-18) stimulation does not induce NK cell degranulation but induce IFN-γ production, with an inhibitory effect of corticosterone incubation. The plots of PMA/iono stimulated samples show that NK cells in the two conditions (vehicle and cort) have the same potential to degranulate and produce IFN-γ in response to a nonspecific stimulus. The plots of IgG2a stimulated samples show that incubation with an irrelevant plate-bound antibody of this isotype is not able to activate NK cell functions, as it is comparable to unstimulated sample (**e**). This control demonstrates that specific triggering of NK1.1 receptor by plate-bound antibodies of IgG2a isotype activates both NK cell degranulation and IFN-γ production

8. Resuspend the cell pellet in 200 μl FACS buffer and acquire samples with the flow cytometer (*see* **Note 9**).
9. Follow the gating strategy described in Fig. 1 to analyze IFN-γ production and CD107a expression by NK cells: use the control tube (splenocytes incubated in complete medium without any activating stimulus) to gate on lymphocytes on the basis of cell morphology (*see* **Note 10**), on live cells, on single cells (*see* **Note 11**), exclude T and B lymphocytes (CD3^{+}CD19^{+}) and gate on NK cells (NKp46^{+}). Use the IFN-γ APC/CD107a FITC plot to set a gate on this sample, that is, the negative

control. Apply these gates to all the other samples to study the effect of corticosterone addition on cytokine and NK1.1 stimulation on IFN-γ intracellular accumulation and degranulation of NK cells.

4 Notes

1. NK cells from C57BL/6 and SJL mouse strains are characterized by NK1.1 expression, while in other mouse strains, such as BALB/c, NK cells display no reaction to the widely used anti-NK1.1 antibody PK136, because of allelic variations in Nkrp1b and Nkrp1c [12]. This must be taken into account in the choice of the stimulating antibody when splenocytes from BALB/c mice are used.
2. When analyzing cells by flow cytometry it is important to exclude dead cells and restrict the analysis on viable cells. To do so, it is recommended to use amine reactive dyes (LIVE/DEAD fixable stains) that allow the distinction between live and dead cells. This distinction is based on the principle that amines on the outside of a living cell will react with the amine-reactive dye and have a dim fluorescence, while dead cells, that have lost cell membrane integrity, will allow the dye into the cell where it will react with amines throughout the cell resulting in a bright fluorescence. The advantage of using amine reactive dyes instead of cell-impermeant DNA binding dye, such as propidium iodide, is that it is compatible with subsequent cell fixation.
3. To confirm that the effect of the neuroendocrine mediators on NK cell stimulation is intrinsic, it is required to isolate NK cells from the mixed splenocytes populations and perform the experiment on pure NK cells. Moreover, this protocol can be extended to other ILC populations isolated from other organs, such as liver and small intestine, and the kind of activating stimuli can be adapted on the subset of interest of ILCs to analyze.
4. Stimulating NK cell surface receptors by coating the plate with antibodies is a useful tool to mimic receptor triggering by ligands. It is possible to use other antibodies in alternative to anti-NK1.1 antibody to activate NK cells, such as anti-NKp46 antibody. Moreover, coating each well with combinations of antibodies for multiple activating and inhibitory receptors allows the study of the effect of their synergy and/or inhibition.

5. NK cell incubation with IL-12 plus IL-18 is the classical stimulus to induce IFN-γ production, and it has been demonstrated that corticosterone is able to antagonize the effect of these cytokines ([3] and Fig. 1). It may be needed to study the effect of other cytokines or combinations of them. For example, the effect of neuroendocrine hormones on NK cell survival and proliferation can be investigated by incubating cells with IL-15 for longer time points. Moreover, the de novo expression of PD-1 on NK cells from the spleen can be induced by incubating cells for 48 h with corticosterone in combination with IL-15 and IL-18 [4].
6. A longer time of incubation may be required to activate ILCs and induce cytokine production. When it is the case, it is important to consider that prolonged incubation with GolgiPlug and GolgiStop may be toxic for the cells. To increase the detectability of intracellular cytokines while preserving cell viability it may be needed to add these protein transport inhibitors only in the last 4 h of stimulation.
7. To analyze the cell type of interest among splenocytes it is needed to stain lymphocytes with antibodies that allow their univocal identification. In the experiment described in this protocol, however, since splenocytes are stimulated with anti-NK1.1 antibody, it is not recommended to use the same antibody for FACS staining and NK cells are identified for their expression of NKp46. Thus, attention must be paid in the choice of the antigens for FACS analysis when cells have been previously stimulated with antibodies binding surface antigens.
8. To exclude from the flow cytometry analysis cell types that are not relevant, cells must be incubated with antibodies that recognize markers specific for these other lineages and not expressed by the cell type of interest. When the analysis is not focused on these cell types and the cell subset of interest is very rare in the sample (like tissue resident ILCs), it is useful to stain cells with antibodies for lineage (Lin) markers all conjugated with the same bright fluorochrome, to be able to easily exclude them. In this protocol Lin^+ cells are T and B lymphocytes ($CD3^+$ and $CD19^+$, respectively). Moreover, it is sometimes useful to include CD11b in the cytometry panel as an additional marker, to be able to exclude myeloid cells from the analysis. In this case, it must be taken into account that mature murine NK cells express this marker too. Therefore, one must check carefully that $NKp46^+$ $CD11b^+$ cells do not get excluded in the associated gating strategy.
9. It is important to have the correct controls to set up the flow cytometer and correctly compensate for any overlap in emission of each of the fluorochromes. The controls that is

recommended to prepare are unstained cells, single color (i.e., each antibody at a time) and fluorescence minus one (FMO) where all antibodies in the panel are added to cells, removing a single antibody in turn.

10. In the cytometer, cells flow past a set of laser light sources one cell at a time. While being interrogated by the lasers, each cell scatters light and the light scatter is measured by two optical detectors. One detector measures scatter along the path of the laser, that is, the forward scatter (FSC). FSC intensity is proportional to the diameter of the cell and is primarily due to light diffraction around the cell, allowing the discrimination of cells by size. The other detector measures scatter at a 90° angle relative to the laser, called side scatter (SSC), which usually correlates with cell complexity (granularity). These are general indications based on light refraction, but FSC and SSC depend also on the kind of sample, the sheath fluid and the laser wavelength. However, when measured in conjunction, these two parameters allow gating on the cell population of interest based on its physical properties end exclusion of debris, as they tend to have lower FSC levels.
11. As cells pass through the laser in the cytometer, photons are converted to current by the photomultiplier tube (PMT) and a voltage pulse is generated. The voltage pulse has three characteristics: area (the overall amount of fluorescence associated with an event), height (the peak signal), and width (the signal duration). When cell clumps pass through the laser it is interpreted as one event with the combined properties of all cells in the clump. To distinguish single events from clumps it is recommended to plot area vs width for forward scatter. In general, it is useful to collect area, height, and width for both forward and side scatter, and identify what is the best way to exclude cell clumps from the analysis depending on the kind of sample and the property of the instrument used.

Acknowledgments

S. Ugolini lab is supported by institutional grants from INSERM, CNRS, Aix-Marseille University and Marseille-Immunopole to the CIML and received funding from the European Research Council (ERC) under the European Union's Horizon 2020 research and innovation program, under grant agreement No. 648768; from the Agence Nationale de la Recherche (ANR) (No. ANR-14-CE14-0009-01) and from the ARC foundation (No. PGA120140200817). L. Quatrini has received funding from AIRC and from the European Union's Horizon 2020 research and innovation programme under the Marie Skłodowska-Curie grant agreement No 800924.

References

1. Weikum ER, Knuesel MT, Ortlund EA et al (2017) Glucocorticoid receptor control of transcription: precision and plasticity via allostery. Nat Rev Mol Cell Biol 18:159–174
2. Cain DW, Cidlowski JA (2017) Immune regulation by glucocorticoids. Nat Rev Immunol 17:233–247
3. Quatrini L, Wieduwild E, Guia S et al (2017) Host resistance to endotoxic shock requires the neuroendocrine regulation of group 1 innate lymphoid cells. J Exp Med 214:3531–3541
4. Quatrini L, Wieduwild E, Escaliere B et al (2018) Endogenous glucocorticoids control host resistance to viral infection through the tissue-specific regulation of PD-1 expression on NK cells. Nat Immunol 19:954–962
5. Vivier E, Tomasello E, Baratin M et al (2008) Functions of natural killer cells. Nat Immunol 9:503–510
6. Shapiro HM (1993) Trends and developments in flow cytometry instrumentation. Ann N Y Acad Sci 677:155–166
7. Assenmacher M, Schmitz J, Radbruch A (1994) Flow cytometric determination of cytokines in activated murine T helper lymphocytes: expression of interleukin-10 in interferon-gamma and in interleukin-4-expressing cells. Eur J Immunol 24:1097–1101
8. Elson LH, Nutman TB, Metcalfe DD et al (1995) Flow cytometric analysis for cytokine production identifies T helper 1, T helper 2, and T helper 0 cells within the human CD4+CD27− lymphocyte subpopulation. J Immunol 154:4294–4301
9. Jung T, Schauer U, Heusser C et al (1993) Detection of intracellular cytokines by flow cytometry. J Immunol Methods 159:197–207
10. Prussin C, Metcalfe DD (1995) Detection of intracytoplasmic cytokine using flow cytometry and directly conjugated anti-cytokine antibodies. J Immunol Methods 188:117–128
11. Sander B, Hoiden I, Andersson U et al (1993) Similar frequencies and kinetics of cytokine producing cells in murine peripheral blood and spleen. Cytokine detection by immunoassay and intracellular immunostaining. J Immunol Methods 166:201–214
12. Carlyle JR, Mesci A, Ljutic B et al (2006) Molecular and genetic basis for strain-dependent NK1.1 alloreactivity of mouse NK cells. J Immunol 176:7511–7524

Chapter 9

Role of G-Protein Coupled Receptors in Chemotaxis of Innate Lymphoid Cells

Brinda Bhatt, Huabin Zhu, Nikhil Patel, and Nagendra Singh

Abstract

Innate lymphoid cells (ILCs) are a recently identified family of immune cells mostly present at barrier surfaces. They play an important role in the induction, regulation, and resolution of inflammatory responses. Environmental signals play an important role in development and function of ILCs. G-protein coupled receptors (GPCRs) sense and mediate cellular responses to the environmental signals. ILCs express several G-protein coupled receptors, which play a critical role in migration of these cells to appropriate sites. Here, we describe a method to test the migration of ILCs toward 7α,25-hydroxycholesterol, which is mediated by cell surface-expressed GPR183. A similar strategy can be employed to test the role of other GPCRs in mediating the migration of ILCs toward other chemotactic ligands.

Key words Innate lymphoid cells (ILCs), Intestine, Migration, G-protein coupled receptors (GPCRs)

1 Introduction

Innate lymphoid cells, or ILCs, are a family of immune cells present primarily at barrier surfaces of the body. While of lymphoid origin like T cells, they lack specific antigenic receptor and are part of the innate response to pathogens [1]. ILCs can be divided into three distinct groups based on expression of master transcription factors and cytokine production that mirror effector T cell lineages: ILC1 express T-bet and produce IFN-γ and TNF-α, ILC2 express GATA3 and produce both IL-5 and IL-13, and ILC3 express RORγt and produce IL-22 and IL-17 [1]. ILC3 play a key role in both promoting and restraining colonic inflammation in part by differential production of the proinflammatory IL-17 and barrier protective IL-22 [2–6]. Furthermore, we have previously shown in our laboratory that IL-17-producing ILC3 promote inflammatory conditions in the colons of *Gpr109a*$^{-/-}$*Rag1*$^{-/-}$ mice, leading to spontaneous colitis [7].

Shoba Amarnath (ed.), *Innate Lymphoid Cells: Methods and Protocols*, Methods in Molecular Biology, vol. 2121, https://doi.org/10.1007/978-1-0716-0338-3_9, © Springer Science+Business Media, LLC, part of Springer Nature 2020

ILC3 have been recently shown to express GPR183, a G-protein coupled receptor that binds to 7α,25-hydroxycholesterol (7α,25-OHC) an oxysterol [8]. Furthermore, 7α,25-OHC and GPR183 interaction play an important role in localization of ILC3 to cryptopatches (CPs) and isolated lymphoid follicles (ILFs) [9]. The following is a protocol for the isolation of ILC3 from the colon and small intestines of mice, followed by a migration assay using 7α,25-OHC.

2 Materials

Prepare and store all reagents at room temperature, unless indicated otherwise. When disposing of waste, follow appropriate biosafety protocols.

1. Complete media: RPMI1640 containing 5% FBS, penicillin (100 U/ml), streptomycin (100 μg/ml) and 10 mM HEPES (pH 7.4).
2. Migration medium: RPMI1640 containing 0.5% fatty acid-free BSA (Sigma Cat. #A8806-5G), penicillin (100 U/ml), streptomycin (100 μg/ml) and 10 mM HEPES (pH 7.4).
3. Epithelial isolation buffer: Hanks balanced salt solution (HBSS) containing 5% FBS, 5 mM EDTA (pH 8.0), 10 mM HEPES (pH 7.4) (*see* **Note 1**).
4. Dulbecco's Phosphate Buffered Saline (PBS).
5. Tea strainer.
6. Collagenase D (Roche Cat. #11088882001): Prepare a stock solution of 100 mg/ml in RPMI1640 and store it in −80 °C.
7. DNase I (Roche Cat. #10104159001): Prepare a stock solution of 1 mg/ml in RPMI1640 and store it in −80 °C.
8. Percoll. Before use, dilute to a working concentration of 40% into RPMI1640 (*see* **Note 2**).
9. Flow cytometry staining buffer: PBS containing 10 mM HEPES (pH7.4) and 1% FBS.
10. Flow cytometry antibodies. Store at 4 °C (*see* **Note 3**).
11. Cell strainer: 70 μm (*see* **Note 4**).
12. 24 mm transwell (Corning Cat. #3421) with 5 μm pore size.
13. 7α,25-Dihydroxycholesterol (Avanti Polar Lipids Cat. #700080P): Dissolve into DMSO at a concentration of 100 μg/ml and store at −20 °C. For assay, dilute to 100 nM into migration medium in a sterile hood. Mix well.
14. Anti-mouse CD11c (clone: N418).
15. Anti-mouse CD11b (clone: M1/70).

16. Anti-mouse TER119 (clone: TER119).
17. Anti-mouse CD3ε (clone: 145-2C11).
18. Anti-mouse CD5 (clone: 53-7.3).
19. Anti-mouse CD19 (clone: 1D3).
20. Fixable viability dye (such as BioLegend Cat. #423111, or other suitable one).
21. Anti-mouse CD45.2 (clone: 104, if using CD45.1 mouse, use the appropriate clone).
22. Anti-mouse Thy1.2 (30-H12, if using Thy1.1 mouse, use the appropriate clone).
23. Anti-mouse CD127 (clone: A7R34).
24. Anti-mouse NKP46 (clone: 29A1.4).
25. Anti-mouse RORγt (clone: 4G419).
26. Anti-mouse GATA3 (clone: 16E10A23).
27. Cell fixation and permeabilization buffer (ThermoFisher Cat. #00-5521-00) (*see* **Note 5**).

3 Methods

3.1 Isolation of Intestinal Lamina Propria (ILPs) Cells

1. Prewarm the complete media, HBSS and PBS buffer to room temperature.
2. Pour 20 ml of PBS into 100 mm petri dishes.
3. Remove the colons and small intestines from mice.
4. Using scissors and tweezers, remove all of the mesenteric fat from the outside of the colons and small intestines. Remove Peyer's patches, which are located at the wall of small intestine.
5. Cut open the intestines and wash with PBS in order to remove luminal contents. Ensure the tissue does not dry by means of the PBS in the petri dishes.
6. Cut the intestines into approximately 1 cm pieces and transfer into 20 ml of epithelial isolation buffer.
7. Shake at 150 rpm in 37 °C for 20 min.
8. Vortex the sample at maximum speed for 1 min, then filter with a tea strainer.
9. Collect the tissue from the filter, as those in the filtrate will be epithelial cells. Put tissues into epithelial isolation buffer. Discard the filtrate.
10. Repeat **steps 7–9**.
11. Wash the tissue with PBS in order to remove EDTA (*see* **Note 6**).

12. Dilute the collagenase D to 1 mg/ml and DNAse to 10 μg/ml in prewarmed complete media, then collect the remaining tissue and transfer it into this solution.
13. Shake at 150 rpm for 30 min, at 37 °C.
14. Vortex the sample at maximum speed for 1 min again, and then filter through a steel strainer.
15. Collect the filtrate and keep it on ice.
16. Repeat **steps 12–15** one more time.
17. Combine the filtrates from **steps 15** and **16** into one.
18. Centrifuge the sample 1300 × *g* for 10 min at room temperature, then decant the supernatant and break the pellet.
19. As the sample is spinning, dilute Percoll into complete media at concentration of 40%. Resuspend the pellet into 5 ml of 40% Percoll.
20. Centrifuge the cell suspension at 800 × *g* for 12 min at room temperature. Once the spin is complete, carefully remove the tube from the centrifuge so as not to disturb the cells. The contaminating epithelial cells will float on the top of Percoll. Carefully remove the epithelial layer and Percoll without disturbing the pellet. The cells pelleted at the bottom of the tube are ILPs which contain ILCs (*see* **Note** 7).
21. Break the pellet and resuspend the cells in 5 ml of complete media. Filter the cells through 70 μm strainer. Spin cells at 650 × *g* for 5 min. Decant the supernatant, and break the pellet.
22. Repeat **step 21** one more time.

At this step cells are ready for migration assay. To preserve cell viability and function it is better to use total ILPs. If sorting is needed, a negative sorting using magnetic method is recommended. Flow cytometry-based sorting causes mechanical stress on cells that affects their activity.

3.2 In Vitro Migration Assay

1. Resuspend the ILPs in migration media.
2. Dilute 7α,25-OHC to a final concentration of 100 nM into the migration media and carefully pipet this solution (0.6 ml) to the lower chamber of 24 mm transwell plate. Be sure to avoid bubbles.
3. At the same time, set up a control well under the same conditions, but with migration media lacking 7α,25-OHC in the lower chamber of 24 mm transwell plate.
4. Pipet the cells (1×10^6 in 0.1 ml) to the upper chambers of 24 mm transwell plate in 500 μl migration media.

5. Incubate the plates for 4 h at 37 °C. During this incubation ILCs will migrate to the lower chamber.
6. Collect the cells in the lower chamber and stain them as following:

   Lineage markers (CD11c, CD11b, Ter119, CD3, CD5, CD19), fixable viability dye, CD45.2, Thy1.2, CD127, NKP46, RORγt, and GATA3. RORγt and GATA3 are transcription factors, and staining for them is performed following fixation and permeabilization with the kit from ThermoFisher (Cat. #00-5521-00) (*see* **Note 8**).

   ILCs are identified as lineage$^-$CD45.2$^+$Thy1.2$^+$CD127$^+$ live cells. Among ILCs, different subsets of ILCs are identified by the following markers: RORγt$^+$ for ILC3, GATA3$^+$ for ILC2, and NKP46$^+$ RORγt$^-$ for ILC1.
7. Analyze the stained cells using a flow cytometer.
8. Determine the frequency and number of different ILCs subsets in cells migrated to lower chamber in the presence or absence of 7α,25-OHC.
9. Calculate the relative migration of different ILCs subsets to 7α,25-OHC as following; Number of a given ILCs subset migrated in the presence of 7α,25-OHC divided by the number of cells that migrated in its absence. Alternatively, frequencies of different ILCs subsets migrated in the presence or absence of 7α,25-OHC may be presented to show their chemotaxis in response to 7α,25-OHC.
10. Intestinal ILC2 and ILC3 express Gpr183 and migrate toward 7α,25-OHC (see **Note 9**).

4 Notes

1. Dilute the Collagenase D/DNase in HBSS fresh the day of the experiment.
2. Make sure the amount of Percoll pipetted is correct, as it can be very viscous. Be sure to pipet slowly and carefully with the serological pipette.
3. When using multiple antibodies for post-migration stains, ensure that the combination of fluorophores have unique excitation/emission spectra in order to avoid compensation issues with the data. Additionally, look up the data sheets corresponding to the antibodies to determine the correct working dilution to apply to the cells.
4. Make sure to use sterile tea strainer that has been autoclaved in order to avoid contaminating your sample.

5. Intranuclear staining often results in loss of some cells during fixation, permeabilization and additional washing steps. Intranuclear staining for RORγt and GATA3 can be completely avoided if using RORγt (Rorγt-GFP) and GATA3 (GATA3-GFP) reporter mice to identify ILCs.
6. This wash step is critical. EDTA is inhibitor of collagenase D and therefore must be removed before adding collagenase D. Wash the tissue with PBS very well.
7. The layers of epithelial and other cells can also be removed by careful vacuuming. However, do not vacuum too far down the tube and risk losing the pelleted immune cells.
8. The key difference between T cells and ILCs is expression of the TCR/CD3 complex by the former. The majority of intestinal lamina propria cells are T and B cells, which sometime represents a challenge to unequivocally identify ILCs, which represent a minority population. Use of $Rag1^{-/-}$ or $Rag2^{-/-}$ mice, which lack T and B cells, is a great tool for the identification and characterization of ILCS.
9. In order to demonstrate the specificity of this migration, use ILPs isolated from both C57BL/6 and $Gpr183^{-/-}$ mice. Inability of knockout cells to migrate in the presence of chemotactic stimuli will provide evidence that this movement requires Gpr183 signaling.

References

1. Artis D, Spits H (2015) The biology of innate lymphoid cells. Nature 517:293
2. Zheng Y et al (2008) Interleukin-22 mediates early host defense against attaching and effacing bacterial pathogens. Nat Med 14:282–289
3. Tumanov AV et al (2011) Lymphotoxin controls the IL-22 protection pathway in gut innate lymphoid cells during mucosal pathogen challenge. Cell Host Microbe 10:44–53
4. Bernink JH et al (2013) Human type 1 innate lymphoid cells accumulate in inflamed mucosal tissues. Nat Immunol 14:221–229
5. Takayama T et al (2010) Imbalance of NKp44 (+)NKp46(−) and NKp44(−)NKp46(+) natural killer cells in the intestinal mucosa of patients with Crohn's disease. Gastroenterology 139:882–892, 892.e881–883
6. Buonocore S et al (2010) Innate lymphoid cells drive interleukin-23-dependent innate intestinal pathology. Nature 464:1371–1375
7. Bhatt B et al (2018) Gpr109a limits microbiota-induced IL-23 production to constrain ILC3-mediated colonic inflammation. J Immunol 200:2905–2914
8. Chu C et al (2018) Anti-microbial functions of group 3 innate lymphoid cells in gut-associated lymphoid tissues are regulated by G-protein-coupled receptor 183. Cell Rep 23:3750–3758
9. Emgard J et al (2018) Oxysterol sensing through the receptor GPR183 promotes the lymphoid-tissue-inducing function of innate lymphoid cells and colonic inflammation. Immunity 48:120–132.e128

Chapter 10

Group 2 ILC Functional Assays in Allergic Airway Inflammation

Lewis J. Entwistle, Franz Puttur, Lisa G. Gregory, and Clare M. Lloyd

Abstract

ILC2s are a rare innate cell population capable of rapidly producing type 2 cytokines prior to the recruitment and expansion of adaptive type 2 T helper cells. As a result, they are implicated in the pathogenesis of many type-2 immune-mediated diseases, including allergic airway inflammation. Here we describe methods for interrogating and analyzing ILC2 biology in the context of allergic airway inflammation, such as flow cytometric analysis of mouse and human ILC2s as well as live imaging of pulmonary ILC2s.

Key words ILC2, Asthma, Allergic airway inflammation, Lung, Flow cytometry, Microscopy, Imaging

1 Introduction

Type 2 immune responses are orchestrated by the type 2 cytokines IL-4, IL-5, and IL-13 and are broadly associated with tissue repair, allergic inflammation and defense against large multicellular pathogens, such as helminths. These cytokines mediate a suite of physiological and immunological changes by acting on a multitude of immune and stromal cells. These changes often include mobilization of granulocytes, M2/alternate polarization of macrophages, B cell expansion and class switching, goblet cell hyperplasia and mucus hypersecretion, and smooth muscle contraction [1, 2].

Since their discovery, ILC2s have been implicated in many different diseases. Their enrichment at mucosal sites, such as the lung, and responsiveness to alarmin cytokines renders them perfectly poised to rapidly respond to epithelial damage and initiate type 2 immunity. With their ability to produce the type 2 effector cytokines, IL-5 and IL-13, ILC2s are strongly implicated in type 2 immune-mediated diseases [3].

Allergic airway inflammatory diseases, such as asthma, exhibit characteristic type 2 pulmonary inflammation induced following

Shoba Amarnath (ed.), *Innate Lymphoid Cells: Methods and Protocols*, Methods in Molecular Biology, vol. 2121, https://doi.org/10.1007/978-1-0716-0338-3_10,

exposure to normally inert external stimuli, such as allergens. ILC2s are potently induced in the lung and required in some mouse models of allergic airway inflammation (AAI) through their production of type 2 cytokines, driving eosinophilia, mucus hyperproduction, and airway hyperresponsiveness [4–8]. In addition, they have been shown to be increased in skin, airways and nose of atopic subjects [9–14]. As a result, ILC2 research in the context of AAI has been of great interest in the field of type 2 disease immunology. Indeed, recent findings have suggested that pulmonary ILC2s drive M2/alternate macrophage polarization in the lungs of eosinophilic asthmatic patients [9]. Furthermore, both mouse and human ILC2s express the neuropeptide receptor NMURI [15] and localize with pulmonary neuroendocrine cells in the lung, which are capable of stimulating ILC2 cytokine production and amplifying allergic responses [16].

Here we describe methods to interrogate and analyze pulmonary ILC2 biology in both mouse models of AAI and human disease. Specifically, we describe methods to analyze the phenotype, location, and dynamics of pulmonary ILC2s in mice as well as isolate and analyze ILC2s from human blood.

2 Methods

2.1 Flow Cytometry of ILC2s

2.1.1 Materials

1. Complete RPMI media (cRPMI): RPMI supplemented with fetal calf serum (10%), penicillin/streptomycin (100 U/mL), and L-glutamine (2 mM).
2. Lung digest media: cRPMI (as above) supplemented with collagenase (0.15 mg/mL; Type D, Roche Diagnostics) and deoxyribonuclease (25 mg/mL; Type 1, Roche Diagnostics).
3. Staining Buffer: PBS supplemented with BSA (1% w/v) and sodium azide (0.05%).
4. Red blood cell lysis buffer: to 1 L of deionized water add 8.29 g ammonium chloride, 1 g of potassium bicarbonate, and 37.2 mg disodium-EDTA. Adjust pH to 7.2–7.4.
5. Stimulation mix: cRPMI supplemented with PMA (20 ng/mL; Sigma-Aldrich), ionomycin-free acid (1.5 μg/mL) from *Streptomyces conglobatus* (Merck), and Brefeldin A (5 μg/mL; Sigma-Aldrich).
6. RosetteSep™ Human ILC2 Enrichment kit.
7. Lymphoprep™.
8. Suggested mouse flow cytometry panel.

Antibody	Clone	Fluorochrome	Manufacturer	Use
CD45	30-F11	AF700	BioLegend	Extracellular
CD11b	M1/70	APC	BioLegend	Lineage
CD11c	N418	APC	BioLegend	Lineage
CD19	ebio ID3	APC	BioLegend	Lineage
CD5	53-7.3	APC	BioLegend	Lineage
F4/80	BM8	APC	BioLegend	Lineage
FCεR1	MAR1	APC	BioLegend	Lineage
GR-1	RB6-8C5	APC	BioLegend	Lineage
TCR-β	H57-597	APC	BioLegend	Lineage
TCR-γδ	ebio GL3	APC	eBioscience	Lineage
TER-119	TER-119	APC	BioLegend	Lineage
CD3e	145-2C11	APC-Cy7	BioLegend	Lineage
IL-5	TRFK5	BV421	BioLegend	Intracellular cytokine
CD25	PC61	BV510	BioLegend	Extracellular
CD90.2	53-2.1	BV605	BioLegend	Extracellular
CD127	A7R34	BV711	BioLegend	Extracellular
Sca-1	D7	BV785	BioLegend	Extracellular
ST2	DJ8	FITC	MD Biosciences	Extracellular
IL-13	ebio 13A	PE	eBioscience	Intracellular cytokine
KLRG1	2F1/KLRG1	PE-Cy7	BioLegend	Extracellular
NKp46	29A1.4	PE-Dazzle	BioLegend	Lineage
Gata3	TWAJ	PerCP-eFluor 710	eBioscience	Intracellular transcription factor
LIVE/DEAD™ Fixable Blue Dead Cell Stain	n/a	UV	Invitrogen	Viability dye

*2.1.2 Isolating Cells from Lungs, Lymph Nodes, and Bronchoalveolar Lavage Fluid of Mice (Section 3, **item 1** and **item 2**).*

1. Following euthanasia by overdose of pentobarbital, adult mice are lavaged with three times 0.4 mL PBS to recover airway lumen cells before isolating and removing the lungs and mediastinal (lung draining) lymph nodes (mLN). Lungs and mLN are collected in cRPMI on ice. (Section 3, **item 4**).
2. BAL cells are centrifuged at 200 × *g* for 5 min at 4 °C. Cells are then resuspended in 500 μL cRPMI and left on ice until required.
3. mLNs are processed into a single cell suspension by maceration through a 70 μM cell strainer into a 50 mL tube. The strainer is then washed 2× with 5 mL cRPMI before centrifugation at 200 × *g* for 5 min at 4 °C. Cells are then resuspended in cRPMI. Single cell suspension of mLN is to be left on ice until required.
4. Lung lobes are finely chopped, resuspended in 5 mL of lung digest media and incubated in a 37 °C water bath under agitation for 1 h.
5. The digested lung tissue is forced through a 70 μM cell strainer. The strainer is washed 2× with 5 mL cRPMI before centrifugation at 200 × *g* for 5 min at 4 °C.
6. The lung cell pellet is then resuspended in 3 mL red blood cell lysis buffer for 4 min at room temperature. Approximately 10 mL of cRPMI is then added to prevent further cell lysis and the cells centrifuged at 200 × *g* for 5 min at 4 °C.
7. The lung cell pellet is resuspended in cRPMI as a single cell suspension.
8. Count all single cell suspensions.

2.1.3 Re-Stimulation for Cytokine Production

If intracellular cytokine staining is not required, proceed to Subheading 2.1.4.

1. Add 2–4 × 10^6 cells for each sample to a round-bottom 96-well plate, adding extra samples for appropriate fluorescence-minus-ones (FMOs) and single-stain controls. Centrifuge at 200 × *g* for 5 min at 4 °C and discard supernatant. (Section 3, **item 6**).
2. Add 200 μL of stimulation mix to each well and resuspend cells. Incubate at 37 °C, 5% CO_2 for 4 h. (Section 3, **item 5**).

2.1.4 Antibody Staining for ILC2s

If re-stimulation not performed please start with i.

i. Add 2-4 × 10^6 cells for each sample to a round-bottom 96-well plate, adding extra samples for appropriate FMOs and single-stain controls.

1. Centrifuge cells at 200 × *g* for 5 min at 4 °C and discard supernatant.

2. Wash cells in PBS by adding 200 μL per well. Centrifuge at 200 × *g* for 5 min at 4 °C and discard supernatant.
3. Stain appropriate wells with 50 μL fixable live/dead marker for 20 min at room temperature (protected from light).
4. Add 150 μL of PBS per well and centrifuge at 200 × *g* for 5 min at 4 °C and discard supernatant.
5. Wash as above in 2.
6. Resuspend cells in appropriate cell surface stain cocktail (50 μL per well) and incubate at 4 °C (protected from light) for 20 min.
7. Add 150 μL of stain buffer per well and centrifuge at 200 × *g* for 5 min at 4 °C and discard supernatant.
8. Wash cells in stain buffer by adding 200 μL per well. Centrifuge at 200 × *g* for 5 min at 4 °C and discard supernatant. *If fluorescence-activated cell sorting live ILC2s, resuspend cells in flow buffer and stop here.*
9. Fix cells using eBioscience intranuclear fix/perm according to the manufacturer's instructions. *See notes for information about pausing here.*
10. Add 150 μL of stain buffer per well and centrifuge at 200 × *g* for 5 min at 4 °C and discard supernatant.
11. Permeabilize cells by suspending in 1× perm buffer (eBioscience), according to the manufacturer's instructions.
12. Resuspend cells in appropriate intracellular stain cocktail (50 μL per well) and incubate at 4 °C (protected from light) for 30 min.
13. Add 150 μL of 1× perm buffer per well and centrifuge at 200 × *g* for 5 min at 4 °C and discard supernatant.
14. Wash cells in 1× perm buffer by adding 200 μL per well. Centrifuge at 200 × *g* for 5 min at 4 °C and discard supernatant.
15. Resuspend cells in 200 μL stain buffer and analyze by flow cytometry.

2.1.5 Gating Strategies for Defining ILC2s

1. Mouse ILC2s for analysis: Single cells, live cells, $CD45^+$ cells, lymphocytes, $Lineage^-$, $Gata3^+$ (Fig. 1a, b). (Section 3, **item** 7).
2. Mouse ILC2s for fluorescence-activated cell sorting:
   - If not using a reporter mouse: Single cells, live cells, $CD45^+$ cells, lymphocytes, $Lineage^-$, $CD90.2^+$, $CD25^+$, $KLRG1^+$, $Sca\text{-}1^+$ (Fig. 1a, c).
   - If using a reporter mouse: Single cells, live cells, $CD45^+$ cells, lymphocytes, $Lineage^-$, *Il13*-$reporter^+$/*Gata3*-$reporter^+$ (Fig. 1a, d).

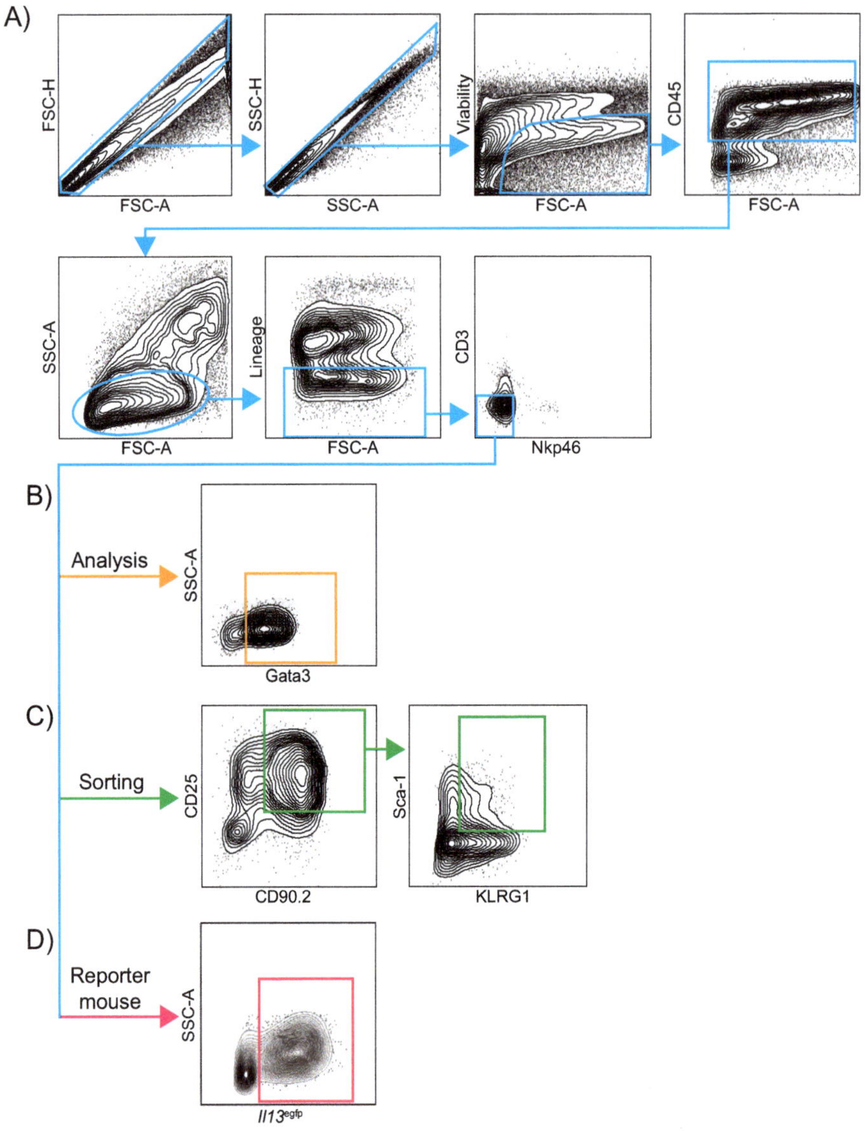

Fig. 1 Mouse pulmonary ILC2 gating strategies. (**a**) Gating Schematic. ILCs are defined as Single, Live, CD45^{+}, Lymphoid, Lineage^{neg} (TCRβ, TCRγδ, CD3e, CD5, CD19, CD11b, CD11c, FCεR1, GR-1, F4/80, and TER-119) negative, CD3^{-}, NKp46^{-}. (**b**) ILC2s are further defined as expressing GATA3. (**c**) If not using a reporter mouse ILC as in (**a**) are sorted with the additional markers CD90.2^{+}, CD25^{+}, KLRG1^{+}, Sca-1^{+}. (**d**) If using an *Il13*-reporter^{+}/*Gata3*-reporter^{+} reporter mouse, ILCs as in (**a**) are sorted with the additional marker, eGFP in this example

Transcription factor staining (Gata3) defines an ILC2. However, we consider "functional" cells to be those with the capacity to secrete type 2 cytokines which drive disease pathology [4, 6, 7].

2.2 Imaging Mouse ILC2s In Situ (PCLS and Intravital Microscopy (IVM))

ILC2s play an essential role in orchestrating immune responses in mucosal tissues such as the lungs. Hence, understanding the spatiotemporal organization of ILC2, visualizing their behavior and dynamic interaction with lung immune and structural cells in a 3D lung microenvironment is vital to evaluate their function in inflammation [17, 18]. In this section we discuss imaging modalities that can be applied to evaluate ILC2 behavior in the inflamed lungs.

2.2.1 Materials Required for PCLS

1. Mice of interest.
2. Respiratory protective equipment (RPE; current standard FFP2/3 or as advised by Occupational Health) appropriate for person carrying out the protocol.
3. Anesthetic (e.g., Euthatal).
4. 70% (v/v) ethanol.
5. Phosphate-buffered saline (PBS).
6. Triton X-100.
7. Normal Goat Serum (NGS).
8. Bovine serum albumin (BSA).
9. Dissection board.
10. Dissection instruments.
11. Suture (e.g., 3-0 silk).
12. 1 and 2 mL sterile syringes.
13. 27 G Hypodermic Needles.
14. Microwave.
15. Low Melting Point (LMP) Agarose.
16. Tissue (Blue Roll).
17. Externally vented Class II (micro) biological safety cabinet (BSC; MSC).
18. 16% methanol free Formaldehyde (sold in 10 mL glass ampules), diluted 1:4 in PBS (v/v) to give 4% working solution.
19. Water bath (37–40 °C); located near externally vented class II BSC).
20. 1.1 mm diameter cannula.
21. 50 mL disposable centrifuge tubes.
22. Wet Ice.
23. Compresstome® VF-300 Tissue Slicer (Precisionary Instruments).
24. Stainless steel blades from World Precision Instruments (Cat. #7550-1-SS).
25. Antibodies (Ab).

26. Sodium Azide (1%; VWR Cat. #786-299).
27. Superglue.
28. Scalpel.

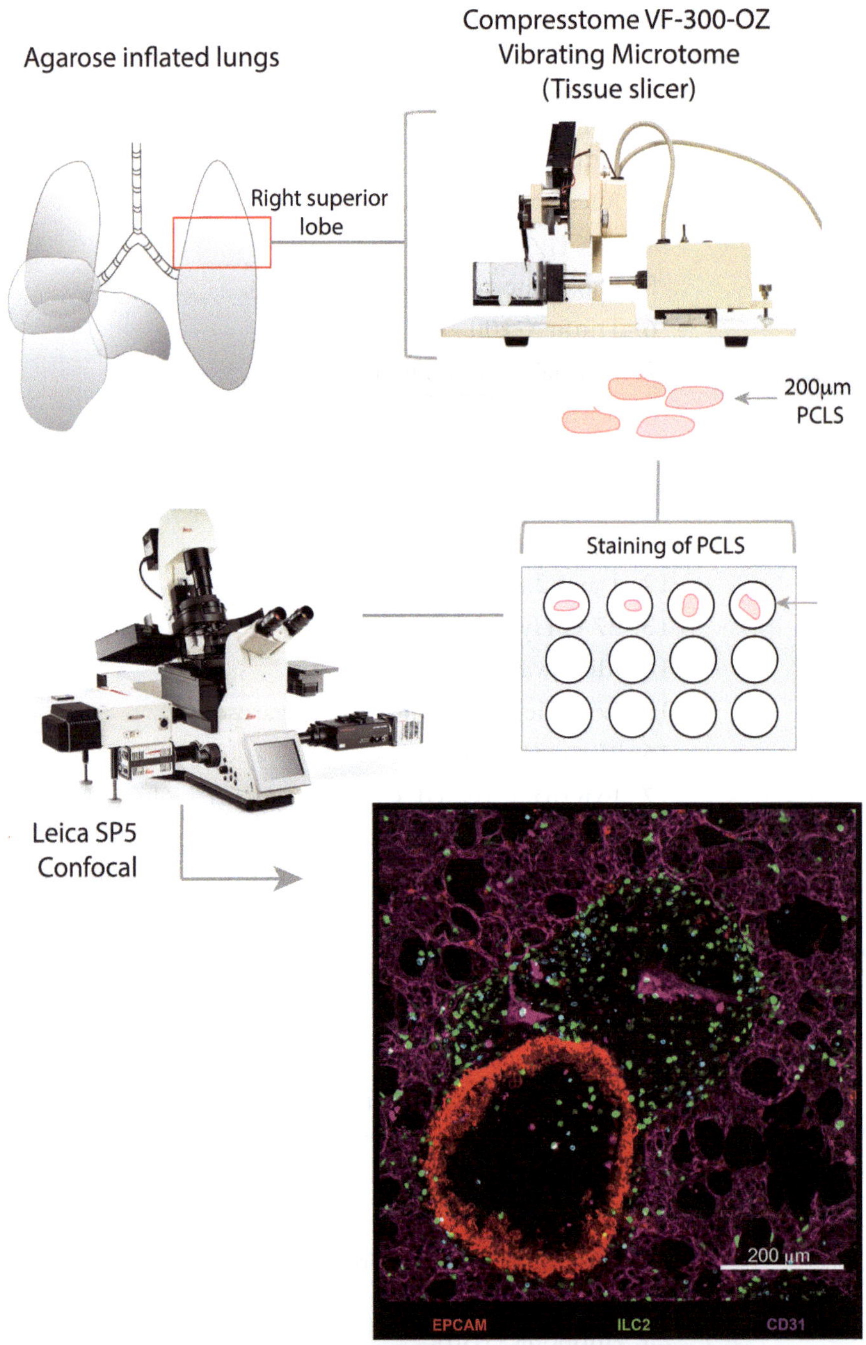

Fig. 2 Methodology for imaging precision cut lung slices

2.2.2 Detailed PCLS Protocol (Depicted in Fig. 2)

Isolating Lung Lobe from Mouse

1. Prepare 2% (w/v) low melting point agarose solution in PBS (microwave to dissolve). Place in water bath at 37–40 °C to prevent gelling and allow temperature to equilibrate.
2. Place cage in safety cabinet. Euthanize mouse by overdose of anesthetic by i.p. injection with an anesthetic agent (e.g. Euthatal) appropriate for the size of mouse (do not use cervical dislocation or CO_2 overdose as this will affect tissue morphology). (Section 3, **item 3**).
3. Ensure that the mouse does not respond to paw pinch and respiratory arrest has occurred.
4. Introduce cannula into trachea via tracheotomy. Seal the tracheotomy site with suture applied around the trachea below the insertion site of the cannula.
5. Slowly but continuously fill lungs with approx. 0.8–1 mL agarose solution, chest wall is still intact so should be easy to feel that you are not overinflating. (Filling the lungs with agarose before dissection, fixing and slicing, helps preserve the lung structure.)
6. Remove cannula and tie suture to stop leakage.
7. Place ice over chest cavity until cool (2–3 min).
8. Dissect to remove lungs heart and trachea "en bloc" and place in HBSS containing 2.5% HEPES (GIBCO, Life Technologies).
9. Carefully remove the left lobe from the lungs and cut the lower quarter of the lobe off with a scalpel to leave a flat surface.
10. Carefully fix the isolated lobe onto the slicing mount with superglue.
11. Once the glue is dry, embed the mounted lobe in LMP agarose.
12. Once the agarose is set, slice lung lobes to desired thickness (usually 200 μm) using the Compresstome® VF-300 Tissue Slicer. Slices are collected in HBSS containing 2.5% HEPES (GIBCO, Life Technologies).

Staining Live Lung Slices

1. Preincubate lung slices in complete RPMI 1640 medium (RPMI + 10% HFCS + 1% Pen/Strep) for 15 min to remove excess agarose. This is performed in a 24-well plate containing 1 mL RPMI medium per well. Up to two slices can be incubated per well.
2. Following the washing step, aspirate the media and directly incubate the lung slices in primary Ab (unconjugated) 1:100–1:500 or fluorochrome antibody (preconjugated) depending on clone and antibody concentration. Incubate slices for 1.5 h at 37 °C and 5%CO_2. This is performed in a

24-well plate containing 1 mL RPMI medium per well. Up to three slices can be stained per well.

3. Aspirate the primary antibody solution and wash with 2 × 1 mL media by gently adding and withdrawing media in the wells containing the lung slices. (Washing steps can be performed in serum free RPMI media.)
4. If fluorochrome labeled secondary antibody is required, then incubate with secondary antibody for 1 h at 37 °C and 5% CO_2.
5. Repeat wash steps as in **item 3** (under heading staining live lung slices) and transfer the stained slices into μ-Plate 24 Well Black staining plates. Cat. #82406 (Thistle Scientific).
6. Imaging is performed using an inverted confocal microscope.
7. Slices can be imaged for up to 24 h post staining. However, to image cell movement slices should be analysed within the first 4 h for optimal cell movement. (Section 3, **item 9**).

Staining Fixed Lung Slices

1. Following the process of slicing immediately fix in 4% formaldehyde in PBS at room temperature (RT) for 20 mins or at 4 °C for 2 h (incubation time is epitope dependent).
2. At this stage further mice may be dissected to provide additional lung lobes for analysis, make note of what time each slice enters the formaldehyde.
3. Wash 2× in PBS for 5 mins on shaker.
4. Wash slices in 1× PBS for 2 h at 37 °C with agitation (to remove agarose).
5. Permeabilize for 1–2 h at RT on shaker in PBS/BSA 1%/NGS 10%/Triton X-100 0.3%/sodium azide 0.05%.
6. Primary Ab 1:100–1:500 (depending on titer of individual antibody clone) in PBS/BSA 1%/NGS 10%/Triton X-100 0.1%/sodium azide 0.05% 24-48 h (incubation time is epitope dependent) on shaker at 4 °C.
7. Rinse once and wash 3× for 15 min on shaker with PBS/BSA 1%/Triton X-100 0.1%/azide 0.05% at RT.
8. Secondary Ab 1:500–1:1000 depending on Ab in PBS/BSA 1%/NGS 10%/Triton X-100 0.1%/sodium azide 0.05% 2 h on shaker at RT (can add DAPI at this stage).
9. Rinse once and wash 3× for 15 min on shaker with PBS/BSA 1%/Triton X-100 0.1%/sodium azide 0.05% at RT.
10. Wash 2× in PBS for 5 mins on shaker.
11. Mount and visualize or store at 4 °C prior to imaging.

2.2.3 Confocal Image Acquisition of PCLS

1. Transfer live or fixed slices to 24-well ibidi imaging plates.
2. Image up to three slices per well of the μ-Plate 24 Well Black staining plates. Cat. #82406 (Thistle Scientific).
3. In our lab, all images are generated by an inverted laser-scanning confocal microscope (SP5, Leica microsystems).
4. Perform imaging using lasers required to excite the fluorochrome of choice. Lasers available on the SP5 are as follows: UV 405 nm, argon-ion 488 nm, Diode 543 nm, and Diode 633 nm.
5. Acquire images as 1024 × 1024 pixel frames in ten scanning steps using an inverted 20× dry objective, pinhole 60 μm.
6. For time-series images generate *z*-stack images over 45 min to 1 h.
7. Videos are quantified using Imaris version 8.1 (BitPlane, Oxford Instruments) and ILC2 track speed, length and displacement are quantified using Motilitylab software.

2.2.4 Intravital Method for Visualizing ILC2

Imaging set-up for IVM on the murine lung was adapted and modified from a previously described protocol [19] *and discussed in* [20].

1. For IVM studies small custom-built circular suction chambers were developed for surgical insertion into the thorax of terminally anesthetized mice similar to previous described [17, 19].
2. Mechanically ventilate the mouse and place on their right lateral side on a 37 °C heat mat.
3. Make an incision between the ribs (mouse's left side) and gently insert the suction chamber. Gentle suction is applied via the vacuum port to stabilize the lung tissue (Fig. 3).
4. Acquire images using a long working distance water immersion objective and an upright laser scanning confocal microscope with Hybrid Detectors (Leica SP5).
5. All of these methods cause an initial pneumothorax; however, positive end-expiratory pressure and gentle suction stabilize

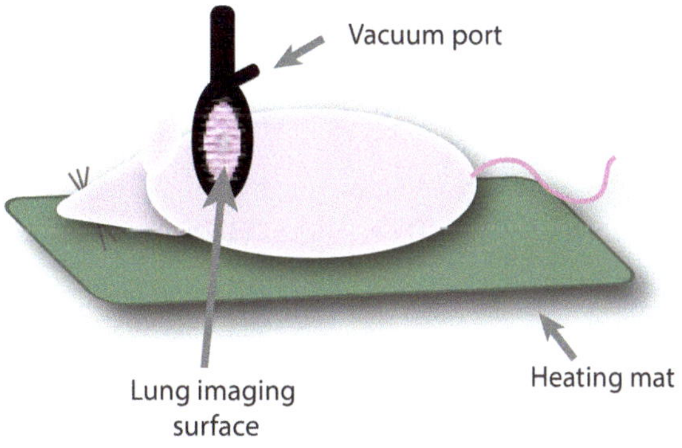

Fig. 3 Mouse setup for intravital microscopy

the lung tissue at the time of surgery, thus restoring normal pressure.

6. Fluorescent labeled anti-CD31 can be administered intravenously to label the endothelium.
7. Time-lapse videos depicting ILC2 movement can be generated through a 0.95 numerical aperture (NA) long working distance water immersion 25× objective.

2.3 Human ILC2 Isolation and Analysis

2.3.1 Materials

1. Stimulation mix: cRPMI supplemented with PMA (20 ng/mL; Sigma-Aldrich), ionomycin-free acid (1.5 μg/mL) from *Streptomyces conglobatus* (Merck), and Brefeldin A (5 μg/mL; Sigma-Aldrich).
2. RosetteSep™ Human ILC2 Enrichment kit.
3. Lymphoprep™.
4. Suggested Human ILC2 flow panel.

Antibody	Clone	Fluorochrome	Manufacturer	Use
CCR6	G034E3	AF700	BioLegend	Extracellular
KLRG1	14C2A07	APC	BioLegend	Extracellular
CD25	M-A251	APC CY7 (APC eFluor 780)	BioLegend	Extracellular
CRTH2	BM16	BV421/ Eflour450	BioLegend	Extracellular
CD90	5E10	BV510	BioLegend	Extracellular
CD127	A019D5	BV605	BioLegend	Extracellular
CD117	104D2	BV711	BioLegend	Extracellular
CD45	HI30	BV785	BioLegend	Extracellular
CD11c	3.9	FITC	BioLegend	Lineage
CD123	6H6	FITC	BioLegend	Lineage
CD14	HCD14	FITC	BioLegend	Lineage
CD16	3G8	FITC	BioLegend	Lineage
CD19	BIB19	FITC	BioLegend	Lineage
CD1a	HI149	FITC	BioLegend	Lineage
CD20	2H7	FITC	BioLegend	Lineage
CD3	OKT3	FITC	BioLegend	Lineage
CD34	561	FITC	BioLegend	Lineage
CD4	OKT4	FITC	BioLegend	Lineage

(continued)

Antibody	Clone	Fluorochrome	Manufacturer	Use
CD5	UCHT2	FITC	BioLegend	Lineage
FCεR1α	AER-37 (CRA-1)	FITC	BioLegend	Lineage
ST2	2A5	PE	MBL	Extracellular
CD161	HP-3G10	PE CY7	BioLegend	Extracellular
CD56	HCD56	PE-Dazzle	BioLegend	Extracellular
GATA3	TWAJ	PerCP eFlour710	Invitrogen	Intracellular transcription factor
LIVE/DEAD™ Fixable Blue Dead Cell Stain	n/a	UV	Invitrogen	Viability dye

2.3.2 Enriching ILC2s from Human Blood

1. Collect 45 mL of blood with 500 μL EDTA to prevent clotting.
2. Proceed with RosetteSep™ Human ILC2 Enrichment kit, following the manufacturer's instructions.

2.3.3 Staining Human ILC2s for Flow Cytometry

Following enrichment of ILC2s from human blood using the RosetteSep™ Human ILC2 Enrichment kit.

1. Centrifuge cells at 200 × *g* for 5 min at 4 °C and discard supernatant.
2. Wash cells in PBS by adding 500 μL per tube. Centrifuge at 200 × *g* for 5 min at 4 °C and discard supernatant.
3. Stain cells with 200 μL fixable live/dead marker for 20 min at room temperature (protected from light).
4. Add 500 μL of PBS and centrifuge at 200 × *g* for 5 min at 4 °C and discard supernatant.
5. Resuspend cells in appropriate cell surface stain cocktail (200 μL per well) and incubate at 4 °C (protected from light) for 20 min.
6. Add 500 μL of stain buffer and centrifuge at 200 × *g* for 5 min at 4 °C and discard supernatant. (Section 3, **item 8**).
7. Wash cells in stain buffer by adding 500 μL per well. Centrifuge at 200 × *g* for 5 min at 4 °C and discard supernatant. *If fluorescence-activated cell sorting live ILC2s, stop here.* Fix cells using eBioscience intranuclear fix/perm according to the manufacturer's instructions. *See notes for information about pausing here.*

8. Add 150 μL of stain buffer per well and centrifuge at 200 × *g* for 5 min at 4 °C and discard supernatant.
9. Permeabilize cells by suspending in 1× perm buffer (eBioscience), according to the manufacturer's instructions.
10. Resuspend cells in appropriate intracellular stain cocktail (50 μL per well) and incubate at 4 °C (protected from light) for 30 min.
11. Add 150 μL of 1× perm buffer per well and centrifuge at 200 × *g* for 5 min at 4 °C and discard supernatant.
12. Wash cells in 1× perm buffer by adding 200 μL per well. Centrifuge at 200 × *g* for 5 min at 4 °C and discard supernatant.

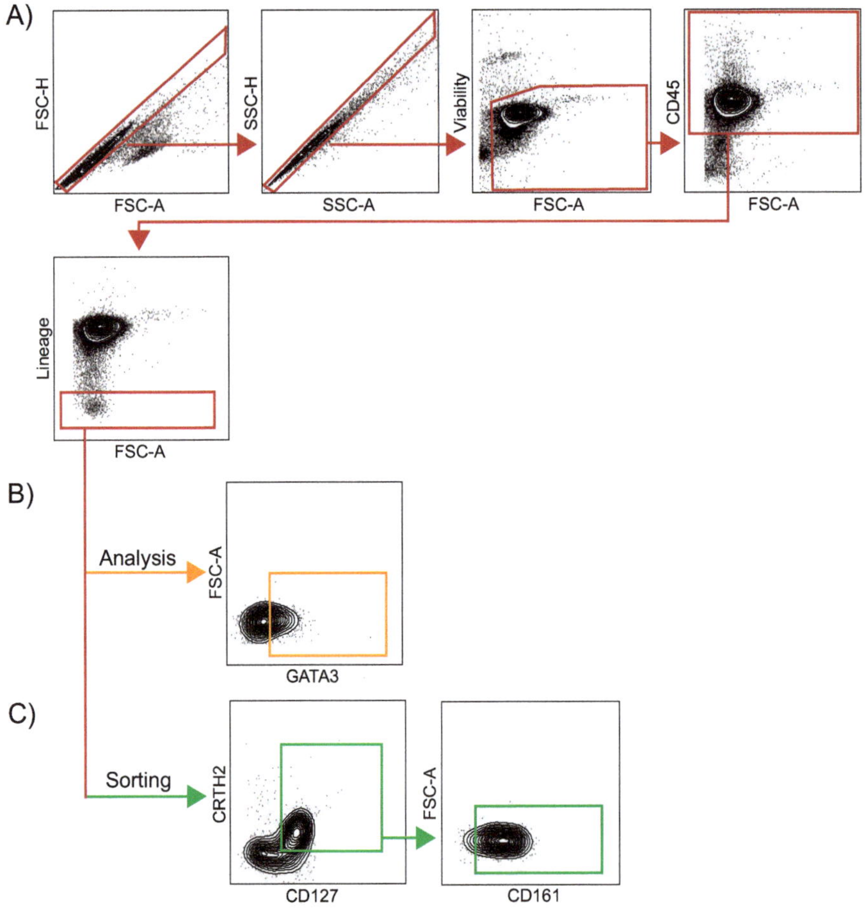

Fig. 4 Human pulmonary ILC2 gating strategies. (**a**) Gating Schematic. ILCs are defined as Single, Live, CD45$^+$, Lymphoid, Lineage^{neg} (TCRβ, TCRγδ, CD3e, CD5, CD19, CD11b, CD11c, FCεR1, GR-1, F4/80, and TER-119) negative. (**b**) ILC2s are further defined as expressing GATA3. (**c**) ILCs are sorted as in (**a**) with the additional markers CRTH2, CD127, and CD161

13. Resuspend cells in 200 μL stain buffer and analyze by flow cytometry.

2.3.4 Human ILC2 Gating Strategies

1. Human ILC2s for analysis: Single cells, live cells, $CD45^+$ cells, lymphocytes, $Lineage^-$, $GATA3^+$ (Fig. 4a, b).
2. Human ILC2s for fluorescence-activated cell sorting: Single cells, live cells, $CD45^+$ cells, lymphocytes, $Lineage^-$, $CRTH2^+$, $CD127^+$, $CD161^+$ (Fig. 4a, c).

3 Notes (Any Hints or Tips to Aid with the Methods)

1. Different mouse strains have altered pulmonary ILC2 phenotypes.
2. Different model allergens and ILC2-inducing stimuli (such as IL-33) alter pulmonary ILC2 phenotype.
3. To cull mice for ILC2 isolation or imaging, do not use cervical dislocation or CO_2 overdose, as this will affect tissue morphology and cellular composition.
4. ILC2 phenotype differs between BAL, Lung, and mLN.
5. Ex vivo stimulation of ILC2s will alter their extracellular marker phenotype, such as downregulation of surface receptors.
6. Utilize FMOs for flow cytometry analysis of ILC2s as some markers have poor separation of the positive population from the negative population.
7. ILC2 flow cytometry analysis requires the acquisition of many cells as they are a rare population; therefore, ensure plenty of cells are stained for each sample. For example, in the lungs of a naïve mouse $Gata3^+$ ILC2s account for approximately 0.6% of all live $CD45^+$ cells. Therefore, to analyze 500 ILC2s, approximately 8.3×10^5 live $CD45^+$ cells need to be acquired and recorded.
8. Perform flow cytometry antibody staining of a large number of cells in 15 mL centrifuge tubes (when sorting). Cell pellets can be lost when performing in 50 mL centrifuge tubes.
9. While performing live PCLS make sure to keep the lung slices hydrated in 100 μL cRPMI while imaging.

References

1. Lloyd CM, Snelgrove RJ (2018) Type 2 immunity: expanding our view. Sci Immunol 3(25). https://doi.org/10.1126/sciimmunol.aat1604
2. Allen JE, Sutherland TE (2014) Host protective roles of type 2 immunity: parasite killing and tissue repair, flip sides of the same coin. Semin Immunol 26(4):329–340

3. Halim TY (2016) Group 2 innate lymphoid cells in disease. Int Immunol 28(1):13–22
4. Halim TY et al (2012) Lung natural helper cells are a critical source of Th2 cell-type cytokines in protease allergen-induced airway inflammation. Immunity 36(3):451–463
5. Halim TY, Takei F (2014) Isolation and characterization of mouse innate lymphoid cells. Curr Protoc Immunol 106:3.25.1–3.25.13
6. Barlow JL et al (2012) Innate IL-13-producing nuocytes arise during allergic lung inflammation and contribute to airways hyperreactivity. J Allergy Clin Immunol 129(1):191–8.e1–4
7. Bartemes KR et al (2012) IL-33-responsive lineage- CD25+ CD44(hi) lymphoid cells mediate innate type 2 immunity and allergic inflammation in the lungs. J Immunol 188 (3):1503–1513
8. Van Dyken SJ et al (2014) Chitin activates parallel immune modules that direct distinct inflammatory responses via innate lymphoid type 2 and gammadelta T cells. Immunity 40 (3):414–424
9. Kim J et al (2019) Innate immune crosstalk in asthmatic airways: innate lymphoid cells coordinate polarization of lung macrophages. J Allergy Clin Immunol 143(5):1769–1782.e11
10. Bartemes KR et al (2014) Enhanced innate type 2 immune response in peripheral blood from patients with asthma. J Allergy Clin Immunol 134(3):671–678.e4
11. Christianson CA et al (2015) Persistence of asthma requires multiple feedback circuits involving type 2 innate lymphoid cells and IL-33. J Allergy Clin Immunol 136(1):59–68.e14
12. Chen R et al (2017) Allergen-induced increases in sputum levels of group 2 innate lymphoid cells in subjects with asthma. Am J Respir Crit Care Med 196(6):700–712
13. Smith SG et al (2016) Increased numbers of activated group 2 innate lymphoid cells in the airways of patients with severe asthma and persistent airway eosinophilia. J Allergy Clin Immunol 137(1):75–86.e8
14. Nagakumar P et al (2016) Type 2 innate lymphoid cells in induced sputum from children with severe asthma. J Allergy Clin Immunol 137(2):624–626.e6
15. Cardoso V et al (2017) Neuronal regulation of type 2 innate lymphoid cells via neuromedin U. Nature 549(7671):277–281
16. Sui P et al (2018) Pulmonary neuroendocrine cells amplify allergic asthma responses. Science 360(6393):eaan8546
17. Puttur F et al (2019) Pulmonary environmental cues drive group 2 innate lymphoid cell dynamics in mice and humans. Sci Immunol 4 (36):eaav7638
18. Dahlgren MW et al (2019) Adventitial stromal cells define group 2 innate lymphoid cell tissue niches. Immunity 50(3):707–722.e6
19. Looney MR et al (2011) Stabilized imaging of immune surveillance in the mouse lung. Nat Methods 8(1):91–96
20. Secklehner J, Lo Celso C, Carlin LM (2017) Intravital microscopy in historic and contemporary immunology. Immunol Cell Biol 95 (6):506–513

Chapter 11

Isolation, Culture, and Induction of Plasticity in ILC2s

Jonathan Silver, Alison A. Humbles, and Yoichiro Ohne

Abstract

Innate lymphoid cells (ILCs) are lymphocytes with critical roles in homeostasis, inflammation, and immunity to pathogens. ILCs are rare relative to other immune cell populations and are primarily defined by lack of expression of markers associated with other immune cell lineages and are predominantly found in mucosal tissues like the gut, lung and skin. They are classified into distinct subsets, ILC1, ILC2, and ILC3, which mirror subsets of $CD4^+$ helper T cells. ILC subsets have distinct cytokine and transcription factor profiles which align with their biological functions, although recently it has emerged that ILC subsets are not phenotypically fixed and exhibit considerable heterogeneity and plasticity in different contexts. Here, we describe protocols for the maintenance, expansion, and induction of plasticity in mouse and human ILC2s. The resulting cells can be used for molecular interrogation of ILC function and biology, both in vivo and in vitro.

Key words Innate lymphoid cells, ILC2s, Cytokines

1 Introduction

Innate lymphoid cells are a recently described subset of innate cells with prominent roles in mucosal homeostasis and defense [1, 2]. ILCs develop from common lymphocyte progenitors in the bone marrow and are present throughout the body; particularly in the lung, liver, gut, and skin. They are phenotypically similar to $CD4^+$ T cells and NK Cells and are classified into three distinct subsets: ILC1, ILC2, and ILC3. These subsets are functionally analogous to Th1, Th2, and Th17 cells, respectively. The similarities between ILCs and other lymphocyte populations, particularly $CD4^+$ T cells and NK Cells, have enabled the rapid derivation of ILC developmental pathways, regulatory circuits, and biological functions. It is now recognized that ILCs play critical roles in

Shoba Amarnath (ed.), *Innate Lymphoid Cells: Methods and Protocols*, Methods in Molecular Biology, vol. 2121,
https://doi.org/10.1007/978-1-0716-0338-3_11,

maintaining barrier function in homeostasis and disease, regulating the type and magnitude of antipathogen inflammation [1]. More recently, altered ILC populations have been identified in a range of diseases including COPD, IBD and asthma [3, 4]. Whether these aberrant responses can be controlled for therapeutic benefit remains unclear.

Despite the importance of ILCs in immunity and inflammation, several factors make the isolation and characterization of these cells much more difficult compared to other lymphocytes. For example, ILCs are rare in peripheral blood, they do not undergo clonal differentiation like B or T cells and are primarily identified by the lack of expression of lineage-specific surface markers. Moreover, while they can be found in lymphoid tissues like lymph nodes and spleen, ILCs are generally more common in mucosal-associated tissues like lung, gut, and skin—all of which require specialized digestion protocols. In addition to their rarity and location in difficult-to-process organs there are many tissue-specific subsets of ILCs; studies have identified a range of ILC1s and ILC3s in the gut tissue for example that express a variety of surface markers [1]. In hindsight, it is now obvious why ILCs evaded detection for so long–they lack specific markers and are very rare.

ILCs are classified into distinct subsets on the basis of transcription factor expression and functional characteristics; however, it has become clear in recent years that these cells are considerably plastic phenotypically. Reports have identified that ILC1s can convert to ILC3s and vice versa [5, 6]. Moreover, expression gradients of lineage-specifying transcription factors have been identified in ILCs; thus, ILC3s can express GATA-3 which has a nonredundant role in regulating their function [7].

Although ILC2 cells were initially thought to be less heterogeneous and plastic compared to other subsets it is now evident that they exhibit similarly plasticity [3, 8–10]. Recent studies using single cell sequencing have also revealed a range of ILC2 populations in the gut suggesting that ILC2s exist on a functional and phenotypic gradient naturally, even in the absence of an ongoing pathogen-induced immune response [11]. In order to understand how ILC2s respond to a range of stimuli and elucidate their contribution to ongoing inflammatory responses, we developed several protocols for isolation and polarization of Type 2 ILCs. Magnetic depletion using a custom cocktail of biotinylated lineage antibodies allows for efficient enrichment of mouse ILC2 populations for in vitro culture and subsequent adoptive transfer. The enrichment method takes advantage of a lack of lineage specific surface markers on ILCs and typically achieves 85–90% purity when performed on mouse lung. Similar techniques for human ILCs enrich the ILC2 population before FACS sorting and expansion in vitro. Although we focus on ILC2s here, these protocols can be easily adapted to other ILC subsets and various tissues. Taken together, these methods allow for very pure isolation, culture of ILCs from both mouse and humans and subsequent interrogation of their lineage potentials.

2 Materials and Methods

2.1 Mouse ILC2s

2.1.1 Materials

1. Isolation buffer (PBS; 2 mM EDTA; 0.5% BSA).
2. Biotinylated Antibodies (*see* Table 1).

Table 1
Mouse antibodies for FACS and enrichment

			Clone	Vendor	Cat#	per 100 μL
Surface markers:	CD3	BUV395	17A2	BD Biosciences	740268	1
	CD49b	PE eFluor 610	DX5	eBiosciences	61-5971-82	1
	CD45	BV510	30-F11	Biolegend	103137	1
	ST2	FITC	DJ8	MDBiosciences	101001F	2.5
	CD90	APC eFluor 780	53-2.1	eBiosciences	47-0902-82	0.5
	CD44	Alexa 700	IM7	Biolegend	103026	0.33
	CD25	BV650	PC61	Biolegend	102037	1
	CD127	PE	A7R24	eBiosciences	12-1271-82	1
Lineage:	CD11c	eFluor 450	N41.8	eBiosciences	48-0114-82	0.5
	CD11b	eFluor 450	M1/70	eBiosciences	48-0112-82	0.5
	FceRI	eFluor 450	MAR_1	eBiosciences	48-5898-82	0.5
	CD19	eFluor 450	1D3	eBiosciences		0.5
	B220	eFluor 450	RA3-6B2	eBiosciences	48-0452-82	0.5
	Gr-1	eFluor 450	RB6-8C5	eBiosciences	48-5931-82	0.5
	TCRab	eFluor 450	H57-597	eBiosciences	48-5961-82	0.5
	TCRgd	eFluor 450	GL3	eBiosciences	48-5711-82	0.5
	CD5	eFluor 450	53-7.3	eBiosciences	48-0051-82	0.5
	CD27	eFluor 450	O323	eBiosciences	48-0279-42	0.5
	CD19	eFluor 450	1D3	eBiosciences		
Cytokines:	IL-5	PE	TRFK5	eBiosciences	12-7052-82	0.2
	IL-13	eFluor 660	eBio13A	eBiosciences	50-7133-82	0.33
	IFN-gamma	PE/Cy7	XMG1.2	Biolegend	505826	0.2
Transcription factors:	GATA3	Alexa Fluor 647	16E10A23	Biolegend	653810	1
	T-bet	PE/Cy7	4B10	Biolegend	644824	1 per 10e6 cells:

(continued)

Table 1 (continued)

			Clone	Vendor	Cat#	per 100 μL
For enrichment	CD11c	biotinylated	N41.8	eBiosciences	13-0114-85	2.5
	CD11b	biotinylated	M1/70	eBiosciences	13-0112-85	2.5
	FceRI	biotinylated	MAR_1	eBiosciences	13-5898-85	2.5
	CD19	biotinylated	1D3	eBiosciences	13-0193-85	2.5
	B220	biotinylated	RA3-6B2	eBiosciences	13-0452-85	2.5
	Gr-1	biotinylated	RB6-8C5	eBiosciences	13-5931-85	2.5
	TCRab	biotinylated	H57-597	eBiosciences	13-5961-85	2.5
	TCRgd	biotinylated	GL3	eBiosciences	13-5711-85	2.5
	CD5	biotinylated	53-7.3	eBiosciences	13-0051-85	2.5
	CD27	biotinylated	O323	eBiosciences	13-0279-85	2.5
	CD326	biotinylated	G8.8	eBiosciences	13-5791-85	2.5
						22.5 μL isolation buffer

3. Cytokines (*see* Table 2).
4. RPMI supplemented with 10% FBS, 1% Pen/Strep (cRPMI).
5. DNase I Enzyme.
6. Liberase TM.
7. 5 ml polypropylene tube
8. BFA.
9. GolgiPlug.
10. PMA.
11. Ionomycin.
12. 32% PFA (electron microscope tech) diluted with PBS to 4%.
13. BD Perm/Wash buffer.
14. FoxP3 staining kit.

2.1.2 Methods for Induction of Plasticity in Mouse ILC2s

1. Prior to intranasal cytokine administration, mice are anesthetized with isoflurane, verified by slow breathing and lack of reflexes.
2. 50 μl of PBS or appropriate cytokine (at 20 μg/ml for 1 μg/dose; *see* Table 3) is administered in each nostril, 50 μl, dropwise. For induction of ILC2 plasticity, mice are dosed intranasally with either PBS, IL-33 or IL-33 + IL-12/IL-18.
3. Mice are placed back in caging and watched for 5–10 min to ensure they wake from anesthesia properly.
4. **Steps 1–3** are repeated 48 and 96 h later for a total of three doses.
5. On Day 5 or Day 6, mice are sacrificed using CO_2 according to institutional guidelines.

Table 2
Human antibodies for FACS staining

Antigen	Fluorochrome	Clone	Vendor	Cat#	100 μl
CD3	PE/Dazzle594	SK7	BioLegend	344827	1
CD4	PE/Dazzle594	RPA-T8	BioLegend	301047	1
CD14	PE/Dazzle594	N5E2	BioLegend	301841	1
CD16	PE/Dazzle594	3G8	BioLegend	302047	1
CD19	PE/Dazzle594	HIB19	BioLegend	302241	1
CD20	PE/Dazzle594	2H7	BioLegend	302347	1
CD34	PE/Dazzle594	581	BioLegend	343527	1
CD123	PE/Dazzle594	6H6	BioLegend	306021	1
HLA-DR	PE/Dazzle594	L243	BioLegend	307653	1
CD45	BUV395	HI30	BD	563792	0.5
CD127	VioBright FITC	RIA614	Miltenyi Biotec	130-120-113	1
CRTH2	AF647	BM16	BioLegend	350104	1
c-Kit	PE-Cy7	104D2	BioLegend	313212	1
CCR6	BV421	G034E3	BioLegend	353408	1
CD56	APC-Cy7	HCD56	BioLegend	318332	0.1
GATA3	BUV395	L50-823	BD	565448	1
T-bet	PE/Cy7	4B10	BioLegend	644823	1
IL-5	Vio515	JES1-39D10	Miltenyi Biotec	130-108-099	5
IL-13	BV711	JES10-5A2	BD	564288	1
IFN-γ	BUV395	B27	BD	563563	1

6. Lungs are perfused with 10 ml of cold PBS by opening the chest cavity, making a small incision in the left ventricle of the heart, followed by inserting the syringe into the right ventricle of the heart and gently pushing the PBS through circulation. The lungs will inflate, then turn white, following proper perfusion.
7. Lung tissue is resected and diced into ~2 cm pieces in 2 ml of complete RPMI and placed on ice.
8. 3 ml of enzyme mixture (2.5 ml of complete RPMI, 500 μl Liberase TM, 5 μl DNase I) is added per sample taking care that each segment of lung tissue is immersed in the enzyme mixture.
9. Each sample is placed on a shaker at 200 rpm at 37 °C for 30 min.

Table 3
Cytokines for expansion and induction of plasticity

	Vendor	Cat. #	Working concentration
Human cytokines			
IL-2	R&D Systems	202-IL	10 U/ml
IL-1β	R&D Systems	201-LB	10 ng/ml
IL-12	R&D Systems	219-IL	10 ng/ml
Mouse cytokines			
IL-33	R&D Systems	3626-ML-010	20 μg/ml
IL-12	BioLegend	577002	20 μg/ml
IL-18	R&D Systems	9139-IL-010	20 μg/ml

10. Samples are gently dissociated using a plunger from a 1 cc syringe over a 70 μM filter before being rinsed with 5 ml of complete RPMI. This step is repeated two more times for a total of 15 ml washes before the filter is discarded.
11. Cells are centrifuged at 365 × *g* for 5 min before media is aspirated and the cells are resuspended in 2 ml of ACK Lysing buffer for 3 min.
12. Each sample is filled with cRPMI up to 30 ml before being centrifuged again and excess media aspirated.
13. Samples are resuspended in 2 ml of cRPMI, filtered again if necessary, and counted on a ViaFlow counter. Cells at this step can be stained for ILCs; a typical stain is shown in Fig. 1A (whole lung).
14. In preparation for ILC isolation, samples from each relevant treatment group are pooled and washed with cold isolation buffer.
15. Each sample is incubated with appropriate biotinylated lineage antibodies at a concentration of 2.5 μl of each antibody +2.5 μl of PBS buffer per ten million total cells and mixed thoroughly before being incubated at 4 °C for 15 min.
16. Samples are washed with 20 ml of the PBS buffer before being incubated with Miltenyi anti-biotin microbeads according to the manufacturer's protocol.
17. Following incubation with anti-biotin microbeads, the cells are washed again with 20 ml of PBS buffer. After this centrifugation step the cell pellet should be brown, indicating successful binding of the anti-biotin microbeads. The cells are filtered and run through a Miltenyi MACS LS column according to the manufacturer's protocol; multiple columns may be necessary depending on how many cells were labeled.

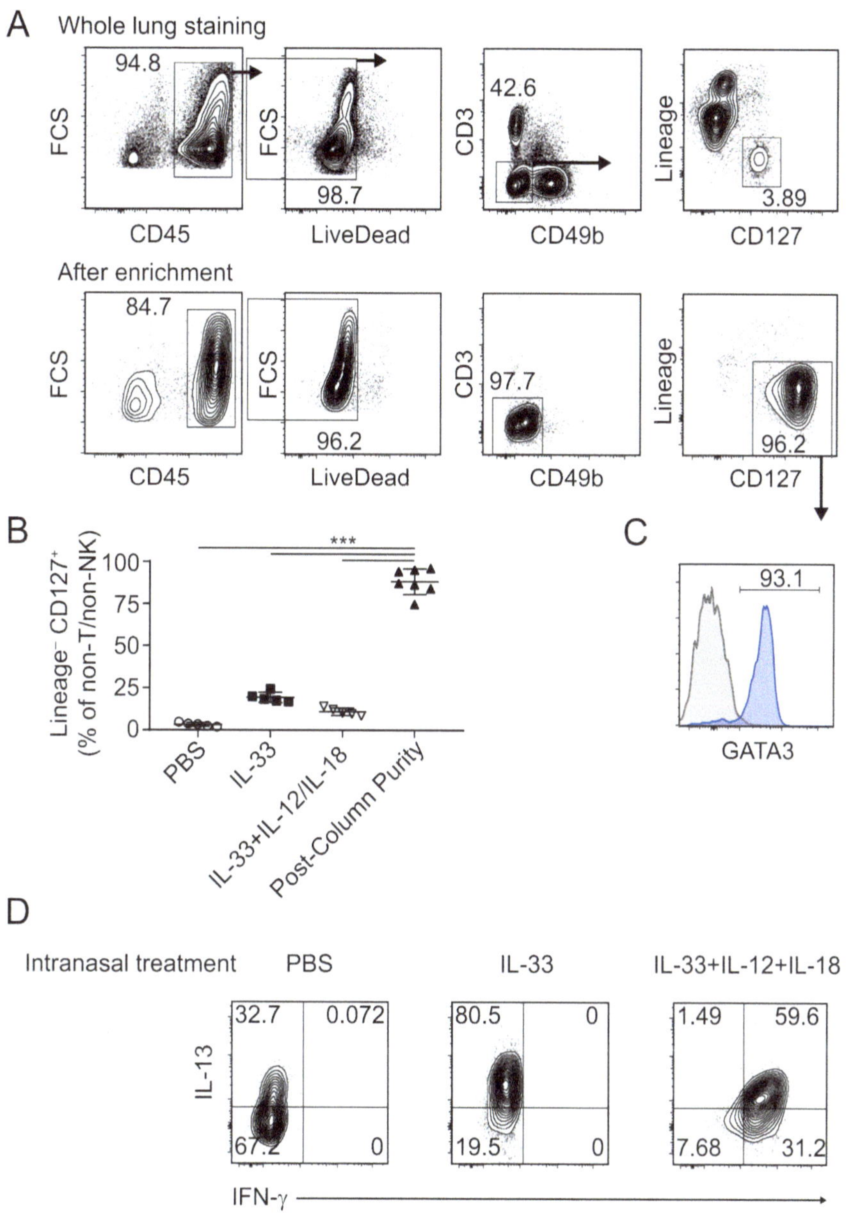

Fig. 1 Gating strategy for mouse ILC2s. (**a**) FACS staining of lung-resident ILCs in whole tissue versus after enrichment. ILCs are $CD45^+$ viable $CD3/CD49b^-$ $Lineage^-$ $CD127^+$ cells. (**b**) Comparison of ILC frequency between different treatment groups vs postcolumn purity, $^{***}p < 0.0001$. (**c**) GATA-3 expression in enriched ILCs from an IL-33-treated mouse, confirming the ILC2 phenotype. (**d**) Intracellular staining of cells isolated from lungs of mice treated as indicated (PBS, IL-33 or IL-33+IL-12+IL-18) demonstrating coexpression of IL-13 and IFN-γ in the IL-33+IL-12+IL-18 group, reflecting the ILC2 → ILC1 plasticity of these cells. Data are representative of >20 different experiments

18. Flow through is collected, centrifuged, and resuspended in 5 ml of cRPMI before cell counting. At this time point the ILCs average 85% purity (see **Notes 1** and **2**) (Fig. 1B, C) and the majority of them express GATA-3, consistent with an ILC2 phenotype (Fig. 1D). The cells treated with IL-33+IL-12+IL-18 are co-producing IFN-γ and IL-13 which indicates successful induction of plasticity (*see* **Notes 3** and **4**).

2.2 Human ILC2s

2.2.1 Materials

1. Staining buffer (PBS; 2% FBS).
2. EasySep human NK enrichment kit.
3. Antibodies (*see* Table 1).
4. Cytokines (*see* Table 2).
5. Ultra eComp beads.
6. Yssel's medium supplemented with 2% AB serum.
7. IMDM supplemented with 10% FBS.
8. Strainer cap tube.
9. 5 ml polypropylene tube.
10. BFA.
11. GolgiPlug.
12. PMA.
13. Ionomycin.
14. 32% PFA (electron microscope tech).
15. Perm/Wash buffer.
16. FoxP3 staining kit.

2.2.2 Instruments and Equipment

1. FACS Fusion.
2. FACS LSRFortessa.
3. Vi-cell cell counter.
4. MACSQuant VYB.
5. Multichannel suction.
6. ELISA shaker.

2.2.3 Methods

1. Isolate PBMCs from 200 ml peripheral blood by density gradient centrifugation using Ficoll.
2. Wash cells twice with Staining buffer. Centrifugation speed is at $365 \times g$ for 5 min at RT for all the procedures.
3. Count cell number using Vi-Cells.
4. Enrich ILC2s using EasySep human NK cell enrichment kit by following the manufacturer's instructions (Fig. 2A) (*see* **Note 5**).
5. Centrifuge and pellet enriched cells.

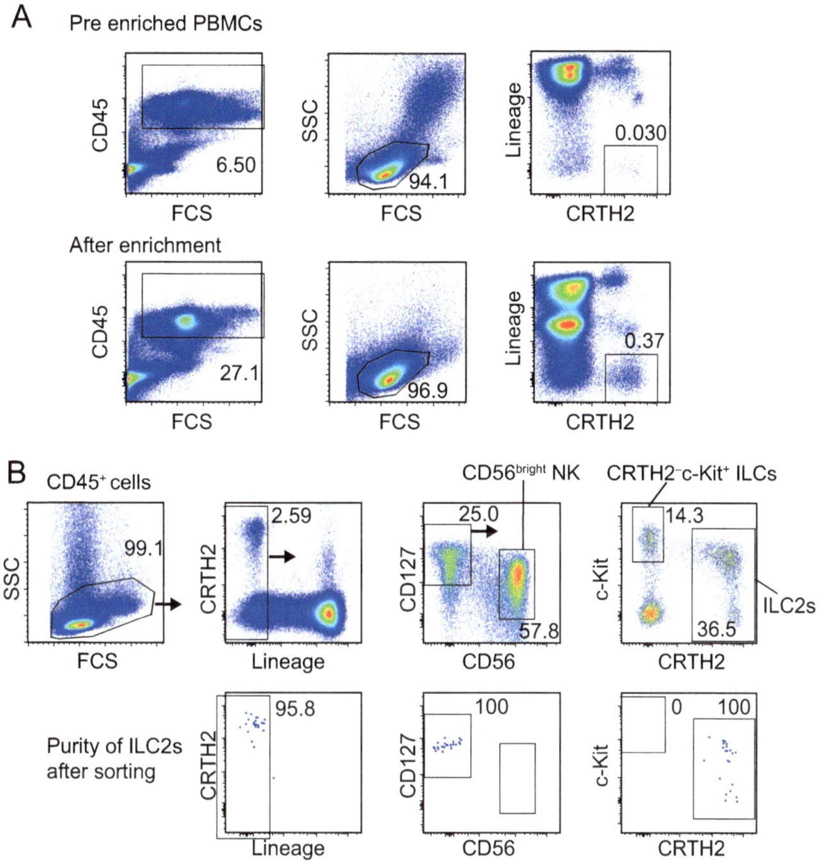

Fig. 2 Gating strategy for human ILC2s. (**a**) Representative FACS plots of staining for human ILC2s in whole PBMCs versus NK-cell-enriched PBMCs. ILC2s are defined as CD45$^+$, lineage negative CRTH2$^+$ cells. (**b**) FACS gates for sorting human ILC2s; which are defined as lineage negative, CD127$^+$, CD56$^-$ expressing CRTH2 and variable for c-Kit expression. Post sort purity is shown at the bottom. Data are representative of >20 independent experiments

6. Add antibody cocktail to cells so that final volume is 100 μl. Incubate for 15 min at RT.
7. Wash cells with 5 ml Staining buffer.
8. Resuspend cells in 500 μl Staining buffer.
9. Remove cell aggregate by passing through strainer cap tube.
10. Prepare collection tube by dispensing 3 ml medium into 5 ml polypropylene tube.
11. Perform cell sorting by FACSFusion (Fig. 2B). Collect ILC2s and other ILCs into collection tubes.
12. Centrifuge collection tubes and aspirate supernatant by 200 μl.
13. Count number of cells by MACSQuant VYB after diluting cell suspension by PBS at 1:10 ratio.

14. Plate certain number of cells (*see* **Note 6**) to 96 well U-bottom plate (*see* **Note 7**).
15. Add IL-1β with or without IL-12 in the presence of IL-2 and fill up with Yssel's medium supplemented with 2% AB serum (*see* **Note 8**) to 200 μl. Cells are cultured for 7–10 days.
16. Half of medium is replaced to fresh one with cytokines at Day 3 or Day 4.
17. (Optional) For ELISA, 100 μl of medium is collected.
18. (Optional) For cell proliferation, collect 10 μl of cells, dilute 1:10 in PBS with 7-AAD and count cells by MACSQuant VYB.
19. Intracellular cytokine staining.
    (a) Add 110 μl of stimulation cocktail (PMA and Ionomycin diluted in medium at final concentration 25 ng/ml and 1 μg/ml, respectively).
    (b) After incubation in CO_2 incubator for 1 h, add Golgi protein transport inhibitors (BFA and GolgiPlug) and incubate for another 4 h.
    (c) Stain dead cells by Fixable LIVE/DEAD Blue in 100 μl volume in the same plate (1:1000 dilution in PBS).
    (d) Add 200 μl Staining buffer, centrifuge plate at 650 × *g* for 1 min and remove supernatant.
    (e) (Optional) Stain cells with antibodies against surface receptors. Incubate for 15 min at RT, wash with 200 μl Staining buffer.
    (f) Fix cells with 100 μl of 4% PFA/PBS for 15 min at 4 °C.
    (g) Wash cells twice with 200 μl Staining buffer.
    (h) Permeabilize cells by adding 200 μl Perm/Wash buffer and incubating for 10 min at RT.
    (i) Centrifuge plate and remove supernatant by multichannel suction.
    (j) Add antibody cocktail for cytokines and incubate for 30 min at RT with continuous shaking on ELISA shaker (Speed 6).
    (k) Wash cells twice with Perm/Wash buffer.
    (l) Resuspend cells in 100 μl Staining buffer for measurement by flow cytometry (Fig. 3A).

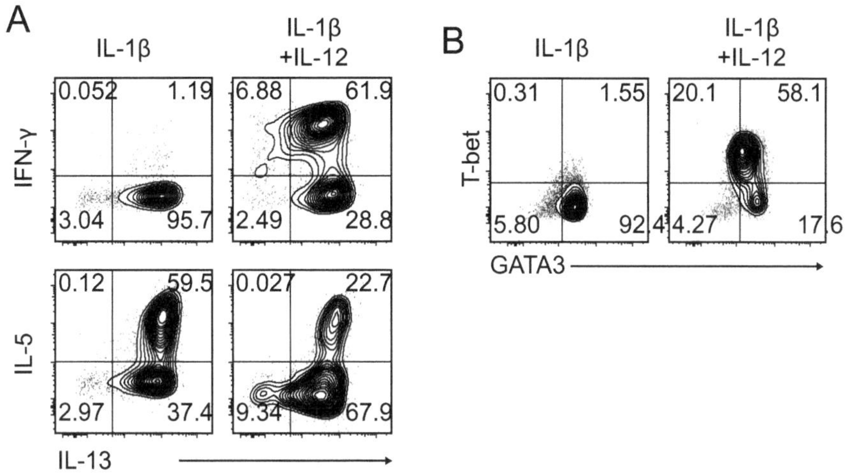

Fig. 3 Intracellular staining of human ILC2s post culture. Representative FACS plots of human ILC2s cultured with IL-1β or IL-1β + IL-12 and stained intracellularly for IFN-γ, IL-5, and IL-13 following stimulation with PMA and ionomycin (**a**), or T-bet and GATA-3 at steady state (**b**). Cells cultured with IL-1β + IL-12 coproduce IFN-γ and IL-13, and co-express T-bet and GATA-3 consistent with their plasticity. All conditions contain IL-2

20. Intranuclear transcription factor staining.
    (a) Stain dead cells by Fixable LIVE/DEAD Blue in 100 μl volume in the same plate (1:1000 dilution in PBS).
    (b) Add 200 μl Staining buffer, centrifuge plate at 650 × *g* for 1 min and remove supernatant.
    (c) (Optional) Stain cells with antibodies against surface receptors. Incubate for 15 min at RT, wash with 200 μl Staining buffer.
    (d) Fix cells with 200 μl of FoxP3 Fixation/Permeabilization working solution for 45 min at 4 °C.
    (e) Centrifuge plate and remove supernatant.
    (f) Wash cells with 250 μl of 1× Permeabilization Buffer.
    (g) Add 200 μl of 1× Permeabilization Buffer, resuspend well, and incubate for 15 min at RT.
    (h) Centrifuge plate and remove supernatant.
    (i) Add antibody cocktail for transcription factor and incubate for 60 min at RT with continuous shaking on ELISA shaker (Speed 6).
    (j) Wash cells three times with 1× Permeabilization Buffer.
    (k) Resuspend cells in 100 μl Staining buffer for measurement by flow cytometry (Fig. 3B).

3 Notes

1. Enriched cells can be further cultured in vitro with IL-2 ± other cytokines or growth factors of interest.
2. Cells can be easily sorted following enrichment for adoptive transfer.
3. The number of cells may be very low in each well, so for FACS analysis after culture the gating strategy may have to be modified. We find starting with FSC versus CD45 gives better resolution to see the cells in culture if the cell numbers are lower than 10,000 per cell.
4. The enrichment protocol works for other organs: spleen, lymph node, and liver.
5. ILC2 numbers can vary between seasons, with a tendency of higher ILC2 frequencies in allergy season.
6. A minimum of 300 cells should be plated per well to ensure cell survival, appropriate cytokine levels, and enough cells for flow cytometric analysis. Cells can be cultured for several weeks if fresh media + IL-2 and IL-1β is added every 3–4 days.
7. The shape of the bottom of the plate is critical, as ILC2s do not grow well in flat-bottom plates.
8. Cell culture media can impact ILC numbers. The conversion to ILC1 can be seen in either Yssel + 2% human serum or IMDM + 10% FBS; however, cells proliferate better in the latter. Additionally, some specifically configured medium (e.g., NK medium from Miltenyi) can change the phenotype of ILC2s in culture [12].

References

1. Artis D, Spits H (2015) The biology of innate lymphoid cells. Nature 517:293–301. https://doi.org/10.1038/nature14189
2. Vivier E et al (2018) Innate lymphoid cells: 10 years on. Cell 174:1054–1066. https://doi.org/10.1016/j.cell.2018.07.017
3. Silver JS et al (2016) Inflammatory triggers associated with COPD exacerbations orchestrate ILC2 plasticity in the lung. Nat Immunol 17(6):626–635
4. Zeng B et al (2019) ILC3 function as a double-edged sword in inflammatory bowel diseases. Cell Death Dis 10:315. https://doi.org/10.1038/s41419-019-1540-2
5. Bernink JH et al (2015) Interleukin-12 and -23 control plasticity of CD127(+) group 1 and group 3 innate lymphoid cells in the intestinal lamina propria. Immunity 43:146–160. https://doi.org/10.1016/j.immuni.2015.06.019
6. Klose CS et al (2013) A T-bet gradient controls the fate and function of CCR6-RORgammat+ innate lymphoid cells. Nature 494:261–265. https://doi.org/10.1038/nature11813
7. Zhong C et al (2016) Group 3 innate lymphoid cells continuously require the transcription factor GATA-3 after commitment. Nat Immunol 17:169–178. https://doi.org/10.1038/ni.3318
8. Ohne Y et al (2016) IL-1 is a critical regulator of group 2 innate lymphoid cell function and plasticity. Nat Immunol 17(6):646–655
9. Bal SM et al (2016) IL-1beta, IL-4 and IL-12 control the fate of group 2 innate lymphoid cells in human airway inflammation in the

lungs. Nat Immunol 17:636–645. https://doi.org/10.1038/ni.3444

10. Lim AI et al (2016) IL-12 drives functional plasticity of human group 2 innate lymphoid cells. J Exp Med 213(4):569–583. https://doi.org/10.1084/jem.20151750
11. Gury-BenAri M et al (2016) The spectrum and regulatory landscape of intestinal innate lymphoid cells are shaped by the microbiome. Cell 166:1231–1246.e1213. https://doi.org/10.1016/j.cell.2016.07.043
12. Camelo A et al (2017) IL-33, IL-25, and TSLP induce a distinct phenotypic and activation profile in human type 2 innate lymphoid cells. Blood Adv 1:577–589. https://doi.org/10.1182/bloodadvances.2016002352

Chapter 12

Bacterial Infection Allows for Functional Examination of Adoptively Transferred Mouse Innate Lymphoid Cell Subsets

Goo-Young Seo, Daniel A. Giles, and Mitchell Kronenberg

Abstract

Innate lymphoid cells (ILCs) are important regulators of the early responses to infection at mucosal barriers, including the intestine. Recently, we have shown that specific ILC3 subsets protect against enteric bacterial pathogens. Here, we describe a mouse model of oral infection by *Yersinia enterocolitica* (*Y. enterocolitica*) and several different methodologies to assess the severity of the infection. We also detail how ILC3 subsets can be isolated from the mouse small intestine and transferred into recipient immune deficient mice to study the function of these ILCs in the small intestine.

Key words Adoptive transfer, Group 3 innate lymphoid cells, Innate immunity, Small intestine lamina propria, *Yersinia enterocolitica*

1 Introduction

ILCs are a heterogeneous lymphoid population that do not express rearranged antigen receptors, and they respond to infection and other insults promptly by secreting effector cytokines. Therefore, ILCs contribute to barrier integrity, homeostasis, and host defense of mucosal surfaces [1]. ILC subsets have been defined by patterns of expression of effector cytokines, transcription factors, and cell surface markers. These subsets can be broadly defined as natural killer (NK) cells, as well as ILC1, ILC2, and ILC3. While natural killer (NK) cells share features with $CD8^+$ cytotoxic T cells [1], ILC1, ILC2, and ILC3 subsets share features with $CD4^+$ T helper (Th)1, Th2, and Th17 subsets, respectively. Shared features include expression of key transcription factors and cytokine production. In mice, ILC3s are further divided into subsets distinguished by CCR6 and NKp46 expression [2–5]. $CCR6^+$ ILC3s, which are lymphoid tissue inducer (LTi) cells, produce IL-22 and IL-17 [6]. $CCR6^{neg}$ ILC3s, which include $NKp46^+$ and $NKp46^-$ cells,

Shoba Amarnath (ed.), *Innate Lymphoid Cells: Methods and Protocols*, Methods in Molecular Biology, vol. 2121, https://doi.org/10.1007/978-1-0716-0338-3_12,

can express T-bet and produce IFN-γ in addition to IL-22 [2]. ILC3s are important for host defense against bacterial infection of the large intestine, including *Citrobacter rodentium* [5, 7, 8], *Helicobacter hepaticus* [9], and *Salmonella typhimurium* [2, 10]. Interestingly, ILC3s are more abundant in the small intestine than in the large intestine [11].

Y. enterocolitica is a facultative intracellular bacterium that targets the small intestine and causes food borne illness. After oral uptake, *Y. enterocolitica* replicates in the small intestine, invades Peyer's patches of the distal ileum and disseminates to the spleen and the liver [12]. Frequently the infection is cleared in 1–2 weeks [13, 14], suggesting that innate immune cells play important roles in protection from *Y. enterocolitica* infection. Our studies recently demonstrated that ILC populations, in particular $CCR6^{neg}$ ILC3s, are important for resistance to *Y. enterocolitica* infection in the small intestine by secreting IFN-γ [15].

In this chapter, we provide a strategy to isolate ILC3 subsets from the small intestine, based on cell surface marker expression, for the purpose of adoptive transfer. Following transfer, we describe the establishment and monitoring of *Y. enterocolitica* infection model and using analysis of infection parameters for study of ILCs.

2 Materials

2.1 Isolation of ILC3 Subsets and Adoptive Transfer of ILC3 Subsets

1. Donor $Rag1^{-/-}$ mice.
2. Recipient $Rag2^{-/-}gc^{-/-}$ mice, age 8–10 weeks.
3. RPMI Wash Buffer 1: 1000 mL RPMI, 25 mL of 1 M HEPES (final 25 mM), 10 mL of 100× P/S/G (Penicillin-Streptomycin-Glutamine), 50 mL of FBS (final 5%).
4. RPMI Wash Buffer 2: 1000 mL RPMI, 25 mL of 1 M HEPES (final 25 mM), 10 mL of 100× P/S/G, 100 mL of FBS (final 10%).
5. EDTA Buffer: 500 mL HBSS, 12.5 mL of 1 M HEPES (final 25 mM), 5 mL of 0.5 M EDTA (final 5 mM).
6. Collagenase Buffer: 100 mL of RPMI Wash Buffer 2, 1 mL of Collagenase Type-8 (1:100 dilution; Stock = 50 mg/mL; Final = 0.5 mg/mL), 50 μL DNase I (1:2000 dilution; Stock = 100 mg/mL; Final = 0.05 mg/mL).
7. 100% Percoll: 45 mL of Percoll + 5 mL of 10× HBSS.
8. 40% Percoll: 30 mL of RPMI Wash Buffer 1 + 20 mL of 100% Percoll (final 40%).
9. 80% Percoll: 10 mL of RPMI Wash Buffer 1 + 40 mL of 100% Percoll (final 80%).
10. Fluorescence-activated cell sorting (FACS) staining buffer: 49 mL of 1× Phosphate buffered saline (PBS) + 1 mL of FBS (final 2%).

11. Dissecting scissors.
12. Forceps.
13. Petri dishes.
14. 70 μm cell strainers.
15. 37 °C incubating orbital shaker.
16. Centrifuge.
17. 50 mL conical tubes, FACS tubes.
18. Flow cytometry.
19. Fluorophore-conjugated antibody cocktail for lineage (lin): CD19 (clone ID3), B220 (clone RA3-6B2), Grl (clone RB6-8C5), CD11c (clone N418), CD11b (clone M1/70).
20. Fluorophore-conjugated antibodies for isolation of ILC3 subsets: CD3 (clone 145-2C11), CD45 (clone 30-F11), CD90.2 (clone 53-2.1), NK1.1 (clone PK136), NKp46 (clone 29A1.4), and CCR6 (clone 140,706).
21. Live/Dead Fixable Yellow Dead Cell Stain Kit.

2.2 Y. enterocolitica Infection Model: Infection

1. Mice, age 8–10 weeks, *Ifng*$^{-/-}$ mice (*see* **Note 1**).
2. *Y. enterocolitica* strain WA-C (pYV::CM) (*see* **Note 2**) [16].
3. Luria broth (LB) broth media; LB agar plates supplemented with chloramphenicol (final 30 μg/mL).
4. 30 °C incubating orbital shaker, 30 °C microbiological incubator.
5. Sterile bacteria culture tubes.
6. Spectrophotometer.
7. Oral gavage needle with bulbous tip (Curved Feeding Needles).
8. 1 mL syringe.

2.3 Y. enterocolitica Infection Model: Monitoring and Quantifying Infection

1. Zinc formalin.
2. 70% isopropanol.
3. Homogenizer.
4. Chloramphenicol (final 30 μg/mL)-containing LB agar plates.

3 Methods

Figure 1 outlines the entire procedure for measuring ILC function against *Yersinia* infection. To determine if there is a role for ILC3 subsets during *Y. enterocolitica* infection, isolated ILC3 subsets from the small intestinal lamina propria lymphocytes (SI-LPLs) of *Rag1*$^{-/-}$ mice, which lack B and T cells, are transferred into

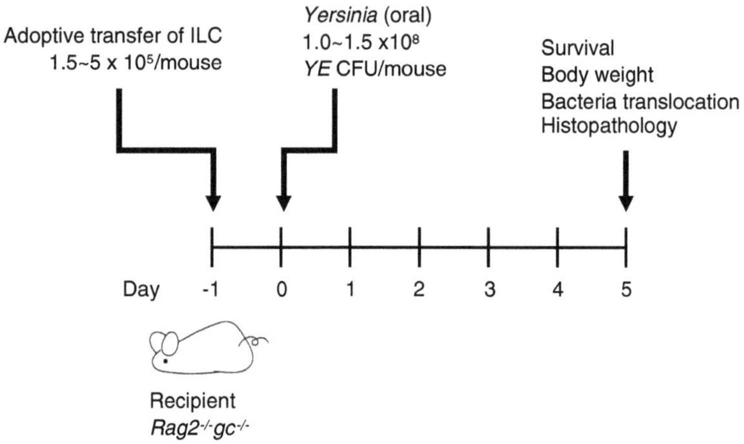

Fig. 1 Measuring ILC function against *Yersinia* infection. Schematic outline of the entire procedure to measure ILC function against *Yersinia* infection. Isolated SI-LPLs from uninfected *Rag1*$^{-/-}$ are stained with fluorophore-conjugated antibodies. ILC3 subsets, such as NKp46^{-}ILC3, NKp46^{+}ILC3, and CCR6^{+}ILC3 cells, are sorted by flow cytometry. Next, ILC3 subsets are transferred by retro-orbital injection into *Rag2*$^{-/-}$*gc*$^{-/-}$ recipients at day −1, and the mice are infected with *Y. enterocolitica* the following day (day 0). Control mice are injected with PBS. After *Y. enterocolitica* infection, mice are monitored by body weight and survival rate daily until days 5–7. At which point, the mice are euthanized and tissues are collected to assess the infection

recipient *Rag2*$^{-/-}$*gc*$^{-/-}$ mice, which lack all ILC subsets, as well as B and T cells. Recipient mice are then monitored to assess the infection. The following description of infection of mice with *Y. enterocolitica* and subsequent monitoring has been approved by the La Jolla Institute for Immunology Institutional Animal Care and Use Committee (IACUC). All procedures must be approved by a relevant institutional animal ethics committee.

3.1 Isolation of ILC3 Subsets and Adoptive Transfer of ILC3 Subsets

3.1.1 Lamina Propria Lymphocyte from Small Intestine (See Note 3)

1. Cut the entire small intestine from *Rag1*$^{-/-}$ mice and open longitudinally (*see* **Note 4**).
2. Wash briefly with RPMI Wash Buffer 1 and cut into 1.5 cm pieces.
3. Transfer the tissue into a 50 mL conical tube containing 25 mL of EDTA Buffer. Incubate the tissue pieces in a 37 °C incubating orbital shaker at 225 rpm for 10–15 min.
4. After incubation, pour off tissue pieces and liquid into a petri dish, collect all tissue pieces, put them back in a 50 mL conical tube, and then add 25 mL of prewarmed (*see* **Note 5**) EDTA Buffer into a 50 mL conical tube. Incubate the tissue pieces in the 37 °C incubating orbital shaker at 225 rpm for 10–15 min. Repeat **steps 3** and **4** several times until the epithelial layer is removed (*see* **Note 6**).

5. Pour off tissue pieces and liquid in the petri dish, collect all tissue pieces, put them back in a 50 mL conical tube. Add 20 mL of RPMI Wash Buffer 2 into a 50 mL conical tube and wash tissue pieces by shaking by hand 20 times.
6. Pour off tissue pieces and liquid in the petri dish, collect all tissue pieces, and transfer tissues to a 50 mL conical tube containing 25 mL of prewarmed Collagenase Buffer. Incubate the tissue pieces in the 37 °C incubating orbital shaker at 225 rpm for 20–25 min.
7. Separate lamina propria lymphocytes from digested tissues by passing cells through a 70 μm cell strainer and add 20 mL of RPMI Wash Buffer 2.
8. Centrifuge the flow-through cell suspension at 400 × *g* at 4 °C for 5 min.
9. Pour off the supernatant. Resuspend pellet well in 2 mL of 40% Percoll and gently load this cell containing fraction over 2 mL of 80% Percoll in an FACS tube (*see* **Note 7**).
10. Centrifuge the gradient at 650 × *g* at room temperature for 10 min (*see* **Note 8**). This step must be completed at room temperature.
11. Remove the upper phase containing fat and cell debris with a suction pump. Collect the ring interphase between the 80% Percoll and 40% Percoll, transfer it into a 50 mL conical tube, add 30–40 mL of RPMI Wash Buffer 2, and invert the tube several times (*see* **Note 9**).
12. Centrifuge the cell suspension at 400 × *g* at 4 °C for 5 min. Pour off the supernatant and resuspend cells in the FACS staining buffer.

3.1.2 Staining, Sorting of ILC3 Subsets by Flow Cytometry for Adoptive Transfer

ILC3s are divided into three subsets distinguished by CCR6 and NKp46 expression in mice. These include the CCR6^{+} ILC3 LTi subset and the CCR6^{neg} populations, which include NKp46^{+} and NKp46^{-} subsets. Previously, Guo et al. showed that intestinal ILC3s from mice are identified by their differential surface expression of CD45 and CD90 [17]. Here, we further show how to sort the individual ILC3 subsets from the SI-LPLs by flow cytometry.

1. Once lamina propria lymphocytes are isolated, resuspend the cells with FACS staining buffer containing anti-mouse CD16/CD32 (Mouse Fc Block diluted in FACS buffer at 1:400 ratio) and incubate for 15 min at 4 °C.
2. Prepare antibody mixture for ILC staining and sorting of ILC3 subsets as follows: For antibody mixture, CD19, B220, Grl, CD11c, CD11b (*see* **Note 10**), CD3, CD45, CD90.2, NK1.1, and NKp46 are used at a 1:400 dilution in FACS buffer corresponding to 0.5 μg/mL. CCR6 (concentration is lot specific) is diluted in FACS buffer at 1:20 ratio.

3. Add antibody mixture to the cells and incubate for 30 min at 4 °C in the dark.
4. Wash the cells with FACS staining buffer, centrifuge the cell suspension at 400 × *g* at 4 °C for 5 min and pour off the supernatant (*see* **Note 11**).
5. Resuspend cells with 1 mL of 1× PBS, add 1 μL of the dissolved LIVE/DEAD Fixable Dead Cell Stain Kits and mix well. Incubate the cells for 30 min at 4 °C, protected from light. Wash the cells with FACS staining buffer, centrifuge the cell suspension at 400 × *g* at 4 °C for 5 min and pour off the supernatant (*see* **Note 11**).
6. Resuspend the cells with FACS staining buffer up to 100×10^6 cells/mL (*see* **Note 12**) and filter the cells using a 70 μm cell strainer to remove cell debris.
7. The stained cells are ready for sorting. Proceed to the cell sorter. ILC3 subsets should be sorted according to the strategy detailed in Fig. 2.
8. Wash the sorted ILC3 subsets with 1× PBS and resuspend the cells in PBS at 1.5–5 × 10^5 cells/200 μL.
9. Transfer 200 μL of sorted cells (1.5–5 × 10^5/mouse) into $Rag2^{-/-}gc^{-/-}$ recipients at day −1 by retro-orbital injection. After the injection, monitor the mouse to ensure it is breathing and behaving normally.

3.2 Y. enterocolitica Infection Model: Infection

1. To recover *Y. enterocolitica* from the glycerol stock (*see* **Note 13**), open the tube and use a sterile loop, or pipette tip to scrape some of the frozen bacteria off of the top. Inoculate *Y. enterocolitica* to 4 mL of sterile LB media (*see* **Note 14**) in the culture tube. Incubate the culture tube in a 30 °C incubating orbital shaker overnight at 200 rpm (*see* **Note 15**).
2. For second culture of *Y. enterocolitica*, add 500 μL of the overnight culture to 50 mL of sterile LB media in a new 200 mL culture flask and place in a 30 °C incubating orbital shaker at 200 rpm for 5–6 h.
3. Measure the optical density of the second culture with a spectrophotometer at 600 nm to ensure that the correct concentration has been attained prior to oral gavage. The OD600 should be approximately 0.6 which is in the log or exponential phase of bacteria growth and is the most reproductive portion of the growth curve. If the OD600 is less 0.6 after second culture, continue incubation until OD600 reaches 0.6. In our hands, OD600 ≈ 0.6 is about 5.0–10.0 × 10^8 CFU/mL.

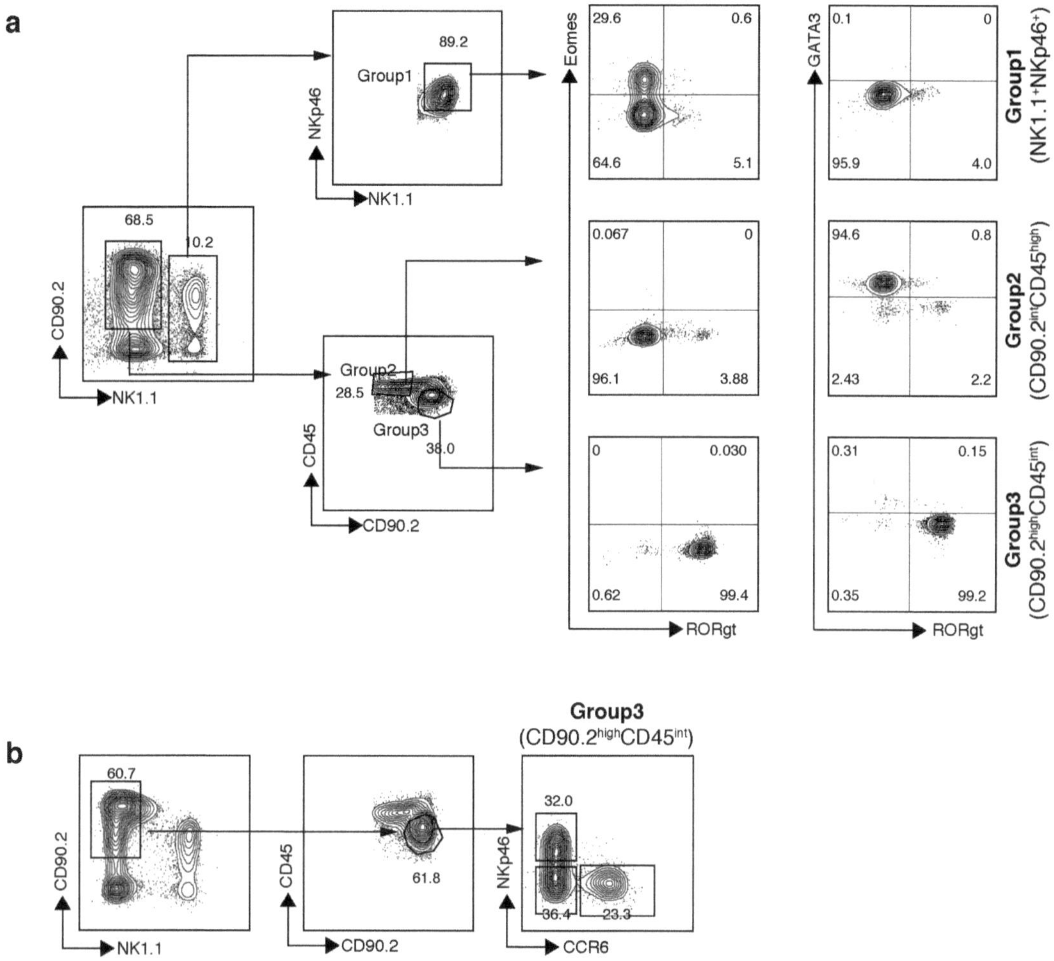

Fig. 2 Gating strategy for isolation of intestinal ILC3 subsets from the small intestine of *Rag1*$^{-/-}$ mice by flow cytometry. (**a**) Strategy for staining of RORγt$^+$ILC3s from the small intestine of *Rag1*$^{-/-}$ mice by their differential surface expression of CD45 and CD90. Gating is performed on lymphocytes and doublets are excluded. Gating is performed on live CD45$^+$ Lineage (Lin)$^-$CD3$^-$ cells. Lineage markers are CD19, B220, CD11c and Gr1. Data show initial flow cytometry plot gated on live CD45$^+$Lin$^-$CD3$^-$NK1.1$^+$cells and CD45$^+$Lin$^-$CD3$^-$CD90.2$^+$NK1.1$^-$cells. Cells are identified as Group 1, Group 2 and Group 3 by their differential surface expression of CD45, CD90, and NK1.1. Group 1 (CD45$^+$Lin$^-$CD3$^-$NK1.1$^+$NKp46$^+$ cells) include NK cells, ILC1, and ILC3. Group 2 (CD45$^+$Lin$^-$CD3$^-$ CD90^{int}CD45^{high} cells) are mainly GATA3$^+$ ILC2, but also contain a few RORγt$^+$ILC3. Group 3 (CD45$^+$Lin$^-$CD3$^-$ CD90^{high}CD45^{int} cells) are 99% RORγt$^+$ILC3. (**b**) Gating strategy for isolation of intestinal ILC3 subsets from *Rag1*$^{-/-}$ mice. Gating is performed on live CD45$^+$Lin$^-$CD3$^-$CD90.2$^+$NK1.1$^-$cells. Then, CD90^{high}C-D45^{int} cells are selected, and divided further based on NKp46 and CCR6 expression as follows: NKp46$^-$ILC3 (CD45$^+$Lin$^-$CD3$^-$NK1.1$^-$CD90.2^{high}CD45^{int}CCR6$^-$NKp46$^-$), NKp46$^+$ILC3 (CD45$^+$Lin$^-$CD3$^-$NK1.1$^-$CD90.2-^{high}CD45^{int} CCR6$^-$NKp46$^+$), and CCR6$^+$ILC3 (CD45$^+$Lin$^-$CD3$^-$NK1.1$^-$CD90.2^{high}CD45^{int}CCR6$^+$NKp46$^-$)

4. To confirm the accuracy of the inoculum preparation colony density, serially dilute 100 μL of the second culture tenfold until a 10^8-fold dilution of the starting material has been reached. Then plate 30 μL of serial dilutions of 10^6–10^8 in duplicate or triplicate onto chloramphenicol-containing LB agar plates to determine the bacterial density of the inoculate. Incubate the plates at 30 °C incubator for 2 days, then count the number of the colonies on the 10^6–10^8 in plates and calculate the CFU/mL.
5. Proceeding from **step 3**, prepare a 1 mL syringe for oral gavage. Swirl the bacterial culture to ensure a homogenous mixture before loading the syringe. Firmly attach the oral gavage needle onto the syringe and load the bacterial liquid (*see* **Note 16**).
6. Weigh the mouse, and then slowly inject approximately 100–200 μL of culture into the esophagus and remove the gavage needle at the same angle. In our hands, 100–200 μL of second culture is about 1×10^8–2×10^8 CFU of *Y. enterocolitica*. After the oral gavage, monitor the mouse to ensure it is breathing and behaving normally.

3.3 *Y. enterocolitica Infection Model: Monitoring and Quantifying Infection*

1. Survival rate and monitoring for other qualitative signs of infection daily: After *Y. enterocolitica* infection, check the mouse survival rate daily and monitor mice including appearance (scruffy, skin tenting as evidence of dehydration, etc.) and behavior and activity levels (reduced activity, hunched, etc.) daily, until days 5–7. Hunched appearance and reduced activity are a clear sign of severe infection, so these mice should be monitored closely and euthanized if severely ill according to standards set by the IACUC.
2. After *Y. enterocolitica* infection, weigh the mice every day until days 5–7 as measure of overall health. Weight loss is also a clear sign of severe infection, so mice experiencing weight loss should be monitored closely as above.
3. When mice are euthanized, approximately at day 5, collect ileum contents, spleen, and liver for quantification of bacterial burdens (CFU/g).
4. Collect ileum contents by longitudinally cutting open the ileum and scraping out contents using the rounded back of forceps. Place contents into preweighed 1.5 mL tube. Weigh and record weight of ileum contents and resuspend in 1 mL 1× PBS. Serially dilute the contents, according to diagram in Fig. 3a, until a dilution of 1:20,000 has been reached. Next, plate 30 μL of the 1:200, 1:2000, and 1:20,000 dilutions in triplicate onto chloramphenicol-containing LB agar plate to determine the bacterial colonies. Incubate the plates in a 30 °C incubator for 2 days, and then count the number of the colonies in plates and calculate as follow; CFU/g = (no. of colonies in 1 mL of PBS × dilution factor)/(grams of tissue) (Fig. 3a).

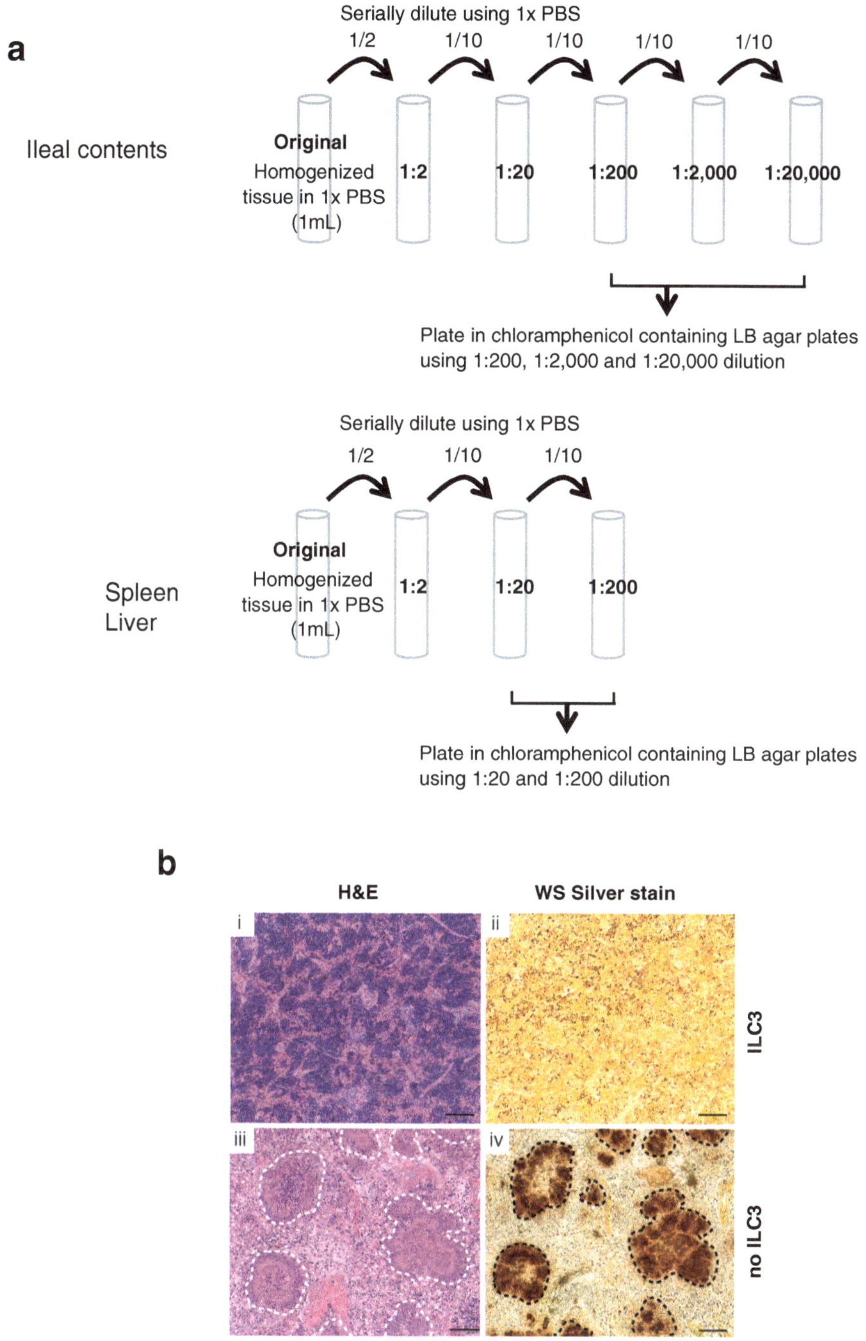

Fig. 3 Quantifying *Y. enterocolitica* infection. (**a**) Serial dilution of bacterial cultures. The CFU/g can be calculated using the formula: CFU/g = (no. of colonies in 1 mL of PBS × dilution factor)/(grams of tissues). (**b**) Representative H&E-stained (i, iii) and WS silver stained (ii, vi) splenic sections of *Rag2*$^{-/-}$*gc*$^{-/-}$ recipients with/without adoptive transfer of ILC3 after *Y. enterocolitica* infection at day 5 p.i. Scale bars, 100 μm. *Rag2*$^{-/-}$*gc*$^{-/-}$ mice without adoptive transfer of ILC3 infected with *Y. enterocolitica* exhibit large areas of splenic necrosis (white dotted lines, ii) that contain *Y. enterocolitica* colonies (black dotted lines, vi) in the spleen. However, *Rag2*$^{-/-}$*gc*$^{-/-}$ recipients infected with *Y. enterocolitica* are rescued from large areas of splenic necrosis that contain *Y. enterocolitica* colonies by transfer of purified ILC3 (i, iii)

5. Collect approximately half of the spleen (reserving the rest for histology) and approximately 100–200 mg of the liver, from the left lateral lobe, into preweighed 1.5 mL tubes and weigh the tissues. Transfer them into an FACS tube (*see* **Note 17**), add 1 mL of PBS into the tube and homogenize the tissue on ice with the homogenizer. Serially dilute the homogenized tissue, according to diagram Fig. 3a, until dilutions of 1:200 have been reached. Next, plate 30 μL of the 1:20 and 1:200 dilutions in triplicate onto chloramphenicol-containing LB agar plate to enumerate bacterial colonies. Incubate the plates in a 30 °C incubator for 2 days, and then count the number of the colonies in the plates and calculate CFU (Fig. 3a).
6. A systemic infection of *Y. enterocolitica* leads to focal abscess formation with necrosis containing *Y. enterocolitica* colonies in the liver and the spleen. To assess a systemic infection by *Y. enterocolitica*, prepare samples for hematoxylin and eosin (H&E) stain or Warthin-Starry (WS) silver stain to allow for assessment of necrosis in the tissues and *Yersinia* colonies, respectively (Fig. 3b).
7. Collect the tissues and fix them with zinc formalin overnight and the next day change out the fixation buffer for 70% isopropanol (*see* **Note 18**).
8. Complete the H&E stain or WS silver stain according to standard protocols.

4 Notes

1. If possible, use *Ifng*$^{-/-}$ mice as a positive control of *Y. enterocolitica* infection to check the virulence of *Y. enterocolitica* at each experiment. *Ifng*$^{-/-}$ mice start losing their body weight and survival diminishes beginning around day 4.
2. This strain is resistant to chloramphenicol and was obtained from Dr. J. Heesemann (Ludwig Maximilian University of Munich, Germany) [16].
3. Based on experience, one male *Rag1*$^{-/-}$ donor mouse (6–8 weeks old) should yield approximately 0.3–0.6 × 10^5, 0.5–1.2 × 10^5, and 0.5–1.2 × 10^5 cells in the CCR6$^+$ ILC3s, CCR6$^-$NKp46$^+$ ILC3s, and CCR6$^-$NKp46$^-$ ILC3s, respectively.
4. Peyer's patches do not develop in *Rag1*$^{-/-}$ mice. Therefore, there is no need to remove Peyer's patches.
5. Before using the buffers, prewarm at 37 °C.

6. While it depends on the condition of the intestine, these steps usually need repeating 5–7 times.
7. Percoll must be kept at room temperature before use, because temperature differences will affect the gradient.
8. Even though the cells are centrifuged with brake , the gradient is not affected.
9. Before centrifuging, be sure to mix the tube well and completely disrupt the remaining Percoll layer.
10. For sorting Group 1 ILCs, do not add CD11b within the lineage cocktail because intestinal NK cells can express CD11b.
11. Wash the cells twice to remove the unbound antibody or LIVE/DEAD Fixable Dead Cell Stain from the cells.
12. Adjust the cell concentration as necessary to achieve a maximum sort rate.
13. For long term storage of *Y. enterocolitica*, make a glycerol stock. After you have second culture, add 700 μL of the *Y. enterocolitica* culture to 300 μL of 50% glycerol in a 2 mL screw top tube or cryovial and gently mix by pipetting. Freeze the glycerol stock tube at −80 °C. Do not let the glycerol stock unthaw.
14. Be sure that LB media looks clear at inoculation to confirm LB media is not contaminated.
15. Ensure that the cap of the culture tube is loose letting oxygen enter and adjust the shaking speed as necessary for full aeration.
16. Ensure removal of air bubbles by flicking the syringe.
17. Do not fix the tissues with zinc formalin for a long time. After 24 h, samples should be placed in 70% isopropanol for long-term storage.
18. A 1.5 mL tube is not suitable for homogenization of spleen and liver. Thus, a bigger tube, such as an FACS tube, should be used.

Acknowledgments

This work was supported by grants from the National Institutes of Health P01 DK46763, R01 AI61516, and MIST U01 AI125955 to M.K.; F32 AI140581 to D.A.G.; and National Research Foundation (NRF) of Korea 2013R1A1A2057931 to G.-Y.S. We thank the staff of the Microscopy & Histology Core (Zbigniew Mikulski and Angela Denn), Flow Cytometry Core and the Department of Laboratory Animal Care (DLAC) at LJI for excellent technical assistance, SDx Histopathology Inc., (Carlsbad, CA) for help with histology analysis.

References

1. Artis D, Spits H (2015) The biology of innate lymphoid cells. Nature 517:293–230
2. Klose CS, Kiss EA, Schwierzeck V et al (2013) A T-bet gradient controls the fate and function of CCR6-RORgammat+ innate lymphoid cells. Nature 494:261–265
3. Luci C, Reynders A, Ivanov I et al (2009) Influence of the transcription factor RORgammat on the development of NKp46+ cell populations in gut and skin. Nat Immunol 10:75–82
4. Sanos SL, Bui VL, Mortha A et al (2009) RORgammat and commensal microflora are required for the differentiation of mucosal interleukin 22-producing NKp46+ cells. Nat Immunol 10:83–91
5. Satoh-Takayama N, Vosshenrich CA, Lesjean-Pottier S et al (2008) Microbial flora drives interleukin 22 production in intestinal NKp46 + cells that provide innate mucosal immune defense. Immunity 29:958–970
6. Eberl G, Marmon S, Sunshine MJ et al (2004) An essential function for the nuclear receptor RORgamma(t) in the generation of fetal lymphoid tissue inducer cells. Nat Immunol 5:64–73
7. Cella M, Fuchs A, Vermi W et al (2009) A human natural killer cell subset provides an innate source of IL-22 for mucosal immunity. Nature 457:722–725
8. Sonnenberg GF, Monticelli LA, Elloso MM et al (2011) CD4(+) lymphoid tissue-inducer cells promote innate immunity in the gut. Immunity 34:122–134
9. Buonocore S, Ahern PP, Uhlig HH et al (2010) Innate lymphoid cells drive interleukin-23-dependent innate intestinal pathology. Nature 464:1371–1375
10. Goto Y, Obata T, Kunisawa J et al (2014) Innate lymphoid cells regulate intestinal epithelial cell glycosylation. Science 345:1254009
11. Song C, Lee JS, Gilfillan S et al (2015) Unique and redundant functions of NKp46+ ILC3s in models of intestinal inflammation. J Exp Med 212:1869–1882
12. Trulzsch K, Oellerich MF, Heesemann J (2007) Invasion and dissemination of Yersinia enterocolitica in the mouse infection model. Adv Exp Med Biol 603:279–285
13. Bottone EJ (1997) Yersinia enterocolitica: the charisma continues. Clin Microbiol Rev 10:257–276
14. Cover TL, Aber RC (1989) Yersinia enterocolitica. N Engl J Med 321:16–24
15. Seo GY, Shui JW, Takahashi D et al (2018) LIGHT-HVEM signaling in innate lymphoid cell subsets protects against enteric bacterial infection. Cell Host Microbe 24:249–260. e244
16. Trulzsch K, Sporleder T, Igwe EI et al (2004) Contribution of the major secreted yops of Yersinia enterocolitica O:8 to pathogenicity in the mouse infection model. Infect Immun 72:5227–5234
17. Guo X, Muite K, Wroblewska J et al (2016) Purification and adoptive transfer of group 3 gut innate lymphoid cells. Methods Mol Biol 1422:189–196

Chapter 13

Isolation, Characterization, and Culture of Intestinal Intraepithelial Lymphocytes

Olivia Jane James, Maud Vandereyken, and Mahima Swamy

Abstract

Intestinal intraepithelial lymphocytes (IEL) comprise distinct groups of innate-like and memory T cells that collectively form one of the largest T cell compartments in the body. IEL are located within the intestinal epithelium and are the first immune cells in the gut to interact with the food, microbiota, and pathogens that the gut is continually exposed to. IEL can respond rapidly to external insults to protect the small intestinal epithelium but are also considered regulatory cells that are important to maintain the homeostasis of the gut. However, the mechanisms of IEL activation and their interactions within the epithelium remain largely elusive. Indeed, IEL are not commonly evaluated even in studies of gut immunology, potentially because they are perceived as being difficult to isolate and study. In this protocol, we present a simplified method to isolate IEL from the murine small intestine and provide representative data for flow cytometric analyses of the different IEL subsets. We also outline two procedures for culturing IEL, which can permit functional studies and coculture with epithelial cells. These strategies should make studies of this large but enigmatic T cell compartment more accessible and open up understanding of homeostatic mechanisms in the intestine, and tissue-associated immunity.

Key words Intraepithelial lymphocytes, IEL, Innate-like lymphocytes, Small intestine, γδ T cells, Unconventional T cells

1 Introduction

The intestinal lumen is home to the diverse network of commensal bacteria that are crucial for intestinal homeostasis, but also to many potentially pathogenic stimuli that could cause the body harm. The main role of the epithelium is nutrient absorption, but it also acts as an important physical barrier between the external environment (intestinal lumen) and the host (lamina propria). This physical barrier is reinforced by resident immune cells, such as intraepithelial lymphocytes (IEL), which are a specialised and extremely heterogeneous population of T lymphocytes that sit between the single-cell layer of epithelial cells. It is estimated that for every ten epithelial cells, there is one IEL, and considering the surface area of the small intestine, IEL account for a substantial population of the

Shoba Amarnath (ed.), *Innate Lymphoid Cells: Methods and Protocols*, Methods in Molecular Biology, vol. 2121, https://doi.org/10.1007/978-1-0716-0338-3_13,

body's lymphocytes [1]. Thus, understanding how these cells function in both health and disease is fundamental to our understanding of intestinal homeostasis.

IEL are T cells and therefore express a T cell antigen receptor (TCR) consisting of either αβ or γδ chains, alongside TCR coreceptors CD8αβ or CD8αα and to a lesser extent CD4(+/−). In mice, the most prevalent IEL subsets within the epithelium of the small intestine are those expressing TCRγδ alongside CD8αα (TCRγδ CD8αα), which account for half the total IEL pool. The high percentage of cells expressing TCRγδ is one of the main phenotypes distinguishing IEL from conventional peripheral blood T cells, where TCRγδ-expressing T cells only account for a small percentage. The remaining IEL subsets are those expressing TCRαβ in addition to either CD8αβ or CD8αα (TCRαβ CD8αβ and TCRαβ CD8αα, respectively). IEL differ in many ways from conventional peripheral T cells, not only in the conformation of the TCR and CD8 coreceptors but also in their constitutive expression of activation markers such as CD44 and CD69. In addition, they express a range of other activating and inhibitory receptors, such as NK receptors, allowing them to rapidly respond to danger and stress signals in their microenvironment [1]. Further, as resident T cells, IEL do not recirculate and are retained in the epithelium by interactions such as αEβ7 (also known as CD103) on their surface binding to E-cadherin on epithelial cells [2]. However, there is still relatively limited research into the functionality of these unconventional T cells, compared to T cells found in the spleen and blood.

Germane to our understanding of IEL is a good protocol to process and isolate these cells. Multiple protocols exist for IEL isolation [3, 4], and various technical improvements have been suggested, including the use of multiple Nylon wool columns [5], multiple Percoll gradients [6], elutriation centrifugation [7], mostly designed to improve the purity of IEL preparations from intestinal epithelial cells. However, we find that for routine analyses, such as evaluating the subsets and numbers of IEL in various relevant knockout mouse models, the simplified protocol presented here is sufficient to get large numbers of IEL (up to ten million per murine small intestine), within 2 h of euthanizing a mouse. This protocol enriches murine small intestinal IEL to about 50–60%; if purer populations are needed, this can be achieved by using a magnetic cell separation technique to quickly enrich for a specific population, for example, CD8 IEL. We have described a straightforward staining panel that allows for not only the identification of total IEL but also distinction between the various subsets. Finally, we describe short-term culture of IEL supported by IL-15 [8], and a longer-term culture protocol for intestinal IEL that involves stimulating IEL through their TCR in the presence of a cocktail of cytokines [9, 10].

2 Materials

2.1 Dissection of Murine Small Intestine

Dissection instruments (Fine Science Tools) consisting of:

Fine scissors (Toughcut, straight).

Reusable Feeding Needle 20 Gauge Straight (referred to as gavage needle in methods).

Bonn artery scissors with Ball tip.

Graefe Extra Fine Forceps Curved Serrated (2×).

10–20 ml disposable syringe.

1× Dulbecco's Phosphate Buffered Saline (PBS), store at 4 °C.

Supplemented RPMI media: RPMI 1640 media, 10% fetal bovine serum (FBS), 1% penicillin/streptomycin antibiotic mix, 1% L-glutamine (all from Thermo Fisher Scientific). Incubate at 37 °C in a water bath prior to IEL isolation.

2.2 Isolation of IEL from Small Intestine

100 μm cell strainers (1× per gut).

DL-Dithiothreitol (DTT): 1 M stock solution in H_2O, sterile filter and store at −20 °C in 500 μl aliquots. Use for IEL isolations at 1 mM in supplemented media.

10× PBS.

Percoll: Prepare a 90% isotonic stock by adding 110 ml of 10× PBS to 1 L Percoll, this stock is kept at 4 °C and used to make fresh working stocks (36/67%) of Percoll solutions for IEL isolations.

36% Percoll: prepare 10 ml by adding 4 ml of 90% Percoll stock to 6 ml 1× PBS.

67% Percoll: prepare 10 ml by adding 7.5 ml of 90% Percoll stock to 2.5 ml 1× PBS.

2.3 Characterization of IEL Subsets by Flow Cytometry

FACS buffer: 1× PBS, 1% FBS, 0.05% sodium azide.

Fc block (Purified anti-mouse CD16/32 Antibody, clone: 93): 1:200 in FACS buffer.

4′,6-Diamidino-2-phenylindole (DAPI): 1:20,000 in FACS buffer.

Antibodies for identification of IEL subsets:

TCRβ-APC (clone: H57-597): 1:200 in FACS buffer.

TCRγδ-PerCP-eFluor710 (clone: GL3): 1:200 in FACS buffer.

CD8β-FITC (clone: H35-17.2): 1:500 in FACS buffer.

CD8α-PE (clone: 53-6.7): 1:400 in FACS buffer.

CD4-PE/Cy7 (clone: RM4-5): 1:400 in FACS buffer.

2.4 Culture of Intestinal IEL

Culture medium: RPMI 1640 media, supplemented with 10% FBS, 1% Penicillin-Streptomycin (stock 10,000U/ml), 2 mM L- glutamine, 1 mM sodium pyruvate, 1 mM β-mercaptoethanol, 2.5 mM HEPES, and 1% non- essential amino acids.

Anti-CD3ε antibody (clone: 2C11).

Cytokines: Human IL-2 (Proleukin, stock 10,000 U/ml), murine soluble IL-15 (Immunotools, stock 100 μg/ml), murine IL-3 (R&D Systems, stock 1 × 10^4 U/ml), murine IL-4 (R&D Systems, stock 2 × 10^4 U/ml), murine IL-15/IL-15R Complex Recombinant Protein (eBioscience, stock 100 μg/ml). Prepare stock concentrations in PBS containing 1% BSA, after dissolving lyophilized cytokines as per the manufacturer's instructions.

3 Methods

3.1 Dissection of Murine Small Intestine

1. Cut intestine from ~1 cm below stomach (proximal duodenum), pulling out the entire length of the small intestine and removing the fat and mesentery using tweezers as you do so. Make the final cut to release the small intestine ~1 cm before caecum (terminal ileum) (*see* Fig. 1; pictures 1–3, also *see* **Notes 1** and **2**).
2. Flush the intestine with 10–20 ml ice-cold PBS using a syringe with a gavage needle attached. Put the gavage needle into the widest opening of the intestine (the end cut from nearest the stomach) and slowly infuse PBS until all the luminal content is removed. Push the gut onto the gavage needle as you flush until it is empty (*see* Fig. 1; pictures 4 and 5, also *see* **Note 2**).
3. [OPTIONAL] Peyer's patches may be removed at this point using curved dissection scissors (*see* **Note 3**).
4. Using ball tip scissors cut the intestine longitudinally to expose the lumen. Insert the ball tip of the scissors into the widest opening of the intestine and continue to push the rest of the intestine onto the scissors while simultaneously cutting until the intestine is cut open along its length (*see* Fig. 1; picture 6).
5. Using forceps, pick up the intestine midway along its length so that it is folded in half, and hold it over a 50 ml falcon tube containing 25 ml pre- warmed supplemented RPMI media (*see* Fig. 1; picture 7, *see* **Note 4**).
6. Cut the small intestine transversely into small 5–10 mm pieces into the media (*see* Fig. 1; picture 7).

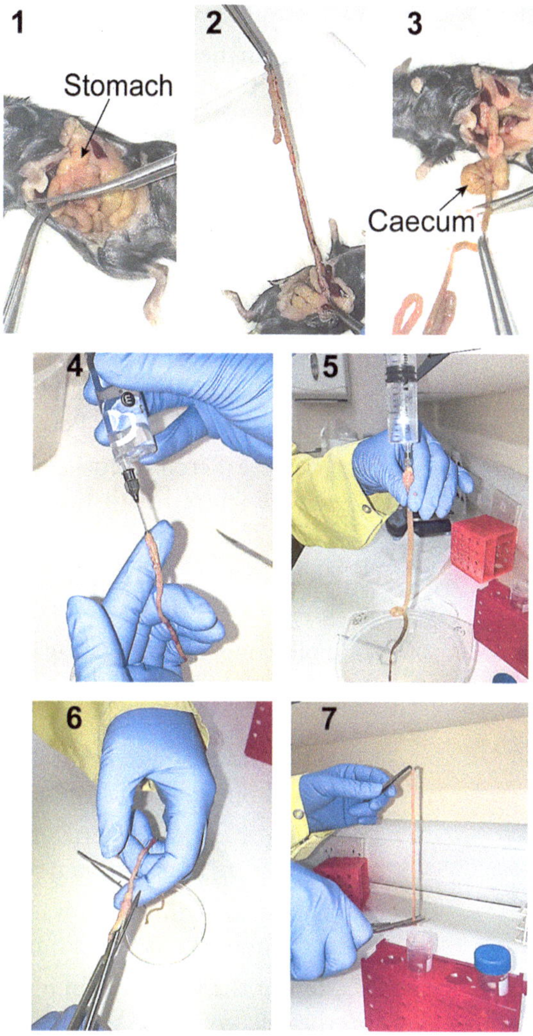

Fig. 1 Dissection of murine small intestine—step-by-step representative images for protocol

3.2 Isolation of IEL from the Small Intestine

1. Add DTT to the media containing the intestinal pieces to a final concentration of 1 mM.
2. Incubate for 40 min with constant shaking at room temperature (*see* **Note 5**).
3. Centrifuge for 5 min at 500 × *g* at room temperature.
4. Discard supernatant and resuspend the pieces in 10 ml warm supplemented media.
5. Vortex at medium speed for 3 min (*see* **Note 6**).
6. Pass the resuspension through a 100 μm cell strainer and collect the flow-through into a new 50 ml Falcon tube (this flow-through contains the IEL).
7. Put the intestinal pieces back into the Falcon tube and resuspend in 10 ml of fresh media.

8. Repeat **steps 9** and **10** for a total of two vortex steps.
9. Use a further 30 ml to wash the cell strainer completely and top up the flow-through to 50 ml.
10. Discard the intestinal pieces.
11. Centrifuge the resulting flow-through for 5 min at 500 × *g* at room temperature.
12. The cells will be in a dense pellet with a red tint, resuspend the pellet in 8 ml 36% Percoll and pipette up and down to thoroughly mix.
13. For good separation of IEL from epithelial cells, we set up two Percoll gradients per gut. Take two fresh 15 ml falcon tubes and add 4 ml of 67% Percoll into each.
14. Create a Percoll density gradient by overlaying 4 ml of 36% Percoll (mixed with the cells) on top of the 67% Percoll (*see* **Note 7**).
15. Centrifuge for 30 min at 700 × *g* at room temperature without brake (*see* **Note 8**).
16. Following centrifugation, the cells of interest appear at the interface between the 36% Percoll top layer and the 67% Percoll bottom layer (*see* Fig. 2).
17. Pipette or aspirate off the epithelial cells and top layer of Percoll containing epithelial cells.
18. Using a p1000 pipette, collect the IEL from the interface, dispensing them into a new 15 ml Falcon tube.
19. Top up to 15 ml with warm media and centrifuge at 700 × *g* for 5 min to wash cells out of Percoll.
20. Repeat the wash step in medium of choice depending on what the cells are going to be used for, for example; if the cells are to be put into culture, do a second wash in culture medium, if the cells are to be used for flow cytometry do a second wash in FACS buffer.
21. Isolated IEL are now ready to use, expected yield is 4×10^6–1×10^7 total live cells per intestine (*see* **Note 9**).
22. If higher purity of IEL is required, it is possible to perform a second Percoll step after the 36/67% density gradient by removing the IEL layer, spinning down the cells and resuspending the cell pellet in 5 ml of 40% Percoll to be centrifuged for 20 min at 500 × *g*. This will further separate some of the remaining epithelial cells from the IEL, giving a purer population. However, for the purest IEL populations, it is advisable that immunomagnetic isolation kits are used, which will incur a decrease in cell numbers but will improve purity (*see* **Note 10**, Fig. 3).

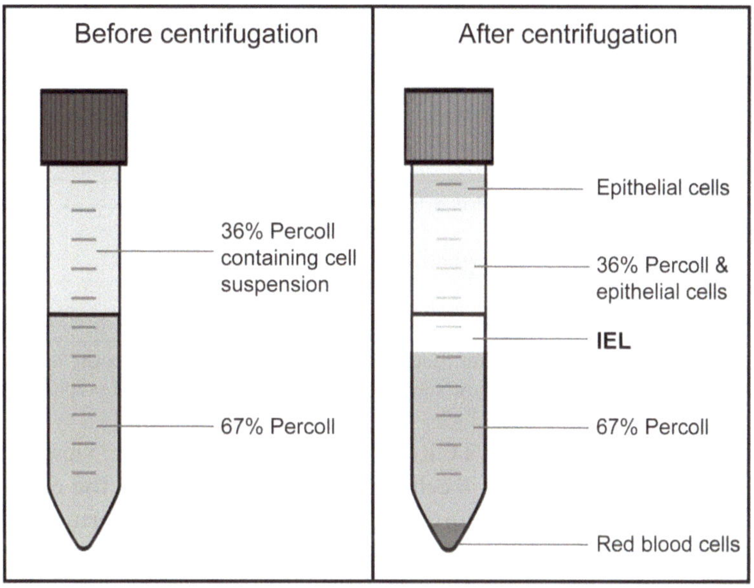

Fig. 2 Diagram of 36/67% Percoll density gradient before and after centrifugation. Before centrifugation, 4 ml of resuspended cells in 36% Percoll is overlaid onto 4 ml 67% Percoll. After centrifugation, the epithelial cells appear as a dense layer at the very top with more epithelial cells diffused amongst the 36% Percoll layer. The IEL are found at the interface between the 36% and 67% Percoll layers as a cloud-like band. Beneath the 67% Percoll layer, at the base of the 15 ml centrifugation tube, a small pellet of red blood cells and dead cells is seen

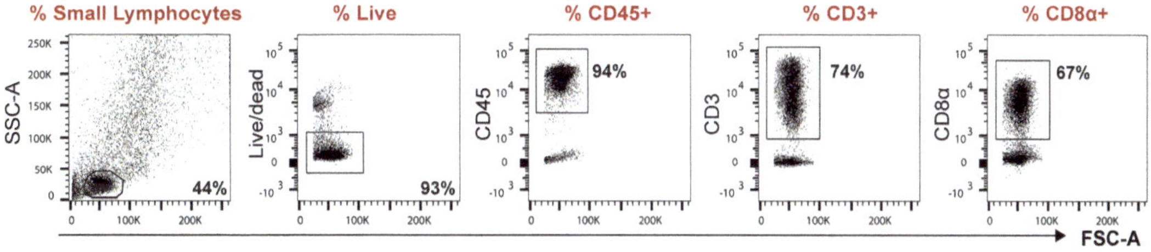

Fig. 3 Flow cytometric data showing the expression of various surface markers on cells isolated via Percoll density gradient. Small lymphocytes account for about half of all the cells isolated using the methods described in this manuscript, and most (>90%) are live as judged by lack of expression of live/dead dye such as DAPI. The percentage of live lymphocytes will vary from mouse to mouse and will also depend on quality of the isolation. Generally, >90% of all live lymphocytes isolated via this method are CD45+, >70% are CD3ε+ and >60% are CD8α+, making these surface receptors useful markers to use for magnetic- or fluorescence-based cell separation techniques to gain purer populations of IEL following a Percoll separation

3.3 Characterization of IEL Subsets by Flow Cytometry

The major IEL subsets routinely retrievable from the murine small intestine are TCRγδ$^+$ CD4$^-$CD8αα$^+$ (TCRγδCD8αα), TCRαβ$^+$ CD4$^-$CD8αα$^+$ (TCRαβ CD8αα), TCRαβ$^+$ CD4$^-$CD8αβ$^+$ (TCRαβ CD8αβ), TCRαβ$^+$ CD4$^+$CD8α$^-$ (TCRαβ CD4) and TCRαβ$^+$ CD4$^+$CD8αα$^+$ (TCRαβ DP). TCRαβ CD8αβ and

Table 1
Staining panel (example) and antibody dilutions

Sample	Antibody–Fluorophore (2× concentrated dilution)
Sample 1: Multistain mastermix for identifying IEL (example)	CD4-PE Cy7 (1:200) TCRβ-APC (1:100) TCRγδ-Percp eFluor710 (1:100) CD8α-PE (1:200) CD8β-FITC (1:250)

This table shows a basic staining panel used to identify IEL isolated via the methods described in this chapter. The fluorophores used are for demonstration purposes and can be substituted as appropriate

TCRαβ CD4 IEL are commonly viewed as tissue-resident memory T cells that have migrated into the epithelium after antigen exposure [11, 12]. Conversely, the former two subtypes are commonly viewed as unconventional or innate-like T cells that constitutively populate the intestinal epithelium after migrating there directly from the thymus or having undergone selective expansion in situ [13]. A basic staining protocol to characterize the presence of these four subsets by flow cytometry is presented below.

1. Count cells and resuspend to a final concentration of 1 million cells/ml in FACS buffer.
2. Plate cells out at 200 μl per stain (per well) in a round-bottom 96-well plate (*see* **Note 11**).
3. Centrifuge the plate at 500 × *g* for 2 min at 10 °C.
4. Flick off supernatant and vortex plate to resuspend cells.
5. Add 25 μl Fc block (1:200 in FACS buffer) and incubate for 5 min on ice.
6. Make a 2× concentrated multistain mastermix of all fluorophore-conjugated antibodies to be used in FACS buffer (*see* Table 1 for antibodies we use to identify IEL subsets).
7. Prepare appropriate compensation controls.
8. Add 25 μl of fluorophore–antibody mastermix to the cells containing the Fc block.
9. Incubate for 15 min on ice and protect from light.
10. Add 150 μl FACS buffer and centrifuge at 500 × *g* for 2 min at 10 °C.
11. Flick off supernatant and vortex the plate.
12. Add 200ul FACS buffer containing a live/dead discrimination dye such as DAPI (*see* **Note 12**).
13. Transfer the stained samples and compensation controls to suitable flow cytometry tubes and acquire on a flow cytometer, such as the BD LSR Fortessa.

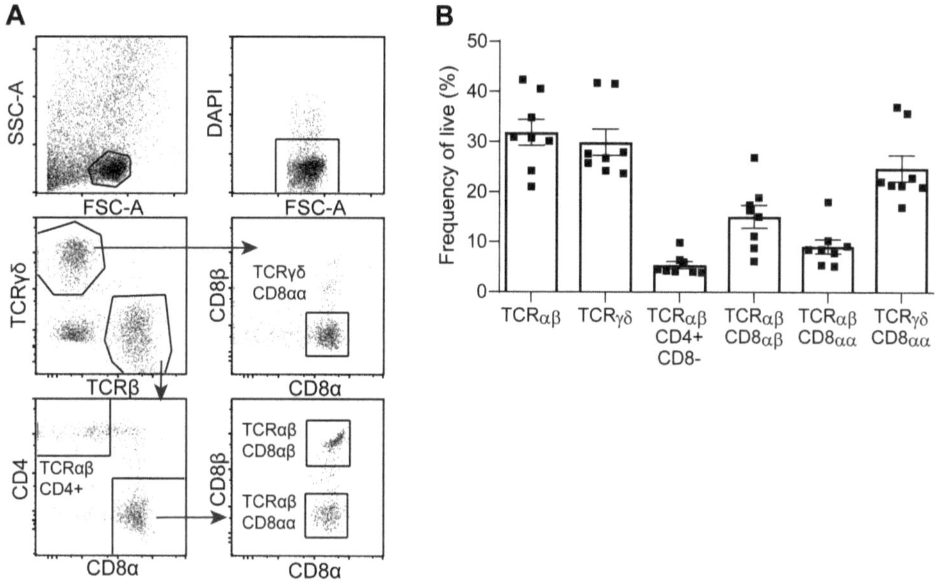

Fig. 4 Identification of IEL subpopulations. (**a**) Gating strategy used to identify distinct IEL subpopulations using flow cytometry. Representative plots show IEL isolated using a 36/67% Percoll density gradient. Cells were gated on size using forward scatter (FSC) vs side scatter (SSC) to separate out cells of interest from debris/non lymphoid cells. Live cells (>90%) were gated on as those that did not express DAPI, as DAPI only penetrates cells with permeabilized cell membranes such as those of dead cells. Live cells were then separated into the various IEL subpopulations based on cell surface marker expression of T cell receptors and T cell receptor-associated proteins; TCRγδ, TCRβ, CD4, CD8α, and CD8β. (**b**) Expected frequencies of IEL subpopulations using the suggested gating strategy, shown as a percentage of live single cells. TCRαβ DP are not depicted in the figure as they typically account for <1% of the total live cells. Each dot represents one murine IEL isolation; $n = 8$

14. After acquisition and appropriate compensation, the basic gating strategy and expected distributions of the different IEL subsets are shown in Fig. 4.

3.4 Culture of Murine Intestinal IEL

IEL can be sorted or plated out immediately after Percoll isolation for further culture. IEL are prone to cell death ex vivo [14], however IL-15 maintains their survival in the short term [8], and we find that IL-15 complexed to the IL-15 receptor α subunit is more efficient than IL-15 on its own. For longer term cultures, and to expand IEL numbers for downstream analyses, TCR stimulation in the presence of a cocktail of cytokines is required.

For short-term culture:

1. Count IEL and resuspend to a final concentration of 1 million cells/ml in culture medium containing 20 ng/ml murine IL-15/IL-15R Complex.
2. Plate cells in a 96-well round-bottom plate at 200 μl per well.
3. Incubate at 37 °C, 10% CO_2. Cells can be maintained in IL-15 for 1–3 days.

For longer term cultures:

4. Coat a 96-well round bottom plate with anti-CD3ε antibody (1 μg/ml in PBS, 50 μl per well) for 2 h at 37 °C.
5. Wash with 1×PBS.
6. Count IEL and resuspend 0.5 million cells/ml in culture medium. Add a cytokine cocktail containing IL-2 (10 IU/ml), IL-3 (100 IU/ml), IL-4 (100 IU/ml), and soluble IL-15 (100 ng/ml).
7. Plate cells in the anti-CD3ε coated 96-well round bottom plate at 200 μl per well.
8. Incubate at 37 °C, 10% CO_2 for 48 h.
9. Wash the cells twice and resuspend in IL-2 (10 IU/ml) containing culture medium at a concentration of 100,000 cells per well in a fresh noncoated 96-well plate (*see* **Note 13**).
10. Replate the cells every 3–4 days in fresh IL-2-containing medium. Cells can be utilized for further downstream analyses after a further 10–12 days in culture.

4 Notes

1. Remove the small intestine as quickly as possible after culling the mouse as this has a significant influence on the number of IEL isolated. It is also important to keep the gut in ice-cold PBS until it is processed into the RPMI, to minimize cell death and losses in cell numbers.
2. There are on average 6–10 Peyer's patches in a healthy adult mouse small intestine, and they tend to be concentrated at the ileal end. We do not routinely remove Peyer's patches, as with the mild isolation protocol above, there is minimal contamination of IEL with Peyer's patch lymphocytes, and these can normally be gated out as they are mainly $CD4^+$ $CD8\alpha^{-ve}$ $CD103^{-ve}$ T cells.
3. Do this within a high-walled beaker as luminal content tends to spray through leaks under tension.
4. A minimum of 10 ml media per gut can be used.
5. This depends on what is available. We use a rotating shaker at room temperature; however, 37 °C is optimal.
6. As a guide, a medium speed on a Vortex Genie is achieved by setting the speed between 7 and 8.
7. Take your time when creating the Percoll density gradients as the layers can easily mix if you are not careful, a good gradient is best achieved using a 3 ml Pastette or a Pipette Boy set at the lowest ejection speed.

8. To achieve a low brake force in a Beckman Coulter Allegra X-12R centrifuge, set acceleration to 6 and deceleration to 3.
9. The yield tends to increase with age and varies dependent on the facility in which the mice are bred. Numbers indicated here are from C57BL6/J mice obtained from Charles River and analyzed between 8 and 12 weeks of age.
10. More than 90% of live cells isolated via the IEL isolation methods described here are CD45+, >80% express CD3ε and >60% are CD8α+ (*see* Fig. 3), making these surface receptors ideal for enrichment of IEL following a Percoll density gradient. We routinely use a CD8α positive selection kit to further purify IEL, such as the EasySep™ CD8α positive selection kit provided by StemCell Technologies.
11. Staining in a 96-well plate allows you to use less cells and a smaller amount of antibody, it can also be faster when washing and more accurate due to similar dead volume in each sample after every wash.
12. There are many dead epithelial cells remaining in the stained cell samples, so adding a live–dead discrimination dye is vital for exclusion of these cells and hence for cleaner gating of IEL subpopulations.
13. During culture of IEL, there is always an initial loss in cell numbers after 1 day, but numbers then stay stable (in short-term cultures) or recover and increase (in long-term cultures).

Acknowledgments

MV and MS are supported by the Wellcome Trust and Royal Society (Sir Henry Dale Fellowship to MS, 206246/Z/17/Z), and OJJ is supported by a Wellcome Trust PhD studentship (215309/Z/19/Z). We are grateful to the members of Prof. Adrian Hayday's lab, King's College London, from whom these protocols were modified, particularly Dr. Sarah Clarke who developed the protocol for long-term culture of IEL.

References

1. Olivares-Villagómez D, Van Kaer L (2018) Intestinal intraepithelial lymphocytes: sentinels of the mucosal barrier. Trends Immunol 39:264–275
2. Schön MP, Arya A, Murphy EA et al (1999) Mucosal T lymphocyte numbers are selectively reduced in integrin alpha E (CD103)-deficient mice. J Immunol 162:6641–6649
3. Lefrancois L, Lycke N (2001) Isolation of mouse small intestinal intraepithelial lymphocytes, Peyer's patch, and lamina propria cells. Curr Protoc Immunol. Chapter 3:Unit 3.19
4. Lundqvist C, Hammarström ML, Athlin L et al (1992) Isolation of functionally active intraepithelial lymphocytes and enterocytes from human small and large intestine. J Immunol Methods 152:253–263
5. Lee Mosley R, Klein JR (1992) A rapid method for isolating murine intestine intraepithelial

lymphocytes with high yield and purity. J Immunol Methods 156:19–26

6. Montufar-Solis D, Klein JR (2006) An improved method for isolating intraepithelial lymphocytes (IELs) from the murine small intestine with consistently high purity. J Immunol Methods 308:251–254
7. Binda E, Erhart D, Schenk M et al (2009) Quantitative isolation of mouse and human intestinal intraepithelial lymphocytes by elutriation centrifugation. J Immunol Methods 344:26–34
8. Lai YG, Gelfanov V, Gelfanova V et al (1999) IL-15 promotes survival but not effector function differentiation of CD8+ TCRalphabeta+ intestinal intraepithelial lymphocytes. J Immunol 163:5843–5850
9. Bas A, Swamy M, Abeler-Dörner L et al (2011) Butyrophilin-like 1 encodes an enterocyte protein that selectively regulates functional interactions with T lymphocytes. Proc Natl Acad Sci U S A 108:4376–4381
10. Swamy M, Abeler-Dörner L, Chettle J et al (2015) Intestinal intraepithelial lymphocyte activation promotes innate antiviral resistance. Nat Commun 6:7090
11. Mucida D, Husain MM, Muroi S et al (2013) Transcriptional reprogramming of mature $CD4^+$ helper T cells generates distinct MHC class II-restricted cytotoxic T lymphocytes. Nat Immunol 14:281–289
12. Schenkel JM, Masopust D (2014) Tissue-resident memory T cells. Immunity 41:886–897
13. Mcdonald BD, Jabri B, Bendelac A (2018) Diverse developmental pathways of intestinal intraepithelial lymphocytes. Nat Rev Immunol 18:514–525
14. Viney JL, Macdonald TT (1990) Selective death of T cell receptor gamma/delta+ intraepithelial lymphocytes by apoptosis. Eur J Immunol 20:2809–2812

Chapter 14

Isolation and Characterization of Innate Lymphoid Cells within the Murine Tumor Microenvironment

Grace Mallett, Warren Patterson, Megan Payne, and Shoba Amarnath

Abstract

Innate lymphoid cells (ILCs) are important for both tissue immunity and tissue homeostasis. They are classified into three groups: Group 1 ILCs include NK cells, which are important in eliciting immunity against intracellular pathogens; group 2 ILCs protect against parasitic helminths; and group 3 ILCs protect against extracellular pathogens. The role of ILCs in cancer immunity remains unclear. In this chapter, we discuss methods for isolating and characterizing tumor-infiltrating ILC subsets within the tumor microenvironment in an experimental murine model of B16 melanoma. The chapter also highlights the expression of PD-1 on the various ILC subsets within the tumor microenvironment.

Key words B16, Melanoma, ILCs, PD-1

1 Introduction

Innate lymphoid cells (ILCs) are an important component of innate tissue immunity and tissue homeostasis. ILCs have also been associated with several different disease pathologies including atopic dermatitis and colitis [1, 2]. The role of ILCs in cancer, however, remains ambiguous and unclear with little evidence to demonstrate a direct contribution of ILCs to tumor biology. Therefore, the function of ILCs within the tumor microenvironment is yet to be defined.

ILCs are tissue resident cells which are predominately associated with mucosal tissue and are often described as the innate counterpart of $CD4^+$ T helper cells. Unlike T Cells, ILCs lack an antigen recognition receptor and instead rely on environmental stimuli from cytokines and alarmins for activation. ILCs have been divided into three distinct subsets, mirroring T helper cell subsets, based on transcription factor expression and cytokine production [3].

Group 1 ILCs (including ILC1s and NK cells), produce proinflammatory cytokines IFNγ and TNFα, in response to IL-15 and

Shoba Amarnath (ed.), *Innate Lymphoid Cells: Methods and Protocols*, Methods in Molecular Biology, vol. 2121,
https://doi.org/10.1007/978-1-0716-0338-3_14,

IL-12, and express the master transcription factor T-bet [4–7]. ILC1s are distinguishable from NK cells through their lack of expression of the T-bet family member EOMES [7]. Group 2 ILCs were first identified by Neill et al. [8] and are defined by the expression of transcription factor GATA3. They have since been divided into additional subsets comprising of "Natural" $ST2^{+}$ ILC2s, present under normal homeostatic conditions and "Inflammatory" ILC2s which upregulate the expression of KLRG1 in response to IL-25 activation [9]. ILC2s produce type 2 cytokines IL-5, IL-9, and IL-13 in response to parasitic helminth infection. Group 3 ILCs were first described in the intestine as an innate source of IL-22 [10]. They are defined by the expression of the transcription factor RORγt and respond to bacterial infection and IL-23 stimulation. There are three populations of group 3 ILC based on the expression of the surface markers NCR and CD4. $NCR^{+}CD4^{-}$ ILC3 secrete IL-22, $NCR^{-}CD4^{-}$ ILC3 express both IL-17 and IL-22, and $NCR^{-}CD4^{+}$ ILC3s, described as LTi4 cells, express IL-17 [2, 11–14].

It is well established that NK cells are required for tumor immune-surveillance (reviewed [15]), but the data on the role of ILC1s in tumor is not clear in spite of their type 1 cytokine phenotype. IFNγ can drive antitumor responses by increasing MHC class I expression on tumor cells, this data has been coupled with increased T cell infiltration and tumor elimination [16]. IFNγ has also been shown to inhibit both tumor angiogenesis and tumor growth [17, 18]. Similarly, TNFα is linked with increased type 1 response, with increased dendritic cell production of IL-12, promoting tumor regression [19]. Unlike group 1 ILCs, ILC2s have traditionally been associated with promoting tumor progression rather than providing tumor defense. ILC2 activation in response to IL-33 promotes tumor growth in murine breast cancer [20]. ILC2s have been shown to drive conversion of monocytes into M2 macrophages; an immunosuppressive subset [21]. ILC3s have been indirectly linked to tumor progression again through our understanding of cytokine such as IL-22 which promotes colon cancer [22]. Furthermore, IL-23 has been linked to increasing tumor progression through the activation of ILC3s and production of IL-17 [23].

In order to fully identify the contribution of ILCs and establish the composition and contribution of each of the subsets, it is beneficial to characterize ILC phenotype and function within the TME. In this chapter we outline a methodology for isolating and identifying the frequency and function of ILCs from the TME using an experimental murine model of melanoma. Using the appropriate adaptions, this model could easily be adapted to alternative cancer models and can be used in conjunctions with therapeutic interventions.

2 Materials

2.1 Tumor Cell Culture

1. *Tumor cell media*: DMEM supplemented with 10% fetal bovine serum (FBS), 1 mM sodium pyruvate, 100 IU of penicillin/ streptomycin antibiotics.
2. *Trypsin:* 10× trypsin diluted in sterile PBS.
3. B16 F10 Melanoma cells (Kindly provided by Dr. Gangaplara, NIH).
4. Cell culture flasks (T175).
5. Trypan blue (1 in 2 dilution in PBS).
6. PBS.
7. Hemocytometer.
8. Incubator (37 °C; 5% CO_2).

2.2 Tumor Cell Isolation

1. *Complete mouse media*: DMEM supplemented with 10% fetal bovine serum, 1 mM sodium pyruvate, 1 mM nonessential amino acids, 1 mM L-glutamine, 100 U of penicillin/streptomycin, and 50 μM 2-merceptoethanol.
2. *Liberase/DNase Solution*: 0.5 mg/ml DNase I (Roche) dissolved in FBS-free media (DMEM supplemented with 1 mM sodium pyruvate, 1 mM nonessential amino acids, 1 mM L-glutamine, 100 IU of penicillin/streptomycin, and 50 μM 2-merceptoethanol) which is supplemented with 0.5 mg/ml of Liberase TL powder. Store at −20 °C in 1 ml aliquots.
3. Fetal bovine serum.
4. Lymphocyte Separation Media.
5. 100 μM cell strainer.
6. 50 ml centrifuge tubes.
7. 15 ml centrifuge tubes.

2.3 Cell Stimulation

1. PMA (Phorbol 12-myristate 13-acetate).
2. Ionomycin.
3. Monensin.
4. Brefeldin A.
5. 48-Well plate.
6. Incubator (37 °C, 5% CO_2).

2.4 Flow Cytometry Staining and Analysis

1. *Flow cytometry buffer*: PBS supplemented with 0.1% sodium azide and 2.5% bovine serum albumin.
2. *Permeabilization solution*: 10× solution purchased from BD Biosciences and diluted in dH_2O.
3. Cytofix™ fixation solution (BD Biosciences).

4. 75 mm FACS Tubes.
5. Flow cytometry antibody master mix for cell surface antibodies (*see* Table 1 and 2).
6. Flow cytometry antibody master mix for intracellular antibodies (*see* Table 3).
7. Live/Dead Fixable Dead Blue Stain (Invitrogen).

3 Methods

3.1 Tumor Cell Culture

1. Thaw frozen B16F10 melanoma tumor cell lines in a water bath (*see* **Notes 1** and **2**).
2. Work inside a biosafety cabinet for the rest of the process. Transfer thawed tumor cell lines containing cryovials inside the cabinet. Pipette cells into a 15 ml falcon and then add 10 ml of complete media gently to the cells. Centrifuge at 400 × *g* for 5 min. Resuspend cells in tumor media at 0.5 million per ml and then transfer to a T175 cm^2 conical tissue culture flask with a filter top. Incubate tumor cells overnight in a 37 °C Incubator (5% CO_2).
3. Start cell harvest an hour prior to inoculating the mice in order to have adequate time to prepare the cells, but also limit cell death from harvesting too early which will impact tumor engraftment.
4. Transfer the tumor cell containing flasks into the biosafety cabinet. Under sterile conditions, remove the tumor media and discard. Next, wash the cells with warm sterile PBS in order to remove any traces of the tumor media.
5. After removing the PBS, add enough trypsin to cover the bottom of the flask and incubate for 3–5 min at 37 °C (*see* **Note 3**).
6. Equal amounts of warmed tumor media should then be added to the flask in order to inactivate the trypsin. Transfer to 50 ml centrifuge tube and centrifuge cells at 400 × *g* for 5 min. Discard the supernatant. Resuspend cells in PBS at a concentration of 1×10^6 cells per ml.
7. Shave the area of tumour inoculation. Then, inoculate mice with 2×10^5 B16 F10 melanoma cells via subcutaneous injection. This equates to a 200 μl injection of the cell suspension into the flank (*see* **Note 4**).
8. It is important to monitor and record tumor growth throughout the experiment. Typically, tumors are a substantial size at day 12 post inoculation (*see* **Note 5**). Calculate the tumor volume using the following equation:

$$(\pi/6) \times 0.5 \times \text{Length} \times \text{Width}^2$$ [24].

3.2 Tumor Tissue Harvest

1. When the endpoint of the experiment has been reached, euthanize mice via cervical dislocation. Dissect the entire tumor with 2–3 mm of extra surrounding skin tissue to help prevent tumor bursting and to ensure all tumor tissue has been collected. Immediately transfer the tissue to 5 ml of complete mouse media and store on ice in order to minimize cell death whilst other samples are being collected (*see* **Note 6**).
2. Transfer the tumor tissue to a petri dish and remove any excess nontumor tissue. At this stage, briefly wash tissue in 2–3 ml of PBS in order to remove any excess media which may inactivate the digestion process.
3. Transfer to 1–2 ml of Liberase TL/DNase solution (depending on tumor size). Once in solution, finely chop the tissue into small pieces in order to aid the digestion (*see* **Note 7**). Incubate the samples at 37 °C for 30 min (*see* **Note 8**).
4. When the digestion is complete, transfer the tissue solution into a 50 ml centrifuge tube containing 5 ml FBS through a 100 μM cell strainer with 20–25 ml of complete mouse media (in 5 ml intervals). The FBS will inactivate the digestion process and prevent over digestion. In order to encourage the cells to pass though the cell strainer, pressure may be applied in a circular motion using the plunger of a 5 ml syringe.
5. Centrifuge at 400 × *g* for 5 min and resuspend the pellet in complete mouse media (10 ml). Add this slowly to lymphocyte separation media (5 ml) at a ratio of 2:1 in a 15 ml falcon tube (*see* **Note 9**). Centrifuge at 400 × *g* for 20 min with the break off.
6. Transfer the interphase into a new 15 ml falcon tube and wash with complete media in order to remove traces of the lymphocyte separation media.
7. Count live cells with trypan blue staining, and resuspend in complete media at a concentration of 1×10^6 cells per ml.

3.3 Cell Stimulation

ILC subsets can be determined by cell surface markers, cell stimulation is required to determine ILC function through cytokine expression.

1. In order to determine cytokine production, tumor samples can be stimulated in vitro prior to flow cytometry analysis. Plate 1×10^6 cells in each well of a 48 well plate. Use extra cells to plate an addition well for an unstimulated control.
2. Add ionomycin, PMA, Brefeldin A, and Monensin to each well (except for the unstimulated control) (*see* **Note 10**). These components will stimulate the cytokine production by the cells as well as blocking protein transport through the endoplasmic reticulum, preventing the release of the cytokines and

Table 1
Lineage panel: all antibodies in the lineage cocktail were used at a concentration of 0.1 μg per 1×10^6 cells in 100 μl

Marker	Fluorochrome	Clone
CD3	PerCP Cy5.5	1452/C11
CD5	PerCP Cy5.5	53-7.3
CD8	PerCP Cy5.5	53-6.7
CD11b	PerCP Cy5.5	M1/70
CD11c	PerCP Cy5.5	N418
CD19	PerCP Cy5.5	1D3/CD19
CD49b	PerCP Cy5.5	HMα2
B220	PerCP Cy5.5	RA3-6B2
Ter 119	PerCP Cy5.5	TER-119
F4/80	PerCP Cy5.5	BM8
NK1.1	PerCP Cy5.5	PK136
Grl	PerCP Cy5.5	RB6-8C5

enable intracellular cytokine measurement. Incubate samples at 37 °C (5% CO_2) for 4 h.

3. After incubation, harvest the cells and split them for the relevant staining panels to measure cytokine production.

3.4 Flow Cytometry Analysis

In order to identify ILC populations, a lineage panel is required to exclude other cell types including macrophages, T cells, B cells, granulocytes, eosinophils, and monocytes (Table 1). In addition, several surface markers (Table 2) and intracellular transcription factor (Table 3) markers are required to identify each of the ILC subsets. When preparing the antibody cocktails, always prepare extra than what is required ($n + 2$), in order to account for pipetting error, particularly with large studies. Since both surface and intracellular staining is required, the cells will need to go through a fixation/permeabilization process (*see* **Note 11**). In addition to the samples, Fluorescence Minus One (FMO) controls should also be stained for each of the antibodies (including the Live/Dead stain). It is recommended to make an antibody cocktail for each of the FMOs prior to harvest. Finally, an unstained sample with no antibody staining will be required.

1. Fresh cell samples that were not stimulated, as well as stimulated cells, should all be washed with PBS in order to remove any traces of media.

Table 2
Antibody cocktail for surface antibodies

Suggested antibody panel for cell surface staining			
Marker	**Fluorochrome**	**Clone**	**Antibody concentration (per 1 × 10⁶/ml)**
Linage (*see* Table 1)	PerCP Cy5.5	*See* Table 1	
CD45	APC Cy7	30-F11	1 µg
CD127	Brilliant Violet 510	A7R34	1.5 µg
CD25	PE Cy7	PC61	1 µg
ST2	FITC	DJ8	1 µg
KLRG1	Brilliant Violet 421	2F1/KLRG1	1 µg
Nkp46	APC	29A1.4	1 µg
PD-1	Brilliant Violet 605	29F.1A12	1 µg
Suggested surface antibody panel for ILC cytokine analysis			
Marker	**Fluorochrome**	**Clone**	**Antibody concentration**
Linage (*see* Table 1)	PerCP Cy5.5	*See* Table 1	
CD45	Alexa Fluor 700	30-F11	1 µg
ST2	FITC	DJ8	1 µg
KLRG1	APC Cy7	2F1/KLRG1	1 µg
NKp46	APC	29A1.4	1 µg

Suggested antibody staining panel for unstimulated samples (top panel) and antibody staining panel for stimulated cells in order to assess cytokine production (bottom panel). Here we describe a panel in which all ILC subsets can be identified, this should be adapted according to the required outcome of each individual experiment. Suggested antibody concentrations are shown but should be optimized. Antibody cocktail should be made up at a concentration such that 100 µl of the cocktail can be added to each sample. Antibody cocktails for surface antibodies should be made up in FACS buffer. Antibody cocktails for intracellular antibodies should be made up in permeabilization buffer

2. Next, cells should be stained with Live/Dead stain (with the exception of the unstained cells and the Live/Dead FMO) at a previously optimized concentration. Ensure cells are mixed thoroughly by pipetting, then incubate in the dark at 4 °C for 30 min. After the incubation, wash samples in PBS in order to remove any excess stain.
3. Resuspend each sample in 100 µl of the prepared cell surface antibody cocktail and mix well by pipetting. Incubate at 4 °C, protected from light; unstained control should be incubated at 4 °C protected from light in 100 µl of FACs buffer alone. Individual cocktails for each of the FMO can also be added to each of the FMO tubes at this stage.

Table 3
Suggested antibody staining panel for intracellular antibodies

Marker	Fluorochrome	Clone	Antibody concentration (per 1 × 10^6/ml)
RORγt	PE	B20	1.5 μg
Suggested antibody panel for ILC cytokine analysis			
Marker	Fluorochrome	Clone	Antibody concentration (per ml)
IL-5	Brilliant Violet 421	TRFK5	1 μg
IL-13	PE	eBio13A	1 μg
IL-17	PE Cy7	eBio17B7	1 μg
IFNγ	Brilliant Violet 510	XMG1.2	1 μg
TNFα	Brilliant Violet 785	MP6-XT22	1 μg
IL-10	Brilliant Violet 605	JES5-16ES	1 μg

4. After 30 min, wash cells with FACS buffer. If cells require fixing, resuspend them in 200 μl of fixation buffer, vortex, and incubate at 4 °C for 30 min. If cells do not require fixing, resuspend the samples in 500 μl of FACS buffer and analyze by flow cytometry immediately.
5. After 30 min of fixation, wash cells twice with 1 ml 1× permeabilization buffer. Resuspend the samples in 100 μl of intracellular antibody cocktail and mix well. Incubate the samples at 4 °C in the dark overnight.
6. Wash samples with 1 ml of permeabilization buffer and then resuspend in 400 μl of FACS buffer. Samples are now ready for flow cytometry analysis (representative flow analysis Fig. 1).

4 Notes

1. For optimal tumour engraftment, use cell lines with a passage lower than 10. In these studies, cell lines of P5–6 were used.
2. In order to establish a syngeneic model, ensure the murine stain of the tumor cell line source matches the experimental mouse strain. Mis-match in strains can result in immune responses and adverse tumor growth.
3. A maximum time of 7 min should be observed as longer incubation may destroy cells. If cells remain attached after the incubation, a cell scraper can be used in order to detach the remaining cells.

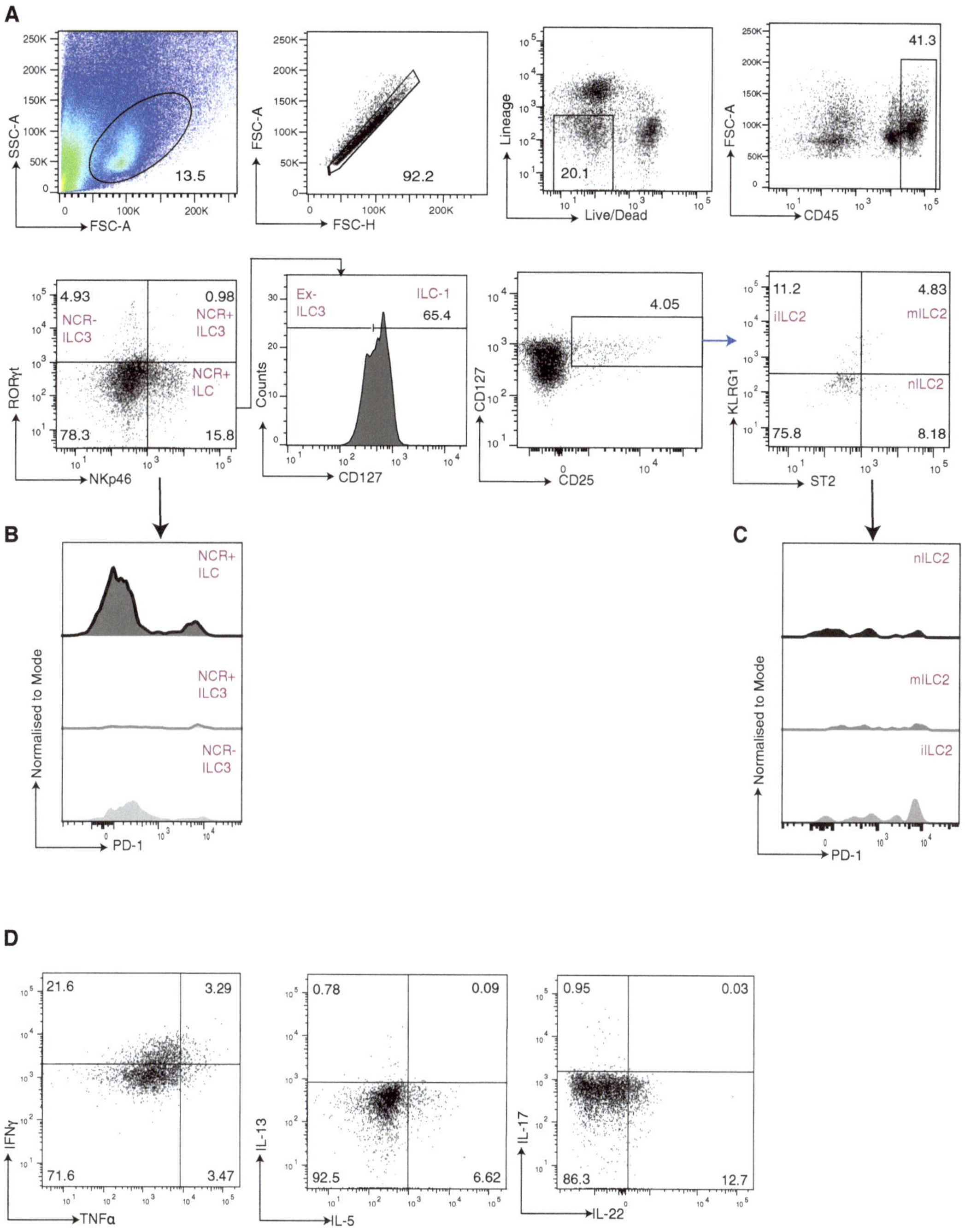

Fig. 1 Representative flow plots of ILC characterization in tumor-infiltrating lymphocytes (TILs). B6 wild-type mouse was inoculated with 2×10^5 B16 melanoma cells via subcutaneous injection. TILs were harvested on day 12 post inoculation and analyzed by flow cytometry. (**a**) Innate lymphoid cells (ILCs) were gated on live cells and were then defined as $Lineage^{-}CD45^{+}$. Lineage was defined as CD3, CD5, CD8, CD11b, CD11c CD19, CD49b, NK1.1, Ter119, B220, F4/80, and Gr1. ILC-1s were defined as $CD127^{+}NKp46^{+}ROR\gamma t^{-}$. Group 2 ILCs were defined as $CD127^{+}CD25^{+}$ and then further subdivided based on expression of KLRG1 and ST2; inflammatory ILC2s were defined as $KLRG1^{+}ST2^{-}$, natural ILC2s were defined as $KLRG1^{-}ST2^{+}$ and mature

4. Tumor placement is important. Tumors placed too low will grow over the lower abdomen and cover the tumor draining lymph nodes; therefore isolating tumor draining lymph nodes will be difficult. Alternatively, if placement is too high, it may increase problems when handling the mice in order to acquire tumor measurement throughout the experiment; tumor ulcerations can affect cell tissue quality.
5. Ensure tumors have reached a large enough volume in order to be able to isolate rare ILC populations; typically, a volume in excess of 400 mm^3 should yield adequate cells for ILC analysis; of note tumors which have grown excessively may have compromised immunobiology so aim for volumes to remain approximately 800 mm^3.
6. Take care to only isolate tumor tissue, take care to not include the tumor draining lymph nodes in the dissection.
7. When processing large studies which use large numbers of mice, Liberase TL solution should be kept on ice until all samples are ready for incubation. This will delay the digestion process in order to avoid over digestion of the first samples.
8. Digestion time is very important, over digestion of samples will destroy cells, under digestion may mean that cells are lost in the tissue. In this model, we propose 30 min is adequate, however for small tumors, digestion time may need to be reduced.
9. Ensure distinct layers between the lymphocyte separation media and the cell suspension is achieved for successful TIL isolation .
10. PMA and Ionomycin can be used at a concentration of 1 μg/ml (PMA) and 0.1 μg/ml (ionomycin); Monensin at a concentration of 2 μM and Brefeldin A at 5 μg/ml. Add Monensin and Brefeldin A only for the last 2hrs of the 4hr incubation .
11. If only surface markers are required for staining, the fixing process is not necessary, instead after surface staining, cells should be washed and analyzed by flow cytometry the same day.

Fig. 1 (continued) ILC2s were defined as KLRG1$^+$ST2$^+$. Group 3 ILCs were subdivided into NCR$^+$ (Nkp46$^+$RORγt$^+$) and NCR$^-$ (Nkp46$^-$RORγt$^+$). Panels (**b**) and (**c**) shows PD-1 expression on the various ILC subsets within the TILs. (**d**) For intracellular cytokine analysis, TILs were stimulated for 4 h with PMA and ionomycin. ILCs were gated on live cells and defined as Lineage$^-$CD45$^+$. ILC cytokine expression was analysed by flow cytometry

Acknowledgments

We would like to thank the Academy of Medical Sciences, Springboard Award [SBF003\1129] supported by Wellcome Trust and Newcastle University Research Fellowship for supporting this work. G.M. is supported by a Medical Research Council [MRC] Discovery Medicine North Ph.D. studentship. M.P. is supported by a Wellcome Trust Vacation Scholarship [216503/Z/19/Z].

References

1. Salimi M, Barlow JL, Saunders SP et al (2013) A role for IL-25 and IL-33-driven type-2 innate lymphoid cells in atopic dermatitis. J Exp Med 210(13):2939–2950
2. Geremia A, Arancibia-Carcamo CV, Fleming MP et al (2011) IL-23-responsive innate lymphoid cells are increased in inflammatory bowel disease. J Exp Med 208(6):1127–1133
3. Spits H, Artis D, Colonna M et al (2013) Innate lymphoid cells—a proposal for uniform nomenclature. Nat Rev Immunol 13 (2):145–149
4. Crellin NK, Trifari S, Kaplan CD et al (2010) Regulation of cytokine secretion in human CD127(+) LTi-like innate lymphoid cells by Toll-like receptor 2. Immunity 33(5):752–764
5. Herberman RB, Nunn ME, Holden HT et al (1975) Natural cytotoxic reactivity of mouse lymphoid cells against syngeneic and allogeneic tumors. II. Characterization of effector cells. Int J Cancer 16(2):230–239
6. Weizman OE, Adams NM, Schuster IS et al (2017) ILC1 confer early host protection at initial sites of viral infection. Cell 171 (4):795–808.e12
7. Klose CSN, Kiss EA, Schwierzeck V et al (2014) Differentiation of type 1 ILCs from a common progenitor to all helper-like innate lymphoid cell lineages. Cell 157(2):340–356
8. Neill DR, Wong SH, Bellosi A et al (2010) Nuocytes represent a new innate effector leukocyte that mediates type-2 immunity. Nature 464(7293):1367–1370
9. Huang Y, Guo L, Qiu J et al (2015) IL-25-responsive, lineage-negative KLRG1(hi) cells are multipotential 'inflammatory' type 2 innate lymphoid cells. Nat Immunol 16(2):161–169
10. Sanos SL, Bui VL, Mortha A et al (2009) RORgammat and commensal microflora are required for the differentiation of mucosal interleukin 22-producing NKp46+ cells. Nat Immunol 10(1):83–91
11. Cella M, Fuchs A, Vermi W et al (2009) A human natural killer cell subset provides an innate source of IL-22 for mucosal immunity. Nature 457(7230):722–725
12. Buonocore S, Ahern PP, Uhlig HH et al (2010) Innate lymphoid cells drive interleukin-23-dependent innate intestinal pathology. Nature 464(7293):1371–1375
13. Sonnenberg GF, Monticelli LA, Ellosi MM et al (2011) CD4(+) lymphoid tissue-inducer cells promote innate immunity in the gut. Immunity 34(1):122–134
14. Takatori H, Kanno Y, Watford WT et al (2009) Lymphoid tissue inducer-like cells are an innate source of IL-17 and IL-22. J Exp Med 206 (1):35–41
15. Morvan MG, Lanier LL (2016) NK cells and cancer: you can teach innate cells new tricks. Nat Rev Cancer 16(1):7–19
16. Martini M, Testi MG, Passetto M et al (2010) IFN-gamma-mediated upmodulation of MHC class I expression activates tumor-specific immune response in a mouse model of prostate cancer. Vaccine 28(20):3548–3557
17. Nakajima C, Uekusa Y, Iwasaki M et al (2001) A role of interferon-gamma (IFN-gamma) in tumor immunity: T cells with the capacity to reject tumor cells are generated but fail to migrate to tumor sites in IFN-gamma-deficient mice. Cancer Res 61(8):3399–3405
18. Beatty G, Paterson Y (2001) IFN-gamma-dependent inhibition of tumor angiogenesis by tumor-infiltrating CD4+ T cells requires tumor responsiveness to IFN-gamma. J Immunol 166(4):2276–2282
19. Larmonier N, Cathelin D, Larmonier C et al (2007) The inhibition of TNF-alpha anti-tumoral properties by blocking antibodies promotes tumor growth in a rat model. Exp Cell Res 313(11):2345–2355
20. Jovanovic IP, Pejnovic N, Radosavljevic GD et al (2014) Interleukin-33/ST2 axis promotes breast cancer growth and metastases by

facilitating intratumoral accumulation of immunosuppressive and innate lymphoid cells. Int J Cancer 134(7):1669–1682
21. Molofsky AB, Nussbaum JC, Liang HE et al (2013) Innate lymphoid type 2 cells sustain visceral adipose tissue eosinophils and alternatively activated macrophages. J Exp Med 210 (3):535–549
22. Huber S, Gagliani N, Zenewicz LA et al (2012) IL-22BP is regulated by the inflammasome and modulates tumorigenesis in the intestine. Nature 491(7423):259–263
23. Chan IH, Jain R, Tessmer MS et al (2014) Interleukin-23 is sufficient to induce rapid de novo gut tumorigenesis, independent of carcinogens, through activation of innate lymphoid cells. Mucosal Immunol 7(4):842–856
24. Schatton T, Murphy GF, Frank NY et al (2008) Identification of cells initiating human melanomas. Nature 451(7176):345–349

Chapter 15

Microbiota Analysis Using Sequencing by Synthesis: From Library Preparation to Sequencing

Andrew Nelson and Christopher J. Stewart

Abstract

The highly parallel nature of sequencing by synthesis (SBS) allows millions of amplicons to be sequenced simultaneously, which has led to enormous interest in the investigation of bacterial communities (often referred to as the microbiota). In this protocol, we describe a method for the 'universal' amplification of the v4 region of the bacterial 16S rRNA gene from genomic DNA and prepare these amplicons so that they can be sequenced using the MiSeq system (Illumina). The protocol provides instruction on sequencing of 188 genomic DNA samples plus PCR positive and negative controls, which can be applied to any sample type where bacterial DNA may be of interest.

Key words DNA sequencing, 16S rRNA, Microbiota, MiSeq, PCR

1 Introduction

The use of massively parallel sequencing has been used extensively to characterise the bacterial populations of the gut, skin and other tissues with each having a unique microbial fingerprint. Innate lymphocyte cells (ILC) are found in peripheral tissues and are thought to play a key role in tissue development and homeostasis. This has led to an interest in the relationship between ILC populations and the local microbiological environment. In this chapter we will cover the use of this technique to determine the microbiota from extracted DNA.

The 16S rRNA gene has been used to investigate bacterial communities (herein referred to as the microbiota) for over 25 years [1]. The appeal of the 16S rRNA gene relates to the nine hypervariable regions (v1–v9), where because it forms part of the structure of the ribosome, flanking regions are well conserved across the kingdom bacteria and large parts of the archaea in order to retain function. Microbial ecologists utilise the highly parallel nature of SBS to investigate the microbiota of hundreds of communities in a single sequencing run (i.e., multiplexing of

Shoba Amarnath (ed.), *Innate Lymphoid Cells: Methods and Protocols*, Methods in Molecular Biology, vol. 2121,
https://doi.org/10.1007/978-1-0716-0338-3_15,

samples). The v4 region of the 16S rRNA gene is the most commonly used of all the hypervariable regions, where is has notably been utilised by the human and earth microbiome projects, as well as other large consortia [2–5]. However, no single hypervariable region can distinguish all species, and each has benefits for the discrimination of particular clades [6]. The v4 region is approximately 250 bp in length, making it a good target for SBS, which has a current maximum length of 300 bp per read. Therefore, it is theoretically possible to sequence fragments up to 600 bp using the MiSeq and combine paired sequences end to end. However, having greater overlap allows each read to act as a QC for the complementary read when processing data downstream, resulting in data that are more accurate and less prone to sequencing artefacts [7, 8].

The requirement for stringent controls for microbiota analysis is becoming more apparent as the technique is employed on samples of low-biomass. Reagent controls for DNA extraction and library preparation are required, as manufacturing practices cannot fully remove contaminating bacterial DNA [9]. New software is being developed to address contamination without reducing data quality, but these are still in their infancy [10]. Mock communities can be used to act as positive controls during library preparation, but also allow researchers to assess the bias introduced during extraction and amplification and are essential for benchmarking these analyses [11]. Quality control of input genomic DNA into microbiota analyses are important as presence of PCR inhibitors carried over from poor extraction practices can prevent optimum amplification [12]. In addition, the number of cycles used for PCR can increase chimera formation, resulting in inflated number of identified taxa which can be problematic especially when working with low biomass samples where increasing cycle number is tempting to improve amplicon concentration [13]. In these cases, running several replicates and pooling then concentrating DNA should be preferred. In all cases, it is necessary to perform identical procedures on both samples and controls.

2 Materials

2.1 PCR

1. Invitrogen™ AccuPrime™ Pfx SuperMix.
2. Barcoded PCR primers.
3. ZymoBIOMICS Microbial Community DNA Standard.
4. 1× Tris-acetate-EDTA (TAE): 40 mM Tris-acetate, 1 mM EDTA.
5. Agarose.
6. 5× Sample Loading Buffer (Bioline).

7. DNA marker: Hyperladder I (Bioline).
8. SYBR safe DNA gel stain (10,000×).
9. 96-Well PCR plates.
10. Heat seal film.
11. Weigh boat.
12. 250 mL conical flask.
13. Casting tray.
14. 15-Well gel comb.
15. Thermal cycler.
16. Heat sealer.
17. Electrophoresis tank.
18. Power pack.
19. Vortex.
20. Plate centrifuge.

2.2 Clean up

1. Agencourt AMPure XP beads.
2. Molecular biology grade water.
3. Molecular biology grade ethanol.
4. 96-Well Midi plate.
5. 96-Well plate magnetic stand.

2.3 Fragment Analysis

1. Agilent High Sensitivity DNA Kit.
   - High sensitivity DNA chip.
   - High sensitivity DNA gel/dye mix.
   - High sensitivity DNA Marker.
   - High sensitivity DNA Ladder.
2. Agilent Bioanalyzer 2100.
3. Chip priming station.
4. IKA vortex mixer.

2.4 Quantification

1. Quant-iT™ PicoGreen® dsDNA Assay Kit.
   - Quant-iT™ PicoGreen® dsDNA reagent.
   - 20× Tris-EDTA (TE); 200 mM Tris–HCl, 20 mM EDTA, pH 7.5.
   - Lambda DNA standard (100 μg/mL).
2. Molecular biology grade water.

3. 96-Well fluorescence microplates.
4. Plate reader with excitation 480 nM/emission 520 nM.

2.5 Normalization

1. Molecular biology grade water.
2. Qubit dsDNA HS Assay Kit.
   - Qubit® dsDNA HS Reagent (200× concentrate).
   - Qubit® dsDNA HS Buffer.
   - Qubit® dsDNA HS Standard #1 (0 ng/μL).
   - Qubit® dsDNA HS Standard #2 (10 ng/μL).
3. 96-Well PCR plate.
4. Heat seal film.
5. LoBind 1.5 mL tube.
6. 500 μL thin-walled tubes.
7. Qubit fluorimeter.
8. Heat sealer.
9. Vortex.
10. Plate centrifuge.

2.6 Sequencing

1. 10 N Sodium Hydroxide.
2. Molecular biology grade water.
3. Sequencing primers (Read 1, Index and Read 2).
4. PhiX Control v3.
5. Illumina V2 500 cycle reagent kit Box 1.
   - Illumina V2 500 cycle reagent cartridge.
   - Hyb buffer.
6. Illumina V2 500 cycle reagent kit Box 2.
   - Flow cell.
   - PR2; Incorporation buffer.
7. LoBind 1.5 mL centrifuge tubes.
8. Lint free wipes.
9. Alcohol wipes.
10. Illumina MiSeq System.
11. Vortex.
12. Centrifuge.

A

	1	2	3	4	5	6	7	8	9	10	11	12
A	SA501 SA701	SA501 SA502	SA501 SA703	SA501 SA704	SA501 SA705	SA501 SA706	SA501 SA707	SA501 SA708	SA501 SA709	SA501 SA710	SA501 SA711	SA501 SA712
B	SA502 SA701	SA502 SA502	SA502 SA703	SA502 SA704	SA502 SA705	SA502 SA706	SA502 SA707	SA502 SA708	SA502 SA709	SA502 SA710	SA502 SA711	SA502 SA712
C	SA503 SA701	SA503 SA502	SA503 SA703	SA503 SA704	SA503 SA705	SA503 SA706	SA503 SA707	SA503 SA708	SA503 SA709	SA503 SA710	SA503 SA711	SA503 SA712
D	SA504 SA701	SA504 SA502	SA504 SA703	SA504 SA704	SA504 SA705	SA504 SA706	SA504 SA707	SA504 SA708	SA504 SA709	SA504 SA710	SA504 SA711	SA504 SA712
E	SA505 SA701	SA505 SA502	SA505 SA703	SA505 SA704	SA505 SA705	SA505 SA706	SA505 SA707	SA505 SA708	SA505 SA709	SA505 SA710	SA505 SA711	SA505 SA712
F	SA506 SA701	SA506 SA502	SA506 SA703	SA506 SA704	SA506 SA705	SA506 SA706	SA506 SA707	SA506 SA708	SA506 SA709	SA506 SA710	SA506 SA711	SA506 SA712
G	SA507 SA701	SA507 SA502	SA507 SA703	SA507 SA704	SA507 SA705	SA507 SA706	SA507 SA707	SA507 SA708	SA507 SA709	SA507 SA710	SA507 SA711	SA507 SA712
H	SA508 SA701	SA508 SA502	SA508 SA703	SA508 SA704	SA508 SA705	SA508 SA706	SA508 SA707	SA508 SA708	SA508 SA709	SA508 SA710	SA508 SA711	SA508 SA712

B

	1	2	3	4	5	6	7	8	9	10	11	12
A	SB501 SB701	SB501 SB502	SB501 SB703	SB501 SB704	SB501 SB705	SB501 SB706	SB501 SB707	SB501 SB708	SB501 SB709	SB501 SB710	SB501 SB711	SB501 SB712
B	SB502 SB701	SB502 SB502	SB502 SB703	SB502 SB704	SB502 SB705	SB502 SB706	SB502 SB707	SB502 SB708	SB502 SB709	SB502 SB710	SB502 SB711	SB502 SB712
C	SB503 SB701	SB503 SB502	SB503 SB703	SB503 SB704	SB503 SB705	SB503 SB706	SB503 SB707	SB503 SB708	SB503 SB709	SB503 SB710	SB503 SB711	SB503 SB712
D	SB504 SB701	SB504 SB502	SB504 SB703	SB504 SB704	SB504 SB705	SB504 SB706	SB504 SB707	SB504 SB708	SB504 SB709	SB504 SB710	SB504 SB711	SB504 SB712
E	SB505 SB701	SB505 SB502	SB505 SB703	SB505 SB704	SB505 SB705	SB505 SB706	SB505 SB707	SB505 SB708	SB505 SB709	SB505 SB710	SB505 SB711	SB505 SB712
F	SB506 SB701	SB506 SB502	SB506 SB703	SB506 SB704	SB506 SB705	SB506 SB706	SB506 SB707	SB506 SB708	SB506 SB709	SB506 SB710	SB506 SB711	SB506 SB712
G	SB507 SB701	SB507 SB502	SB507 SB703	SB507 SB704	SB507 SB705	SB507 SB706	SB507 SB707	SB507 SB708	SB507 SB709	SB507 SB710	SB507 SB711	SB507 SB712
H	SB508 SB701	SB508 SB502	SB508 SB703	SB508 SB704	SB508 SB705	SB508 SB706	SB508 SB707	SB508 SB708	SB508 SB709	SB508 SB710	SB508 SB711	SB508 SB712

Fig. 1 Layout of barcoded primers for dual indexing approach. (**a**) Barcodes SASA for use with first 94 samples and controls. (**b**) Barcodes SBSB for use with first 94 samples and controls. Full sequences can be found in Table 1

3 Methods

It is important to keep DNA free from environmental contamination. To minimise this risk, only open samples and stock reagents in a PCR hood or Class 2 cabinet. All pipetting steps should be performed using DNA/RNA free LoBind filter tips. To prevent degradation, thaw all reagents on ice and aliquot stock reagents where necessary to avoid repeated freeze/thaw cycles. This method details the sequencing of 188 genomic DNA samples plus two PCR positive and two PCR negative controls, whereby a 96-well plate will contain a one positive and negative. A total of two 96-well plates will be used during this example protocol, termed "DNA sample plate 1" and "DNA sample plate 2".

Table 1
Primer sequences required for amplification of the v4 region of the 16S rRNA gene and sequencing

Name	Sequence (5′–3′)
SA501	AATGATACGGCGACCACCGAGATCTACACATCGTACGTATGGTAATTGTG TGCCAGCMGCCGCGGTAA
SA502	AATGATACGGCGACCACCGAGATCTACACACTATCTGTATGGTAATTGTG TGCCAGCMGCCGCGGTAA
SA503	AATGATACGGCGACCACCGAGATCTACACTAGCGAGTTATGGTAATTGTG TGCCAGCMGCCGCGGTAA
SA504	AATGATACGGCGACCACCGAGATCTACACCTGCGTGTTATGGTAATTGTG TGCCAGCMGCCGCGGTAA
SA505	AATGATACGGCGACCACCGAGATCTACACTCATCGAGTATGGTAATTGTG TGCCAGCMGCCGCGGTAA
SA506	AATGATACGGCGACCACCGAGATCTACACCGTGAGTGTATGGTAATTGTG TGCCAGCMGCCGCGGTAA
SA507	AATGATACGGCGACCACCGAGATCTACACGGATATCTTATGGTAATTGTG TGCCAGCMGCCGCGGTAA
SA508	AATGATACGGCGACCACCGAGATCTACACGACACCGTTATGGTAATTGTG TGCCAGCMGCCGCGGTAA
SA701	CAAGCAGAAGACGGCATACGAGATAACTCTCGAGTCAGTCAGCCGGACTACH VGGGTWTCTAAT
SA702	CAAGCAGAAGACGGCATACGAGATACTATGTCAGTCAGTCAGCCGGACTACH VGGGTWTCTAAT
SA703	CAAGCAGAAGACGGCATACGAGATAGTAGCGTAGTCAGTCAGCCGGACTACH VGGGTWTCTAAT
SA704	CAAGCAGAAGACGGCATACGAGATCAGTGAGTAGTCAGTCAGCCGGACTACH VGGGTWTCTAAT
SA705	CAAGCAGAAGACGGCATACGAGATCGTACTCAAGTCAGTCAGCCGGACTACH VGGGTWTCTAAT
SA706	CAAGCAGAAGACGGCATACGAGATCTACGCAGAGTCAGTCAGCCGGACTACH VGGGTWTCTAAT
SA707	CAAGCAGAAGACGGCATACGAGATGGAGACTAAGTCAGTCAGCCGGACTACH VGGGTWTCTAAT
SA708	CAAGCAGAAGACGGCATACGAGATGTCGCTCGAGTCAGTCAGCCGGACTACH VGGGTWTCTAAT
SA709	CAAGCAGAAGACGGCATACGAGATGTCGTAGTAGTCAGTCAGCCGGACTACH VGGGTWTCTAAT
SA710	CAAGCAGAAGACGGCATACGAGATTAGCAGACAGTCAGTCAGCCGGACTACH VGGGTWTCTAAT
SA711	CAAGCAGAAGACGGCATACGAGATTCATAGACAGTCAGTCAGCCGGACTACH VGGGTWTCTAAT

(continued)

Table 1
(continued)

Name	Sequence (5′–3′)
SA712	CAAGCAGAAGACGGCATACGAGATTCGCTATAAGTCAGTCAGCCGGACTACH VGGGTWTCTAAT
SB501	AATGATACGGCGACCACCGAGATCTACACCTACTATATATGGTAATTGTG TGCCAGCMGCCGCGGTAA
SB502	AATGATACGGCGACCACCGAGATCTACACCGTTACTATATGGTAATTGTG TGCCAGCMGCCGCGGTAA
SB503	AATGATACGGCGACCACCGAGATCTACACAGAGTCACTATGGTAATTGTG TGCCAGCMGCCGCGGTAA
SB504	AATGATACGGCGACCACCGAGATCTACACTACGAGACTATGGTAATTGTG TGCCAGCMGCCGCGGTAA
SB505	AATGATACGGCGACCACCGAGATCTACACACGTCTCGTATGGTAATTGTG TGCCAGCMGCCGCGGTAA
SB506	AATGATACGGCGACCACCGAGATCTACACTCGACGAGTATGGTAATTGTG TGCCAGCMGCCGCGGTAA
SB507	AATGATACGGCGACCACCGAGATCTACACGATCGTGTTATGGTAATTGTG TGCCAGCMGCCGCGGTAA
SB508	AATGATACGGCGACCACCGAGATCTACACGTCAGATATATGGTAATTGTG TGCCAGCMGCCGCGGTAA
SB701	CAAGCAGAAGACGGCATACGAGATAAGTCGAGAGTCAGTCAGCCGGACTACH VGGGTWTCTAAT
SB702	CAAGCAGAAGACGGCATACGAGATATACTTCGAGTCAGTCAGCCGGACTACH VGGGTWTCTAAT
SB703	CAAGCAGAAGACGGCATACGAGATAGCTGCTAAGTCAGTCAGCCGGACTACH VGGGTWTCTAAT
SB704	CAAGCAGAAGACGGCATACGAGATCATAGAGAAGTCAGTCAGCCGGACTACH VGGGTWTCTAAT
SB705	CAAGCAGAAGACGGCATACGAGATCGTAGATCAGTCAGTCAGCCGGACTACH VGGGTWTCTAAT
SB706	CAAGCAGAAGACGGCATACGAGATCTCGTTACAGTCAGTCAGCCGGACTACH VGGGTWTCTAAT
SB707	CAAGCAGAAGACGGCATACGAGATGCGCACGTAGTCAGTCAGCCGGACTACH VGGGTWTCTAAT
SB708	CAAGCAGAAGACGGCATACGAGATGGTACTATAGTCAGTCAGCCGGACTACH VGGGTWTCTAAT
SB709	CAAGCAGAAGACGGCATACGAGATGTATACGCAGTCAGTCAGCCGGACTACH VGGGTWTCTAAT
SB710	CAAGCAGAAGACGGCATACGAGATTACGAGCAAGTCAGTCAGCCGGACTACH VGGGTWTCTAAT

(continued)

Table 1
(continued)

Name	Sequence (5′–3′)
SB711	CAAGCAGAAGACGGCATACGAGATTCAGCGTTAGTCAGTCAGCCGGACTACH VGGGTWTCTAAT
SB712	CAAGCAGAAGACGGCATACGAGATTCGCTACGAGTCAGTCAGCCGGACTACH VGGGTWTCTAAT
Read 1	TATGGTAATTGTGTGCCAGCMGCCGCGGTAA
Index	ATTAGAWACCCBDGTAGTCCGGCTGACTGACT
Read 2	AGTCAGTCAGCCGGACTACHVGGGTWTCTAAT

3.1 PCR

1. Dispense 17 μL of Invitrogen™ AccuPrime™ Pfx SuperMix into each well of the 96 well plate Note-multichannel pipette (*see* **Note 1**).
2. Add 2 μL of the SASA barcoded PCR primers to each well following the layout in Fig. 1a and mix by pipetting (*see* **Note 2**).
3. Add 1 μL of sample DNA to each well with the exception of H11 and H12.
4. Add 1 μL of ZymoBIOMICS Microbial Community DNA Standard to well H11.
5. Place the microplate sealing foil on top of the 96 well plate and seal using the heat sealer (*see* **Note 3**).
6. Briefly vortex the plate and pulse centrifuge to bring liquid back to the bottom of the well.
7. Place the plate into the thermal cycler and initiate the program 95 °C 2 min followed by 30 cycles of 95 °C for 20 s, 55 °C for 15 s, 72 °C for 5 min, and then a final extension of 72 °C for 10 min (hold the thermal cycler at 4 °C thereafter).
8. Repeat **steps 1–7** for the second plate containing the sample DNA using the SBSB (Fig. 1b) primer layout and a different thermal cycler or wait until the first plate is complete if only one thermal cycler is available (in which case, collect the plate from **step 7** and place in the fridge while performing this step).
9. While PCR is cycling prepare 1% w/v agarose gel in 1× TAE.
10. Set up casting tray and add water to ensure an effective seal is present. Pour out water and dry casting try.
11. Weigh 0.5 g agarose in a weigh boat and add to a conical flask.
12. Add 50 mL 1× TAE and swirl by hand to mix.

13. Put in the microwave and set to full power and heat for 1 min. Once agarose is fully solubilised remove from microwave using a heatproof glove and allow to cool to ~50 °C.
14. Add 5 μL SYBR safe DNA gel stain (1 μL/10 mL) and swirl by hand to mix.
15. Pour solution into the casting tray and add two combs with 15 wells.
16. Leave to set for 10 min or until the gel appears translucent.
17. Place the gel in the electrophoresis tank.
18. Cover with 1× TAE.
19. Remove 96 well plate containing amplicons from the thermal cycler.
20. To a LoBind 1.5 mL tube add 2 μL of amplicon and 2 μL of 5× sample loading buffer and mix by pipetting. Use wells A1, B2, C3, D4, E5, F6, G7, H8, A9, B10, H11, H12 from DNA sample plate 1 (*see* **Note 4**).
21. Leave the first well in the agarose gel empty and to the next well add 5 μL of Hyperladder I.
22. To the next 12 well add the 4 μL amplicon and 1 μL 5× sample loading buffer mix.
23. Repeat **steps 20–22** for the second DNA sample plate.
24. Place the lid on the tank and plug into the powerpack and run at 120 mA for 20 min.
25. View under a UV light and should see clear bands at ~400 bp (Fig. 2). This can be stored at −20 °C until you are ready to proceed with the next step (*see* **Note 5**).

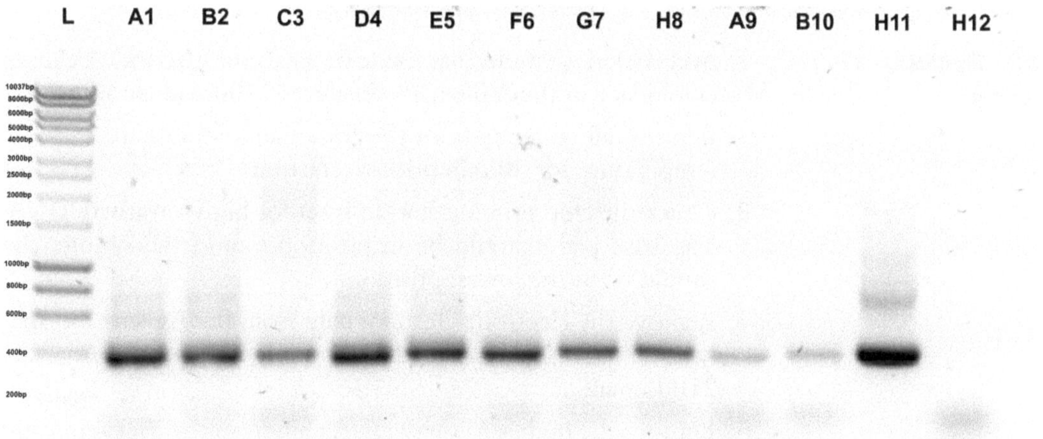

Fig. 2 Agarose gel electrophoresis confirmation of successful PCR amplification. *L* ladder, *A1-B10* samples, *H11* ZymoBIOMICS Microbial Community DNA Standard, *H12* PCR negative control

3.2 Clean up

1. Prepare a 70% solution of molecular biology grade ethanol in molecular biology grade water making 500 μL per amplicon. This solution can be used for no longer than 24 h.
2. Dispense 10 μL of AMPure XP beads into a 96 well midi plate using the same layout as the amplicon 96 well PCR plate.
3. Using a multichannel pipette add 10 μL of amplicon and mix by pipetting up and down ten times (*see* **Note 6**).
4. Leave the midi plate to incubate at room temperature for 5 min.
5. Place the midi plate onto the magnetic stand and leave for 1 min until the beads have aggregated and the solution appears clear. Remove the liquid from the well and keep until the fragment analysis confirms the PCR product was successfully bound to the beads.
6. Add 200 μL of 70% ethanol to each well and leave for 30 s then remove using a multichannel pipette while being careful not to disturb the pellet (*see* **Note 7**).
7. Repeat the previous step again then leave the bead pellet to dry for 5 min.
8. Remove the midi plate from the magnet and add 20 μL of molecular biology grade water to each well and mix by pipetting up and down ten times.
9. Leave the plate to incubate at room temperature for 5 min.
10. Place the midi plate on the magnetic stand and leave for 1 min until the beads have aggregated and the solution appears clear.
11. Pipette the liquid into a new 96 well PCR plate. This can be stored at −20 °C until you are ready to proceed with the next step.
12. Repeat **steps 1–11** for DNA sample plate 2.

3.3 Fragment Analysis

1. Add 350 μL of molecular grade water to the electrode cleaning chip, place in the Agilent BioAnalyzer 2100 and close the lid.
2. Remove all reagents from the fridge and equilibrate to room temperature for 30 min prior to starting.
3. Check the chip priming station is set for high sensitivity DNA. The base plate should be in position C and the syringe clip should be in the lowest position.
4. Remove the electrode cleaning chip from the Agilent BioAnalyzer 2100. Leave the lid of the Agilent BioAnalyzer 2100 open.
5. Remove the high sensitivity DNA chip from the packaging and place it into the chip priming station (*see* **Note 8**).
6. Add 9 μL of gel/dye mix into position 'G' of the high sensitivity DNA chip making sure to pipette into the bottom of the

well, not the sides or air above. If any bubbles are present at stage remove with a clean pipette tip before proceeding.

7. Adjust the syringe plunger so it is in line with the 1 mL graduation.
8. Close the chip priming station making sure it clicks then depress the syringe plunger until it is secured under the syringe clip.
9. Leave for 1 min then release the plunger from the syringe clip and allow the plunger to rise unaided for 5 s then slowly return to the 1 mL position.
10. Open the chip priming station and add 9 μL of gel/dye mix to positions the other positions labelled 'G' of the high sensitivity DNA chip.
11. Add 5 μL of the high sensitivity DNA marker to positions 1–11 and position 'Ladder' of the high sensitivity DNA chip.
12. Add 1 μL of high sensitivity DNA Ladder to position 'Ladder'.
13. Add 1 μL of samples, A1, B2, C3, D4, E5 from DNA sample plate 1 and wells F6, G7, H8, A9, H11 and H12 from sample plate 2 (*see* **Note 9**).
14. Remove the high sensitivity DNA chip from the chip priming station and place in the IKA vortex mixer and vortex at 2000 rpm for 1 min.
15. Place the chip in the Agilent BioAnalyzer 2100 and close the lid.
16. In the assay selection section select Assays>dsDNA>High Sensitivity DNA.xsy.
17. Input sample names in the chip summary section of the 2100 Expert software in the following format "YYYYMMDD_Surname_16S_Well".
18. Click the start button and leave to run (*see* **Note 10**).
19. Once complete, remove the high sensitivity DNA chip (*see* **Note 11**) from the Agilent BioAnalyzer 2100 and replace with the electrode cleaning chip containing the 350 μL of molecular biology grade water, close the lid and leave for 15 min. After the 15 min remove the electrode cleaning chip and close the lid of the Agilent BioAnalyzer 2100. Discard the water from the electrode cleaning chip.
20. Check the fragment patters from the samples should be free from primer contamination and contain the same single peak as the positive control (*see* **Note 12**).

3.4 Quantification

1. Remove Quant-iT™ PicoGreen® dsDNA reagents from the freezer and allow to equilibrate to room temperature on the bench.

2. Prepare 1× TE from 20× TE stock (*see* **Note 13**).
3. Using the 1× TE buffer prepared in **step 2** and the lambda DNA standard provided (100 μg/mL) prepare standard curve ranging from 1000 ng/mL to 0 ng/mL in a final volume of 150 μL.
4. Add 95 μL of 1× TE to each well of the 96-well fluorescence microplates followed by 5 μL of cleaned amplicon and mix by pipetting. Repeat for DNA sample plate 2.
5. In column 1 of a separate 96-well fluorescence microplate add 100 μL of the prepared DNA standard in descending order.
6. Prepare the Quant-iT™ PicoGreen® dsDNA working dye solution by diluting 1:200 with 1× TE buffer. A volume of 100 μL per sample is required. Make 24 mL that is enough working dye for 240 samples to allow for the 188 samples, 4 controls and 8 standards and dead volume to use a multi-channel pipette.
7. Add 100 μL of the working dye solution to each well of the 96-well fluorescence microplates containing the samples, controls and the standards and mix by pipetting.
8. Place the lid onto the 96 well plates and incubate in the dark at room temperature for 10 min.
9. While the plate is incubating turn on the plate reader and set to read with an excitation wavelength of 480 nm and an emission wavelength of 520 nm.
10. Take the 96-well fluorescence microplate containing the DNA standards, remove the lid, place in the plate reader, and read the plate. Save the data in workbook format remove the plate and discard.
11. Repeat **step 10** for the 96-well fluorescence microplates containing the samples and turn off the plate reader.
12. Plot the standard curve for the DNA standards with the intercept set to 0. Use the equation to calculate the concentration (ng/mL) of the samples. Convert this value to nM using the following equation: nM = (concentration/(660 × size in bp)) × 1000. This can be stored at −20 °C until you are ready to proceed with the next step (*see* **Note 14**).

3.5 Normalization

1. Remove Qubit™ dsDNA HS Assay Kit from the fridge and allow to equilibrate to room temperature.
2. Calculate the amount of each amplicon required to achieve a concentration of 4 nM in 10 μL using the following equation: 4 nM = (4 nM × 10 μL)/concentration of sample in nM. If the negative control is below this concentration then proceed and add 10 μL of the negative control (*see* **Note 15**).

3. In a new 96 well PCR plate add the amount of molecular biology grade water required to each well to achieve a concentration of 4 nM.
4. Add the required volume of amplicon to each well of the 96 well PCR plate containing the molecular biology grade water.
5. Seal the 96 well plate containing the 4 nM amplicons using heat seals and sealer and mix using a vortex for 10 s then pulse centrifuge.
6. Remove the seal from the 96 well plate and add 5 μL of each sample to a LoBind 1.5 mL tube.
7. Prepare the Qubit™ dsDNA HS working dye solution by diluting 1:200 with Qubit® dsDNA HS Buffer (provided). In a LoBind microfuge 1.5 mL tube add 1094.5 μL of 1× TE buffer and 5.5 μL of Qubit™ dsDNA HS dye concentrate and mix using a vortex then pulse in the centrifuge.
8. Add 190 μL of working dye solution to two 500 μL thin-walled tubes for the DNA standards and 195 μL of working dye solution to three 500 μL thin-walled tubes for the pooled amplicon.
9. Vortex the DNA standards and pooled amplicon tubes. Add 10 μL of standard 1 to a 500 μL thin-walled tube containing 190 μL of working dye solution then add 10 μL of standard 2 to a 500 μL thin-walled tube containing 190 μL of working dye solution. Add 5 μL of the pooled amplicon to each of the tubes containing 195 μL of working dye solution.
10. Vortex the tubes for 10 s and pulse in a centrifuge then incubate in the dark for 5 min.
11. Turn on the Qubit fluorimeter and select the hsDNA assay.
12. The Qubit fluorimeter will prompt to read new standards, select yes and follow the on-screen instructions to read the standards.
13. Once the standards are accepted, insert the first tube containing the pooled amplicon and select 'Read' on the screen. Select the units and this will take you to a new screen. Change the assay volume to 5 μL and the units to ng/mL and record the value.
14. Repeat **step 10** for the other two replicates.
15. Calculate the mean concentration then convert the concentration in ng/mL value to nM using the following equation: nM = (concentration/(660 × size in bp)) × 1000.
16. Adjust the concentration to 2 nM using the equation: 2 nM = (2 nM × 10 μL)/concentration of sample in nM. This can be stored at −20 °C until you are ready to proceed with the next step.

3.6 Sequencing

1. Thaw MiSeq V2 500 reagent cartridge at room temperature. Thaw 100 μM Read 1, Index and Read 2 primers (Table 1), Hyb buffer and PhiX Control V3 on ice (*see* **Note 16**).
2. Prepare a 1 N solution of sodium hydroxide by adding 25 μL of 10 N sodium hydroxide (grade) to 225 μL of molecular biology grade water. Vortex for 10 s to mix then pulse in a centrifuge.
3. Add 200 μL of 1 N sodium hydroxide to 800 μL of molecular biology grade water to make a 0.2 N sodium hydroxide solution (*see* **Note 17**).
4. To a separate LoBind 1.5 mL tube, add 5 μL of 2 nM pooled amplicon. In a separate tube, add 3 μL Hyb buffer and 2 μL PhiX control v3 and mix by pipetting.
5. Add 5 μL of 0.2 N sodium hydroxide to the two LoBind 1.5 mL tubes containing the pooled amplicon and PhiX control v3 mix by pipetting and leave to incubate at room temperature for 5 min.
6. During the incubation, pipette 3.4 μL of Read 1, Index and Read 2 primer into separate LoBind 1.5 mL tubes and place on ice.
7. Once the 5 min incubation is complete, add 990 μL of Hyb buffer to the pooled amplicon and PhiX control v3 tubes, vortex and pulse centrifuge then place on ice. This step has created a 10 pM pooled amplicon library and 20 pM PhiX control v3.
8. Add 750 μL of Hyb buffer to a new LoBind 1.5 mL tube and place on ice. Add 250 μL of 20 pM PhiX to the 750 μL of Hyb buffer and vortex then pulse centrifuge and place on ice.
9. To a new LoBind 1.5 mL tube add 240 μL Hyb buffer, 60 and 300 μL 10 pM pooled amplicon and vortex then pulse centrifuge and place on ice.
10. Mix the thawed MiSeq V2 500 reagent cartridge by gently rocking.
11. Using a 200 μL pierce the foil covering tubes 12, 13, 14 and 17 on the cartridge.
12. Remove 50 μL of the solution in well 12 using a 200 μL pipette and add to the tube containing 3.4 μL of read 1 primer and mix by gently pipetting.
13. Dispense the entire solution into the LoBind 1.5 mL tube, then set the pipette to a volume of 70 μL. Aspirate the solution, and then dispense the entire solution into well 12 of the cartridge (*see* **Note 18**).
14. Repeat **steps 12** and **13** for the index primer using well 13 and the read 2 primer using well 14.

15. Pipette 600 μL of the solution prepared in **step 9** to well 17 ensuring all of the liquid was dispensed and place in the fridge until ready to use.
16. Remove box 2 of the MiSeq V2 500 kit from the fridge, take out the flow cell container and discard the storage buffer keeping the flow cell in place.
17. Fill the container with molecular biology grade water and invert gently several times.
18. Discard the molecular biology grade water from the flow cell container, remove the flow cell and dry using lint free wipes Note-dry under edges, etc.
19. Take a fresh alcohol wipe and gently wipe the glass part of the flow cell. Note-keep off the gasket.
20. Prepare the sample sheet as outlined in Fig. 3 and save as a .csv file in the MiSeq control folder>SampleSheets.
21. Shutdown the MiSeq by selecting the shutdown function from the menu then turn off the MiSeq at the wall and wait for 1 min. Turn on the MiSeq and wait for the start-up tests to finish (*see* **Note 19**).
22. Select the 'Sequence' function on the MiSeq software and then log in to your Basespace account and click 'Continue'. It is also possible to do this offline.
23. Replace the existing flow cell in the MiSeq with the new flow cell, wait for the RFID to be accepted then press 'Continue' Note-smash.
24. Raise the sipper in the MiSeq and remove the wash solution. Take the PR2 buffer from the box 2 of the MiSeq V2 500 kit, remove the lid and place into the MiSeq. Empty the waste bottle if necessary lower the sipper then press 'Continue'.
25. Remove the wash tray from the MiSeq and replace with the cartridge. Select 'change sample sheet' then select the appropriate sample sheet for the run then select 'Open'. Press 'Continue'.
26. The screen should now show the assay parameters. If correct press 'Continue' (*see* **Note 20**).
27. The MiSeq will now run some pre run checks, once these are passed press 'Sequence'. The sequencing reaction will now run and takes approximately 44 h.

```
[Header],,,,,,,,,
IEMFileVersion,4,,,,,,,,
Investigator Name,Andrew Nelson,,,,,,,,
Experiment Name,YYYYMMDD_NAME_16Sv4,,,,,,,,
Date,07/13/2010,,,,,,,,
Workflow,GenerateFASTQ,,,,,,,,
Application,FASTQ Only,,,,,,,,
Assay,16Sv4,,,,,,,,
Description,Insert_Description,,,,,,,,
Chemistry,Amplicon,,,,,,,,
,,,,,,,,,
[Reads],,,,,,,,,
251,,,,,,,,,
251,,,,,,,,,
,,,,,,,,,
[Settings],,,,,,,,,
Adapter,CTGTCTCTTATACACATCT,,,,,,,,
,,,,,,,,,
[Data],,,,,,,,,
Sample_ID,Sample_Name,Sample_Plate,Sample_Well,I7_Index_ID,index,I5_Index_ID,index2,Sample_Project,Description
SASA1,,,,SA701,CGAGAGTT,SA501,ATCGTACG,,
SASA2,,,,SA702,GACATAGT,SA501,ATCGTACG,,
SASA3,,,,SA703,ACGCTACT,SA501,ATCGTACG,,
SASA4,,,,SA704,ACTCACTG,SA501,ATCGTACG,,
SASA5,,,,SA705,TGAGTACG,SA501,ATCGTACG,,
SASA6,,,,SA706,CTGCGTAG,SA501,ATCGTACG,,
SASA7,,,,SA707,TAGTCTCC,SA501,ATCGTACG,,
SASA8,,,,SA708,CGAGCGAC,SA501,ATCGTACG,,
SASA9,,,,SA709,ACTACGAC,SA501,ATCGTACG,,
SASA10,,,,SA710,GTCTGCTA,SA501,ATCGTACG,,
SASA11,,,,SA711,GTCTATGA,SA501,ATCGTACG,,
SASA12,,,,SA712,TATAGCGA,SA501,ATCGTACG,,
SASA13,,,,SA701,CGAGAGTT,SA502,ACTATCTG,,
SASA14,,,,SA702,GACATAGT,SA502,ACTATCTG,,
SASA15,,,,SA703,ACGCTACT,SA502,ACTATCTG,,
SASA16,,,,SA704,ACTCACTG,SA502,ACTATCTG,,
SASA17,,,,SA705,TGAGTACG,SA502,ACTATCTG,,
SASA18,,,,SA706,CTGCGTAG,SA502,ACTATCTG,,
SASA19,,,,SA707,TAGTCTCC,SA502,ACTATCTG,,
SASA20,,,,SA708,CGAGCGAC,SA502,ACTATCTG,,
SASA21,,,,SA709,ACTACGAC,SA502,ACTATCTG,,
SASA22,,,,SA710,GTCTGCTA,SA502,ACTATCTG,,
SASA23,,,,SA711,GTCTATGA,SA502,ACTATCTG,,
SASA24,,,,SA712,TATAGCGA,SA502,ACTATCTG,,
SASA25,,,,SA701,CGAGAGTT,SA503,TAGCGAGT,,
SASA26,,,,SA702,GACATAGT,SA503,TAGCGAGT,,
SASA27,,,,SA703,ACGCTACT,SA503,TAGCGAGT,,
SASA28,,,,SA704,ACTCACTG,SA503,TAGCGAGT,,
SASA29,,,,SA705,TGAGTACG,SA503,TAGCGAGT,,
SASA30,,,,SA706,CTGCGTAG,SA503,TAGCGAGT,,
SASA31,,,,SA707,TAGTCTCC,SA503,TAGCGAGT,,
```

Fig. 3 Example Illumina MiSeq SampleSheet for generation of demultiplexed fastq files

4 Notes

1. A volume of 220 μL of AccuPrime™ Pfx SuperMix is required per 12 wells if using a multichannel pipette.
2. The primer sequences used in this protocol can be found in Table 1 [3]. Order duplexed primers in 96 well plate format at 20 nmol scale. Select the option that gives a set concentration of primer per well to allow for easy resuspension. Using molecular biology grade water make up a stock plate at 100 μM. Dilute in molecular biology grade water to 10 μM in 96 well PCR plates and store at −20 °C.

3. We have seen a reduction in crossover/contamination when using heat seal film rather than sticky film to seal 96 well plates. We believe this may be due to adherence of primer/DNA/amplicon if the plates are inverted accidentally during storage or are dropped.
4. When successful PCR is confirmed in these wells this shows that reagents were added successfully to each row and column of the 96 well PCR plate.
5. Assuming a successful positive control amplification and no amplification in the negative control, a negative result at this stage for some samples may not necessarily mean that the amplification was unsuccessful. The limit of detection of the gel is higher than the MiSeq system so an extra QA step such as Qubit (*see* Subheading 3) should be used to confirm rather than repeat the assay.
6. The AMPure XP PCR clean up protocol calls for a 1:1.8 sample to bead ratio to remove excess PCR primers. However, we find that this ratio tends to leave some primer in many instances and a 1:1 ratio has a very high success rate first time without reducing the amplicon concentration.
7. Prior to pipetting the 70% ethanol solution into the 96-well Midi plate, you should aspirate and dispense the 70% ethanol using your pipette ten times into its container. This will fill the headspace of the tip and prevent dripping when adding the 70% ethanol to the Midi plate.
8. Although the Agilent High Sensitivity DNA Kit is supplied with Agilent High Sensitivity DNA chips there is excess reagent supplied. It is possible to use any Agilent BioAnalyzer chip to run this assay if you follow the layout of the Agilent High Sensitivity DNA chip and use the Agilent High Sensitivity DNA assay, not the one from the chip you are using [14].
9. These wells are a suggestion to confirm that the amplicon was retained during the AMPure XP clean up and the primer has been successfully removed. We always recommend that you run the positive and negative control wells at this step.
10. The Agilent BioAnalyzer 2100 shows the traces in real time, if the first few look successful then you may wish to proceed to the quantification step.
11. It is possible to reuse the Agilent High Sensitivity DNA chips several times [15]. Remove the chip from the BioAnalyzer 2100 and discard excess reagents by tapping onto a paper towel. Add 15 μL of 1× TAE to each well of the Agilent High Sensitivity DNA chip and vortex using the IKA vortex adapter at 2400 rpm for 1 min. Discard the 1× TAE by tapping onto a paper towel and repeat. Store the chip by filling the wells

with 15 μL 1× TAE, cover using parafilm and store at 4 °C. To use the chip remove 1× TAE if necessary and add 9 μL of gel/dye mix to position III and depress the syringe plunger and leave for 1 min. Release the plunger back to the 1 mL position then remove the gel/dye mix from position III. Add 15 μL of 1× TAE to each well of the Agilent High Sensitivity DNA chip and vortex using the IKA vortex adapter at 2400 rpm for 1 min. Discard the 1× TAE by tapping onto a paper towel. Add the gel/dye mix back to position III then proceed with the chip loading as described above.

12. If there is amplicon present at the expected size (~390 bp) but still primer present, then repeat the Clean up protocol (Fig. 4a). If the amplicon is present at the expected size and there is no sign of primer remaining (50–100 bp) then continue the protocol (Fig. 4b). If there is no primer and no amplicon present, check the solution from Clean up-Step 5 using the Agilent BioAnalyzer 2100. If the amplicon is present, then QA AMPure XP bead stock using Hyperladder I to confirm this is working as expected. If successful, then proceed with Clean up-Step 1. If unsuccessful purchase a new batch of AMPure XP bead solution.
13. Prepare 22 mL of 1× TE by adding 20.9 mL of molecular biology grade water to a sterile 50 mL tube followed by 1.1 mL of 20× TE.
14. The size in bp was determined using the Agilent BioAnalyzer during the Fragment Analysis protocol. If the mean amplicon size was different between samples use the mean for samples that were not assayed.
15. The concentration used here is not prescriptive but is required to dilute the sodium hydroxide concentration below the 1 mM maximum concentration for sequencing. If working with low biomass samples then the sodium hydroxide must first be neutralised [16].
16. To fully thaw the MiSeq V2 500 reagent cartridge at room temperature can take up to 8 h. Therefore, we recommend thawing the cartridge at 4 °C overnight so that the cartridge is ready to use. The cartridge can be stored at 4 °C for up to 1 week.
17. Make sodium hydroxide up fresh each time to ensure successful denaturation of the library.
18. The volume is set higher than the volume of liquid to 'blow out' the entire primer volume into the MiSeq V2 500 reagent cartridge. Make sure to place the tip fully into the tube without applying any force before dispensing. If force is applied, the tip may become lodged in the reagent cartridge and can be difficult to remove.

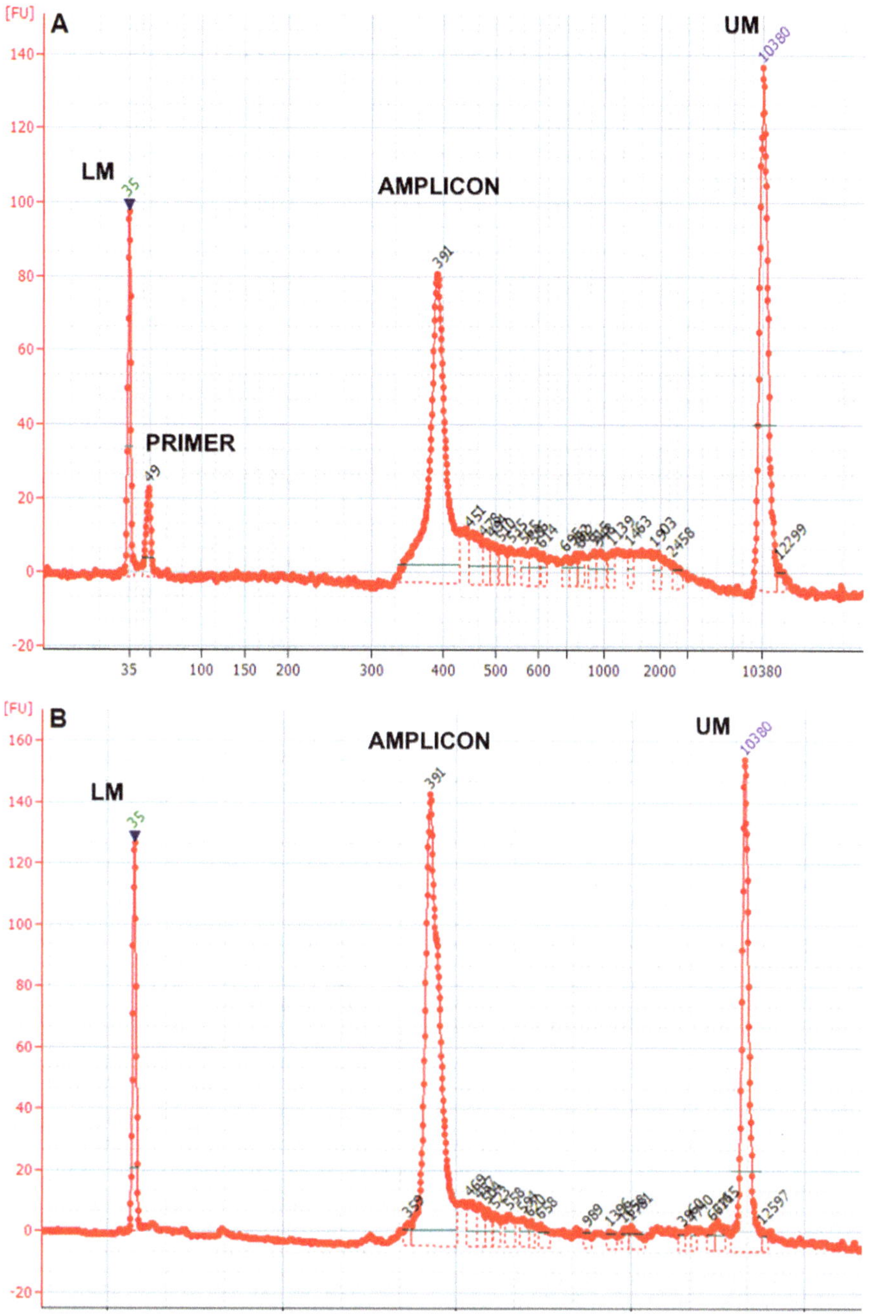

Fig. 4 Agilent Bioanalyzer hsDNA trace showing. (**a**) Successful amplification but remaining primer meaning unsuccessful Clean-up. (**b**) Successful amplification and Clean-up

19. Power cycle the MiSeq system before each run to refresh the MiSeq Control Software to reduce run failure rate.
20. The MiSeq software does not run a 'standard protocol' for each set of reagents the SampleSheet is used to give the MiSeq all of

the information on number of bases to call for each read and index. Therefore, it is important to check this screen as it is the only screen that shows this information.

Acknowledgements

This work was supported by the European Union's Horizon 2020 research and innovation programme under the Marie Skłodowska-Curie Grant Agreement No. 794771 to C.J.S.

References

1. Muyzer G, de Waal EC, Uitterlinden AG (1993) Profiling of complex microbial populations by denaturing gradient gel electrophoresis analysis of polymerase chain reaction-amplified genes coding for 16S rRNA. Appl Environ Microbiol 59(3):695–700
2. Caporaso JG, Lauber CL, Walters WA, Berg-Lyons D, Huntley J, Fierer N et al (2012) Ultra-high-throughput microbial community analysis on the Illumina HiSeq and MiSeq platforms. ISME J 6(8):1621–1624
3. Kozich JJ, Westcott SL, Baxter NT, Highlander SK, Schloss PD (2013) Development of a dual-index sequencing strategy and curation pipeline for analyzing amplicon sequence data on the MiSeq Illumina sequencing platform. Appl Environ Microbiol 79 (17):5112–5120
4. Stewart CJ, Ajami NJ, O'Brien JL, Hutchinson DS, Smith DP, Wong MC et al (2018) Temporal development of the gut microbiome in early childhood from the TEDDY study. Nature 562 (7728):583–588
5. Vatanen T, Franzosa EA, Schwager R, Tripathi S, Arthur TD, Vehik K et al (2018) The human gut microbiome in early-onset type 1 diabetes from the TEDDY study. Nature 562 (7728):589–594
6. Chakravorty S, Helb D, Burday M, Connell N, Alland D (2007) A detailed analysis of 16S ribosomal RNA gene segments for the diagnosis of pathogenic bacteria. J Microbiol Methods 69(2):330–339
7. Schirmer M, Ijaz UZ, D'Amore R, Hall N, Sloan WT, Quince C (2015) Insight into biases and sequencing errors for amplicon sequencing with the Illumina MiSeq platform. Nucleic Acids Res 43(6):e37
8. Edgar RC, Flyvbjerg H (2015) Error filtering, pair assembly and error correction for next-generation sequencing reads. Bioinformatics 31(21):3476–3482
9. Salter SJ, Cox MJ, Turek EM, Calus ST, Cookson WO, Moffatt MF et al (2014) Reagent and laboratory contamination can critically impact sequence-based microbiome analyses. BMC Biol 12:87
10. Davis NM, Proctor DM, Holmes SP, Relman DA, Callahan BJ (2018) Simple statistical identification and removal of contaminant sequences in marker-gene and metagenomics data. Microbiome 6(1):226
11. Bokulich NA, Rideout JR, Mercurio WG, Shiffer A, Wolfe B, Maurice CF et al (2016) Mockrobiota: a public resource for microbiome bioinformatics benchmarking. mSystems 1(5)
12. Schrader C, Schielke A, Ellerbroek L, Johne R (2012) PCR inhibitors – occurrence, properties and removal. J Appl Microbiol 113 (5):1014–1026
13. Wang GC, Wang Y (1997) Frequency of formation of chimeric molecules as a consequence of PCR coamplification of 16S rRNA genes from mixed bacterial genomes. Appl Environ Microbiol 63(12):4645–4650
14. Davies J, Denyer T, Hadfield J (2016) Bioanalyzer chips can be used interchangeably for many analyses of DNA or RNA. Biotechniques 60(4):197–199
15. Nguyen T, Kwak S, Karpowicz SJ (2014) Re-use of commercial microfluidics chips for DNA, RNA, and protein electrophoresis. Biotechniques 57(5):267–271
16. Quail MA, Kozarewa I, Smith F, Scally A, Stephens PJ, Durbin R et al (2008) A large genome center's improvements to the Illumina sequencing system. Nat Methods 5 (12):1005–1010

Chapter 16

Establishing Human Intestinal Enteroid/Organoid Lines from Preterm Infant and Adult Tissue

Christopher J. Stewart, Mary K. Estes, and Sasirekha Ramani

Abstract

Self-organizing mini-intestines cultured ex vivo from intestinal biopsy/resected samples, termed intestinal organoids or enteroids, present a unique opportunity for mechanistic investigation of health and disease of the intestinal epithelium. These patient-derived epithelial cultures are nontransformed, retain the genetic background of the patient, maintain regional specificity, differentiate into all major cell types of the intestinal epithelium, and are physiologically active. The biological relevance of human intestinal enteroids also circumvents the need for animal models for studies on the human gastrointestinal epithelium. Coculture with human endogenous microbes allows for exciting new studies on microbial–host interactions.

While the popularity of organoids/enteroids for human research has risen drastically over the past decade, existing work and published methods are primarily limited to adult tissue. Here, we describe a concise and effective method for the establishment neonatal enteroids (including preterm and term) from surgically resected tissue or biopsy material. While the protocol works on adult tissue/biopsies, it has been specifically adopted and optimised for neonatal tissue. We detail the procedure at each stage ranging from human tissue collection and extraction of stem cells from the tissue, to passaging and general maintenance of organoid/enteroid lines, and how to freeze and revive lines as needed.

Key words Stem cell, Crypt, Intestine, Enteroid, Organoid, Preterm, Infant, Adult, Cell culture, Media

1 Introduction

Innate lymphoid cells (ILCs) are tissue resident populations that occupy specific niches in peripheral tissues including the intestine. Here they form a first line of defense against pathogens and are responsible for expressing cytokines critical for maintaining the integrity of the intestinal mucosa. Intestinal ILC interact with the gut microbiome disruption, which has been implicated in a wide variety of diseases including inflammatory bowel disease and intestinal cancer. Therefore, it is important understand how ILCs modulate the gut microenvironment in humans and one methodology

Shoba Amarnath (ed.), *Innate Lymphoid Cells: Methods and Protocols*, Methods in Molecular Biology, vol. 2121, https://doi.org/10.1007/978-1-0716-0338-3_16,

that might provide insight into ILC function within the gut microenvironment in human is through enteroid cultures.

Epithelial cells line all surfaces of the human body and are of major importance to health and disease. The epithelium provides a barrier from exogenous material coming into contact with the external (e.g., skin) or internal (e.g., intestinal lumen) surfaces. Thus, the epithelial barrier is positioned at the interface where host and non-host meets and, as such, is central for absorption, secretion, susceptibility to pathogens, and immune regulation. Of particular importance to human health and disease are epithelial cells of the intestine, which consists of four sequential sections: duodenum, jejunum, ileum, and the colon. Self-organizing mini-intestines cultured ex vivo from intestinal biopsy/resected samples, termed intestinal organoids or enteroids, present a unique opportunity for mechanistic investigation into health and disease at the epithelial level [1]. As per the 2012 guidelines on nomenclature for intestinal in vitro cultures, an enteroid is defined as a multilobulated structure with a lumen that develops from an enterosphere by formation of budding crypts; thus, we will use the term enteroid from herein [2].

Human intestinal enteroids are generated from isolated intestinal crypts that contain Lgr5+ stem cells and established cultures retain segment-specific properties (e.g., ileal enteroids will be distinct from colonic enteroids) [3, 4]. Enteroids are composed of all the major cell types of the intestinal epithelium (enterocyte, Paneth, goblet, neuroendocrine, and stem) and are physiologically active (e.g., secrete mucin; demonstrate fluid secretion and swelling in response to enterotoxins) [5, 6]. Furthermore, human intestinal enteroids retain the genetic and potentially the epigenetic susceptibility and immune programming of the host. Enteroids are therefore a robust and relevant model to systematically explore host responses to stimuli (e.g., microbial colonization) and drug responses. To this end, recent developments have been made in coculturing bacteria with human intestinal enteroids, allowing mechanistic studies of host–microbial interactions [7, 8]. The use of human intestinal enteroids further facilitates human specific studies of the biology of the gastrointestinal epithelium and thus circumvent the need for animal models while supporting the 3Rs framework (Replacement, Reduction and Refinement) [1]. Of particular note, mechanistic investigation of preterm infants has been challenging due to the inability to recapitulate the extent of human prematurity (e.g., equivalent to 23 week gestation) in animal models. Therefore, enteroids derived from preterm infants may facilitate important and potentially paradigm shifting research.

The use of enteroids for scientific investigation continues to rise and the technology is widely applicable to a range of research. The methods described in this chapter detail the procedure at each stage of enteroid generation, starting from human tissue collection,

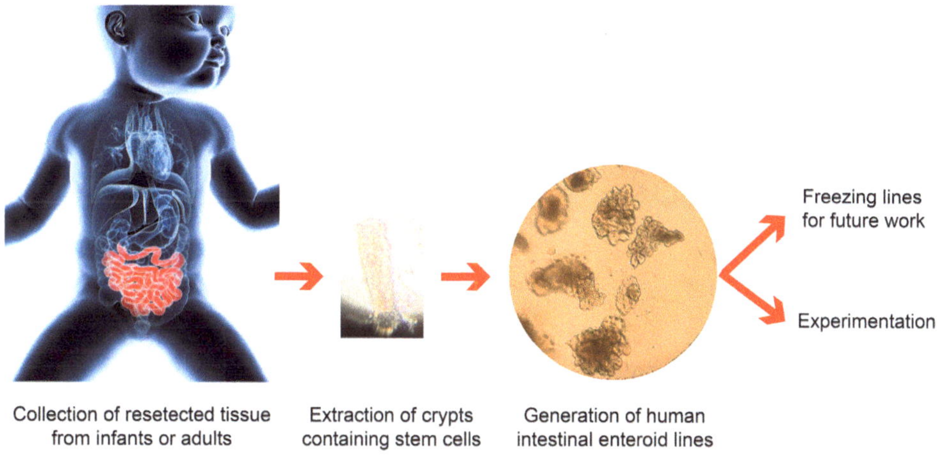

Fig. 1 Schematic showing the main stages in the generation of enteroid lines from intestinal tissue

extraction of crypts containing stem cells from the tissue, passaging and general maintenance of the lines, and methods to freeze and revive lines as needed (Fig. 1). This protocol has been optimized for working with neonatal tissue (including preterm and term) but is also effective on adult tissue/biopsies. These protocols can be performed on both resected tissue and biopsies, but for the purposes of this book chapter we discuss the generation of enteroids from surgically resected tissue. Additionally, this protocol should work with tissue from any region of the intestine.

2 Materials

Ensure all solutions and reagents are tissue culture grade. Perform all steps inside of a tissue culture laboratory hood. Use filter tips where possible. Diligently follow all waste disposal regulations when disposing of waste materials. When making up the media, each stock can be made ahead of time and frozen at −80 °C. In our experience, it is best to avoid storing media for more than 4 months where possible, although media stored beyond this timeframe may still be used.

2.1 Enteroid Media

1. Complete Media Growth Factor negative (CMGF−): 500 ml Advanced DMEM/F12, 5 ml GlutaMAX 100×, 5 ml HEPES 1 M. Aliquot ~50 ml per sterile 50 ml falcon tube.
2. Complete Media Growth Factor positive (CMGF+) in a volume of 500 ml: 78 ml CMGF−, 250 ml Wnt3A-conditioned media produced from ATCC CRL-2647 cells (ATCC,), 100 ml R-spondin-conditioned media (Trevigen Inc.), 50 ml Noggin-conditioned media [9], 10 ml B27 (50×), 5 ml N2 (100×), 5 ml nicotinamide (10 mM), 1 ml N Acetylcysteine (1 mM),

500 μl Gastrin (10 nM), 500 μl A83 (500 nM), 166 μl SB202190 (10 μM), 50 μl EGF (50 ng/mL) [4]. Vacuum filter (0.2 μm) and aliquot into 15 ml and/or 50 ml falcon tubes of varying volumes depending upon ultimately needs and throughput.

3. High Wnt media in a volume of 500 ml: 250 ml CMGF+, 250 ml Wnt3A-conditioned media. Vacuum-filter (0.2 μm) and aliquot into 15 ml and/or 50 ml falcon tubes of varying volumes depending upon ultimately needs and throughput.
4. Differentiation media in a volume of 500 ml: 458 ml CMGF−, 25 ml Noggin conditioned media, 10 ml B27 (50×), 5 ml N2 (100×), 1 ml *N* Acetylcysteine (500 mM), 500 μl Gastrin (100 μM), 500 μl A83 (500 μM), 50 μl EGF (500 μg/mL).
5. Enteroid freezing media: 10% DMSO in CMGF+ containing Y-27632 (10 μM). Alternatively can use Dulbecco's modified Eagle medium/F12, 10% fetal bovine serum, and 10% dimethyl sulfoxide [10].
6. Matrigel Growth Factor Reduced Basement Membrane Matrix.

2.2 Extraction of Stem Cells from Tissue

1. Incomplete chelating solution (ICS) in a volume of 500 ml: Dissolve 2.49 g $Na_2HPO_4{\cdot}2H_2O$ (sodium phosphate dibasic dihydrate), 2.7 g KH_2PO_4 (potassium dihydrogen phosphate), 14 g NaCl (sodium chloride), 0.3 g KCl (potassium chloride), 37.5 g sucrose, and 25 g D-sorbitol in 500 ml tissue culture grade water. Vacuum-filter (0.2 μm) and aliquot 50 ml per sterile 50 ml falcon tube.
2. Complete chelation solution (CCS) in a volume of 50 ml: Prepare the CCS by combining 40 ml sterile water, 10 ml 5× ICS, and 60 μl 1 M DTT. The addition of DDT to ICS should be done immediately prior to use.

3 Methods

Perform all steps inside of a tissue culture laboratory hood. Use filter tips where possible. Diligently follow all waste disposal regulations when disposing of waste materials. All reagents should be on ice and centrifuges should be cooled to 4 °C, unless otherwise stated.

3.1 Collection and Freezing of Resected Tissue

1. Resected tissue should be collected in ice-cold CMGF− or PBS (enough to cover the entire sample) and placed in ice. The resected tissue should be stored on ice for as short a time as possible and ideally stem cells will be isolated within a few hours (*see* **Note 1**).

2. If same day extraction is not possible, resected tissue can be frozen, but this may reduce the crypt yield (*see* **Note 1**).
3. To freeze the resected tissue for later enteroid generation, first place the tissue in a small sterile dish (e.g., 35 mm diameter) containing enough PBS to cover the base of the dish and cut the tissue into ~5 mm^2 fractions.
4. Place each ~5 mm^2 tissue fraction into a cryovial with 500 μl freezing media.
5. Put each vial into a CoolCell (or other freezing containers that allow standardized controlled rate of −1 °C/min) and place the CoolCell in −80 °C overnight.
6. Move each cryovial to liquid nitrogen storage the next day.

3.2 Extraction of Stem Cells from Tissue

For the below protocol, use wide-bore 1 ml low-retention filter tips. All tubes and solutions should be kept on ice throughout the procedure, unless otherwise stated. Set the centrifuge to 4 °C before beginning the protocol.

1. Before starting this protocol, it is important to ensure reagents have been given sufficient time to thaw and that a 24-well plate has been incubated at 37 °C (*see* **Note 2**).
2. Place the resected tissue in a puddle of CMGF− (or PBS) in small sterile dish (e.g., 35 mm diameter). If the resected tissue is fresh and plentiful, use sterile scissors to take a ~5 mm^2 specimen. If the resected tissue is frozen in freezing media, remove the cryovial from liquid nitrogen and place in a 37 °C incubator for 2–5 min to thaw, then proceed as below.
3. Mince tissue into small pieces (*see* **Note 3**) using sterile scissors (i.e., cleaned with 70% ethanol). Sterile forceps can also be used if needed.
4. Transfer minced tissue sample into a 15 ml falcon containing 10 ml CCS and wash tissue sample by inverting tube ~10 times (*see* **Note 4**).
5. Allow tissue sample to settle down to the bottom of the falcon tube (typically ~3–5 min) and then remove supernatant carefully using a pipette (*see* **Note 5**).
6. Add 10 ml of CSS containing 100 μl of Fungizone (1:100) + 100 μl Pen/strep (1:100) and invert the falcon ~10 times.
7. Allow minced tissue to settle down to the bottom of the falcon tube (typically ~3–5 min) and then remove supernatant carefully using a pipette (*see* **Note 5**).
8. Transfer minced tissue to a new 15 ml tube containing 10 ml CCS and wash tissue sample by inverting ~10 times.

9. Allow minced tissue to settle down to the bottom of the falcon tube (typically ~3–5 min) and then remove supernatant carefully using a pipette (*see* **Notes 5** and **6**).
10. Pour CCS to 3 ml mark on the 15 ml falcon tube.
11. Transfer the entire contents (i.e., minced tissue sample in 3 ml CSS) to a single well of a 6-well plate and add 180 μl of 0.5 M EDTA.
12. Place the 6-well plate on a rotary (orbital) shaker in 4 °C room (~500 g) for 30 min (*see* **Note** 7).
13. Transfer the entire contents to a 15 ml falcon tube and add 5 ml CCS and 2 ml FBS (FBS should be at RT) to the falcon tube. The 10 cm dish should be set aside for use in **step 16**.
14. Pulse-vortex for ~5 s.
15. Allow tissue to settle and transfer supernatant to a 15 ml tube. This supernatant should contain the crypt cells. Spin down supernatant at 1000 × *g*, for 5 min, at 4 °C.
16. While waiting for the centrifuge, pour CCS to the 3 ml mark on the 15 ml falcon containing the minced tissue and transfer the entire contents of the falcon back to the same single well of a 6-well plate. Add 250 μl of 0.5 M EDTA and place back on orbital shaker in 4 °C room for 30 min (*see* **Note** 7).
17. When centrifugation is complete, carefully remove supernatant and resuspend the pellet in 2 ml CMGF− (*see* **Note 8**). The pellet contains the crypt cells. Keep the resuspended crypts on ice while doing the additional EDTA step.
18. Once the minced tissue from the second EDTA incubation (**step 16**) has been on the shaker for 30 min, transfer the entire contents to a 15 ml falcon tube.
19. Vortex vigorously for ~10 s.
20. Add 5 ml CCS and 2 ml FBS (FBS should be at RT) to the 15 ml falcon tube.
21. Pulse-vortex for ~5 s.
22. Allow minced tissue to settle and transfer supernatant to a new 15 ml falcon tube and spin down supernatant at 1000 × *g*, for 5 min, at 4 °C.
23. Biopsy samples can now be discarded through the appropriate waste disposal regulations.
24. When centrifugation is complete, carefully remove supernatant and resuspend the pellet (containing crypt cells) in 1 ml CMGF− (*see* **Note 8**). Then, using the same pipette tip, transfer the resuspended crypts to the 15 ml falcon tube on ice containing the first crypt factions (i.e., from **step 17**).

25. Add 5 ml of CMGF− to the falcon tube and pipet up and down ~10 times to mix.
26. Centrifuge the falcon tube at 1000 × *g*, for 5 min, at 4 °C.
27. Carefully remove as much supernatant as possible, ensuring the pellet (containing the crypt cells) is not disturbed.
28. Using cold tips from the fridge, prelube tip with ice-cold CMGF− (i.e., pipet and expel CMGF− into the tip) and then add ~100 μl thawed Matrigel to the pellet in the 15 ml falcon tube (*see* **Note 9**). Gently mix the Matrigel up and down ~10 times (avoiding air bubbles), then add ~30 μl per well to a new prewarmed 24-well plate (*see* **Note 10** and Fig. 2).
29. Incubate at 37 °C 5% CO_2 for 20–30 min to polymerize the Matrigel.
30. Carefully add 500 μl of prewarmed High Wnt containing 10 μM Y-27632 + Pen/Strep (1:100) + Fungizone (1:100) (*see* **Note 11**).
31. Place plate back into 37 °C incubator and check for enteroid growth daily (Fig. 3). *See* **Note 12** for a detailed description on managing the initial phases of enteroid growth.

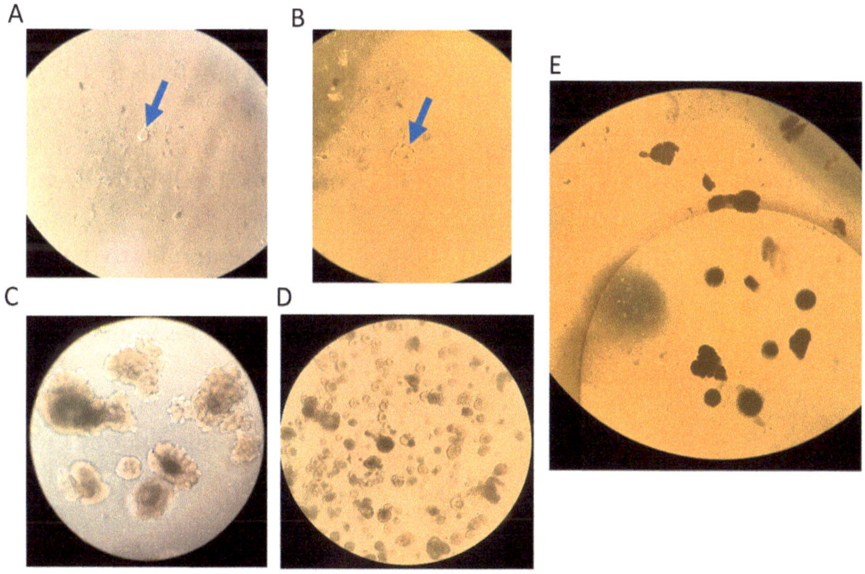

Fig. 2 Images taken down a microscope lens showing enteroids at various stages of growth. (**a**) Enteroid growing from previously frozen tissue after 7 days under 10× magnification (blue arrow points to the enteroid). (**b**) Enteroid growing from previously frozen tissue after 11 days under 10× magnification (blue arrow points to the enteroid). (**c**) Healthy Enteroids prior to passaging after 7 days of growth. (**d**) Unhealthy Enteroids beginning to turn black, signifying they are dying. (**e**) Contaminated well with microbial growth in both the Matrigel and in the media (4× magnification)

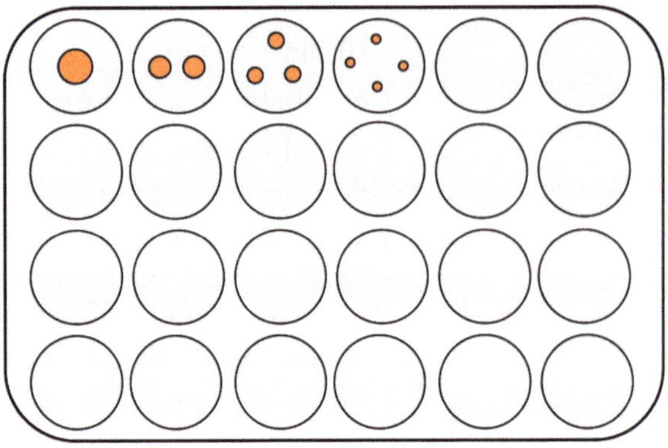

Fig. 3 Schematic representation of a 24-well tissue culture plate showing four possibilities for plating the Matrigel containing enteroids. Each well will have a total volume of ~30 μl. It is advantageous to have multiple Matrigel dots to maximize the contact surface area between the Matrigel and the media. When plating multiple dots, it is important to leave sufficient space between each dot to ensure they do not join together

3.3 Maintenance/ Passaging

1. Remove all media from wells, taking care not to dislodge the Matrigel plug.
2. Add 500 μl cold CMGF− to the first well and mechanically break up Matrigel by pipetting using a P1000 up and down ~10 times.
3. Move from well-to-well collecting the contents as you go.
4. Transfer the CMGF− containing enteroids to a 1.5 ml or 2 ml Eppendorf on ice (*see* **Note 13**).
5. With a new tip, add another 500 μl of cold CMGF− to the last well and work in reverse order from well-to-well, collecting the contents from all wells.
6. Transfer the CMGF− containing enteroids to the same Eppendorf as before.
7. Collect any residual CMGF−/enteroids from each well into to the Eppendorf.
8. Using a 1 ml syringe containing a 25 G × 5/8 needle, draw up 1 ml from the Eppendorf containing enteroids in CMGF−. Then, with moderate force, expel the contents of the syringe back into the Eppendorf.
9. Repeat the previous **steps 1–3** times as necessary (*see* **Note 14**).
10. Spin down the passaged enteroids at 1000 × *g*, for 5 min, at 4 °C.
11. Being careful not to disrupt the enteroid pellet, remove all CMGF− (*see* **Note 15**).

12. Using a cold pipette tip previously rinsed with CMGF−, add Matrigel to the Eppendorf containing the enteroids (*see* **Note 16** for guidance on how to calculate the volume of Matrigel and how best to add this).
13. Set a p200 pipette to 180 μl (i.e., the volume required to fill a row of a 24-well tissue culture plate, assuming ~30 μl per well) and add ~30 μl per well to a new prewarmed 24-well plate (*see* **Note 10** and Fig. 2).
14. Incubate at 37 °C 5% CO_2 for 20–30 min to polymerize the Matrigel.
15. Carefully add 500 μl of prewarmed High Wnt (*see* **Note 11**).
16. Place plate back into 37 °C incubator and check for enteroid growth daily (Fig. 3). After 5–7 days the enteroids are usually ready to be passaged.
17. As a general rule, enteroid media should be replenished 500 μl on Monday, 500 μl on Wednesday, and 600 μl on Friday (extra media on Friday overcomes the need to replenish media over the weekend).

3.4 Freezing

1. As detailed in **Note 12**, it is important to make regular enteroid stocks, particularly during the early passages.
2. Remove all media from wells, taking care not to dislodge the Matrigel plug.
3. Add 500 μl cold CMGF− to the first well and mechanically break up Matrigel by pipetting P1000 up and down ~10 times.
4. Move from well-to-well collecting the contents as you go.
5. Transfer the CMGF− containing enteroids to a 1.5 ml or 2 ml Eppendorf on ice (*see* **Note 13**).
6. With a new tip, add another 500 μl of cold CMGF− to the last well and work in reverse order from well-to-well, collecting the contents from all wells.
7. Transfer the CMGF− containing enteroids to the same Eppendorf as before.
8. Collect any residual CMGF−/enteroids from each well into to the Eppendorf.
9. Centrifuge the passaged enteroids at 1000 × *g*, for 5 min, at 4 °C.
10. Being careful not to disrupt the enteroid pellet, remove all CMGF− (*see* **Note 15**).
11. Resuspend enteroids into freezing medium (each cryovial using 500 μl) at original ratio of 1–2 wells into 1 cryovial (*see* **Note 17**).

12. Put each vial into a CoolCell (or other freezing containers that allow standardized controlled rate of −1 °C/min) and place the CoolCell in −80 °C overnight.
13. Transfer cryovials into liquid nitrogen for long-term storage.

3.5 Thawing and Regrowing Cultures

1. Remove cryovials from liquid nitrogen.
2. Place in 37 °C 5% CO_2 incubator until partially thawed (*see* **Note 18**).
3. Add 500–1000 μl of prewarmed (37 °C) High Wnt to the cryovial of enteroids and pipet up and down to fully defrost the cells and dilute the DMSO.
4. Centrifuge the enteroids at 1000 × *g*, for 5 min, at 4 °C.
5. Being careful not to disrupt the enteroid pellet, remove all CMGF− (*see* **Note 15**).
6. Using a cold pipette tip prelubed with CMGF−, add 120 μl Matrigel to the Eppendorf containing the enteroids and add ~30 μl per well to a new prewarmed 24-well plate (*see* **Note 10** and Fig. 2).
7. Incubate at 37 °C 5% CO_2 for 20–30 min to polymerize the Matrigel.
8. Carefully add 500 μl of prewarmed High Wnt + Y-27632 (*see* **Note 11**).
9. Place plate back into 37 °C incubator and check for enteroid growth daily (Fig. 3).
10. As a general rule, enteroid media should be replenished 500 μl on Monday, 500 μl on Wednesday, and 600 μl on Friday (extra media on Friday overcomes the need to replenish media over the weekend).

4 Notes

1. It is important that the tissue is completely submerged in the collection solution (either CMGF− or PBS) and remains on ice/cold packs or in the fridge until transferred to the lab. Notably, even in cold conditions the crypt yield will reduce over time and so it is important to process the tissue as soon as possible after surgery. Tissue processed within 3 h and without being previously frozen will yield the greatest enteroid number. If sufficient tissue allows, it is suggested to perform immediate extraction on a fresh 5 mm^2 section, as well as freezing down additional 5 mm^2 sections in enteroid freezing media in case the fresh extraction fails. Excess tissue can also be used for other analyses, including histology, metagenomic sequencing, transcriptomics, and others.

2. Matrigel should be thawed at least the night before use where possible. If tissue arrives unexpectedly and no thawed Matrigel is available, it is possible to use Matrigel that has been in the fridge for ~4 h. A 24-well tissue culture plate should be incubated overnight in the 37 °C 5% CO_2 incubator, but there is no harm in having multiple sealed 24-well plates in the 37 °C 5% CO_2 incubator at any given time (if space permits) to ensure there is always a warm plate on hand. A 200 μl tip box should also be left in the fridge as will be needed for plating the Matrigel. Before beginning the protocol, setup a centrifuge to contain the necessary rotor at set at 4 °C, remove a premade aliquot of 4 ml FBS from the freezer to thaw, set the water bath to 37 °C, and thoroughly clean the tissue culture hood with 70% ethanol and other detergents as needed. The media can be placed in the 37 °C water bath at any point so long as sufficient time is given for the entire volume of media to reach at least 30 °C.
3. It is important that the minced tissue is small enough to easily fit into a wide-bore pipette tip.
4. Use a dedicated 15 ml falcon tube of FBS for lubing pipette tips (i.e., suck up and expel FBS before handling solutions containing tissue). This will avoid minced tissue sticking to the inside of the pipette tip.
5. Although effort should be made to avoid bubbles where possible, inevitably bubbles will form. These can be problematic because some minced tissue will stick to bubbles and therefore not settle to the base of the falcon tube, resulting in these bits of tissue being lost when removing the supernatant after each wash step (i.e., ultimately reducing enteroid yield). In order to dislodge pieces of minced tissue from the bubbles and to encourage them to sink, repeatedly tap the base of the falcon tube onto the hood. Additionally, it is not necessary to remove all of the supernatant; it is better to leave a small volume of liquid in the falcon tube rather than risk losing minced tissue.
6. If after the 3× wash steps the supernatant is still dirty (i.e., discolored), repeat additional wash steps with CSS until the supernatant is clear.
7. If using a rotary (orbital) shaker located within a 4 °C room/fridge is not possible, an alternative is to place the 6-well plate on top of an ice bucked filled with ice and use a rotary (orbital) shaker on the bench.
8. The supernatant at this point contains the stem cell containing crypts. After centrifugation, a faint pellet is usually visible; this is the crypt cells, and it is important not to disturb this pellet when removing supernatant. Therefore, it is better to leave a small amount of the supernatant (less than 1 ml) rather than risk losing crypt cells.

9. As a general rule, a fresh (i.e., where the tissue sample was collected within 3 h postsurgery) ~0.5 mm^2 sized piece of tissue will yield ~4 wells in a 24-well plate. The yield will be reduced in tissue that has been left on ice overnight and/or frozen; therefore, less Matrigel (~70 μl) should be added, aiming for ~2–3 wells in a 24 well plate. It is important that the Matrigel is not diluted (at least a Matrigel ratio of 3:1), so consider using a 200 μl pipette tip to carefully remove as much CMGF− as possible. If there is some residual CMGF− in the falcon tube, then use less Matrigel to account for this. For example, if 120 μl is to be plated into 4 well, but ~30 μl of CMGF− remains in the falcon tube, then add ~90 μl of Matrigel.
10. It is important to plate around 30 μl per well in a 24-well tissue culture plate. For optimal enteroid growth it is suggested to plate 2× ~15 μl or 3× ~10 μl dots per well. It is important to avoid getting bubbles in the Matrigel, so take extra care when resuspending the enteroid pellet into the Matrigel solution.
11. When adding enteroid media, slowly pipet down side of well taking care not to disturb the Matrigel plug. Forcefully expelling media onto the Matrigel may cause the plug to dislodge. If the lines are precious (i.e., no spare tissue in frozen storage) or within the initial weeks of growth, add 10 μM Y-27632 + Pen/Strep (1:100) + Fungizone (1:100) as required.
12. Wells that are contaminated will usually appear yellow and cloudy. It might be possible to directly observe fungal growth on the surface of the media. To limit the risk of contamination, High Wnt media containing 10 μM Y-27632 + Pen/Strep (1:100) + Fungizone (1:100) can be used until passage 2. Continue adding Y-27632 if enteroids are not growing well or appear to be dying (Fig. 1). If an isolated well is contaminated, remove all media from the well and wash the well in question with 700 μl 10% bleach and/or 700 μl 70% ethanol, then wash twice with PBS. Leave well empty and mark on the plate the well that was contaminated so it is not used again. If enteroids show widespread contamination (Fig. 1) and the no stocks have been generated previously, it might be possible to save the line by passing the enteroids by adding Normocure. Specifically, passage the enteroids and incubate with 2 ml High Wnt media containing 10 μM Y-27632 and 4 μl Normocure for 30 min on ice. Centrifuge the Eppendorf tube at 1000 × *g*, for 5 min, at 4 °C, remove supernatant, and suspend the enteroid pellet into ~120 μl Matrigel (assuming 4 wells to be plated). Spike ~2 μl of Normocure into the Matrigel and add 500 μl per well of High Wnt media containing 10 μM Y-27632 + Pen/Strep (1:100) + Fungizone (1:100) + 100 μg/ml Normocure. Treat contaminated

enteroids for three passages. The first passage should be performed when many enteroids can be observed and are showing good growth, which is typically 5–20 days following enteroid extraction. It is important to regularly freeze enteroids, particularly during the early passages; freeze down around 50% of the wells in the early passages. This will ensure a plentiful supply of enteroid stocks at early passages and will help to future-proof the line.

13. If there are more than 12 wells of enteroids, use two Eppendorf tubes. This will avoid issues with having too much Matrigel in the CMGF− and ensure (1) a robust passage and (2) that a large Matrigel pellet does not occur after centrifugation.
14. The optimal number of times to pass enteroids through the syringe and the force to apply will take practice and can vary from line to line. It is important to break up enteroids, but over-passaging (i.e., to many passages through the syringe or using too much force) will damage enteroids and cause them to die. As a general rule, it is better to err on the side of caution, passing through the syringe two times and with moderate pressure.
15. If a layer of Matrigel can be seen directly above the enteroid pellet, remove as much CMGF− as possible then add another 1 ml of ice-cold CMGF−. Repeat the centrifugation and remove the CMGF− supernatant. This will avoid loss of enteroids that are stuck in the Matrigel layer and would otherwise be lost when removing the supernatant.
16. Typically we want to at least double the number of wells after each passage, not accounting for wells that are frozen. For example, if you have 8 wells of healthy/dense enteroids and wish to freeze down 3 wells, then the remaining 5 wells will be passaged. Thus, these 5 wells will be split into ~10–15 wells. This estimate will need to be adjusted according to the enteroid density (i.e., dense enteroid wells should be more than doubled and vice versa). To calculate the amount of Matrigel needed, decide how many wells you intend to seed and assuming 30 μl per well, simply multiply 30 μl by the number of wells to be seeded. There will be some residual CMGF− in the Eppendorf so consider subtracting ~20–50 μl from the amount of Matrigel to add. If the volume of Matrigel to be added is <200 μl, use the same 200 μl pipette tip to immediately mix the enteroids into the Matrigel and then plate ~30 μl per well in a prewarmed tissue culture plate (Fig. 2). If the volume of Matrigel to be added is >200 μl, use a 1000 μl pipette tip to add the Matrigel to the Eppendorf containing enteroids and then use a 200 μl pipette tip set at 180 μl to mix the enteroids into the Matrigel, before plating ~30 μl per well in a prewarmed tissue culture plate (Fig. 2).

17. To calculate how much enteroid freezing media to add to the enteroids, decide how many cryovials will be generated and then resuspend the enteroids in 500 μl multiplied by the number of cryovials. As a general rule, if the enteroids are healthy and very dense, then aim for 1 well into 1 cryovial, otherwise a ratio of 1.5 or 2 wells per cryovial should be used.
18. Instead of the 37 °C 5% CO_2 incubator, it is also possible to quickly thaw the enteroids by holding the cryovial under running distilled or warm water. As a rule of thumb, "freeze slow and thaw fast."

Acknowledgments

This work was supported by the European Union's Horizon 2020 research and innovation program under the Marie Skłodowska-Curie Grant Agreement No. 794771 to C.J.S. and Public Health Service Grant U19AI116497 from the National Institutes of Health to M.K.E. We thank Xi-lei Zeng for her expertise and help with all enteroid cultures.

References

1. Blutt SE, Crawford SE, Ramani S et al (2017) Engineered human gastrointestinal cultures to study the microbiome and infectious diseases. Cell Mol Gastroenterol Hepatol 5:241–251
2. Stelzner M, Helmrath M, Dunn JCY et al (2012) A nomenclature for intestinal in vitro cultures. Am J Physiol Gastrointest Liver Physiol 302:G1359–G1363
3. Sato T, Vries RG, Snippert HJ et al (2009) Single Lgr5 stem cells build crypt-villus structures in vitro without a mesenchymal niche. Nature 459:262–265
4. Sato T, Stange DE, Ferrante M et al (2011) Long-term expansion of epithelial organoids from human colon, adenoma, adenocarcinoma, and Barrett's epithelium. Gastroenterology 141:1762–1772
5. Foulke-Abel J, In J, Kovbasnjuk O et al (2014) Human enteroids as an ex-vivo model of host-pathogen interactions in the gastrointestinal tract. Exp Biol Med (Maywood) 239 (9):1124–1134. https://doi.org/10.1177/1535370214529398
6. Kovbasnjuk O, Zachos NC, In J et al (2013) Human enteroids: preclinical models of non-inflammatory diarrhea. Stem Cell Res Ther 4(Suppl 1):S3
7. Jalili-Firoozinezhad S, Gazzaniga FS, Calamari EL et al (2019) Complex human gut microbiome cultured in anaerobic human intestine chips. Nat Biomed Eng 3(7):520–531
8. Fofanova TY, Stewart C, Auchtung JM, et al (2019) A novel human enteroid-anaerobe co-culture system to study microbial-host interaction under physiological hypoxia. *bioRxiv*: 555755
9. Heijmans J, van Lidth de Jeude JF, Koo B-K et al (2013) ER stress causes rapid loss of intestinal epithelial stemness through activation of the unfolded protein response. Cell Rep 3:1128–1139
10. Tsai Y-H, Czerwinski M, Wu A et al (2018) A method for cryogenic preservation of human biopsy specimens and subsequent organoid culture. Cell Mol Gastroenterol Hepatol 6:218–222.e7

Chapter 17

Characterizing Innate Lymphoid Cell Phenotype and Function in Human Inflammatory Bowel Disease

Omer S. Omer, Nick Powell, and Graham M. Lord

Abstract

Innate lymphoid cells (ILCs) are emerging as important effectors of innate immunity and play a critical role in maintaining intestinal immune homeostasis. They are tissue-residing immune cells that can be subdivided based on master transcription factor and cytokine expression, bearing striking resemblance to their $CD4^+$ T helper (Th) cell counterparts. ILCs are increasingly recognized as potential mediators of inflammatory bowel disease (IBD) providing a need to explore their functional and phenotypic differences in health vs. disease. In this chapter we outline protocols for the characterization of human ILCs and intracellular cytokine expression using flow cytometry. We include protocols for isolating human peripheral blood and colonic lamina propria mononuclear cells essential for evaluating human IBD specimens.

Key words Innate lymphoid cell, Inflammatory bowel disease, Cell isolation, Flow cytometry, Gating

1 Introduction

IBD is a chronic inflammatory disorder of the gastrointestinal tract, chiefly comprising ulcerative colitis and Crohn's disease. Its pathogenesis is complex, involving interplay between environmental factors, genetic susceptibility loci, abnormalities in the gut microbiota and dysregulated immune responses. It is widely recognized that $CD4^+$ T helper (Th) 1 cells (expressing IFN-γ and the transcription factor T-bet) and Th17 cells (expressing IL-17A, IL-22, and the transcription factor RORγt) play a central role in driving intestinal inflammation and accumulate in the inflamed colonic mucosa of patients with Crohn's disease and ulcerative colitis [1, 2]. Following the discovery of innate lymphoid cells (ILCs), there has been growing interest in whether these mucosal-dwelling innate lymphocytes contribute to the pro-inflammatory environment in IBD. ILCs appear to be crucial players in mediating mucosal immune homeostasis acting as a bridge between the innate and adaptive immune systems. Defects in the ILC compartment arising from the loss of T-bet in mice lacking an adaptive immune system results

Shoba Amarnath (ed.), *Innate Lymphoid Cells: Methods and Protocols*, Methods in Molecular Biology, vol. 2121,
https://doi.org/10.1007/978-1-0716-0338-3_17,

in a transmissible form of ulcerative colitis [3]. In this model, IL-17A-producing ILCs were potent promoters of disease [4]. In bacteria-driven innate immune-mediated colitis, ILCs are an important source of IFN-γ and of IL-23-inducible proinflammatory cytokines, such as IL-17A and IL-17F [5]. Building on these animal studies, work in humans have identified IL-23-responsive innate mucosal immune cells, which produce IL-22 in mucosa-associated lymphoid tissue including Peyer's patches [6]. Furthermore, in Crohn's disease, a subset of IFN-γ-producing ILCs appear to accumulate at sites of intestinal inflammation suggesting a role for these cells in the pathogenesis of gut mucosal inflammation [7].

ILCs are defined by their lack of recombination activating gene (RAG)-dependent rearranged antigen receptors (such as those found on T cells and B cells); an absence of "classical" immune cell lineage markers; and their lymphoid morphology [8]. They develop from a common lymphoid progenitor and require the common IL-2 cytokine receptor γ-chain and the transcriptional repressor inhibitor of DNA binding 2 (ID2) [9]. Several distinct ILC populations have been identified with lineage-defining transcription factor and cytokine-secreting profiles which bear striking resemblance to their T helper cell counterparts [10]. Like Th1 cells, ILC1s express the transcription factor T-bet and secrete IFN-γ and TNF-α, mounting immune responses to intracellular pathogens. ILC2s mirror Th2 cells, expressing GATA3 and secreting IL-4, IL-5 and IL-13, essential in protecting against parasites and involved in allergic reactions. Finally, ILC3s express RORγt and secrete IL-17 and IL-22, important in defense against extracellular pathogens, similar to Th17 cells.

In general, ILCs can be identified by their expression of IL-7Rα (CD127) and CD161 but lack of lineage defining surface markers for T cells (CD3, TCRαβ, and TCRγδ), B cells (CD19), NK cells (CD16 and CD94), myeloid cells (CD1a, CD11c, CD14, and CD123), granulocytes (FcεRI and CD123), hematopoietic stem cells (CD34), and plasmacytoid dendritic cells (BDCA2 and CD123). This pool of lineage$^-$ CD127$^+$ CD161$^+$ ILCs can be subdivided based on their surface expression of CRTH2 (CD294) or c-kit (CD117) [11]. ILC2s and ILC3s express CRTH2 and c-kit, respectively. ILC1s, however, have no specific defining surface marker. ILC1s share features in common with NK cells, namely the expression of T-bet and secretion of IFN-γ. Distinguishing features of NK cells include the expression of surface markers CD16, CD56, and CD94, and the demonstration of cytotoxic activity through the production of perforin/granzyme B [12]. It is important to note that CD56 is also found on human IL-22-producing ILCs and therefore is not an exclusive marker for human NK cells [6, 13].

This chapter provides detailed methodology for flow cytometric analysis of peripheral blood or colonic lamina propria mononuclear cells to identify ILC populations and evaluate intracellular cytokine expression following activation. This approach will enable the user to characterize changes in ILC compartments in disease models and assess functional status of these populations.

2 Materials

2.1 Isolation of Lamina Propria Mononuclear Cells (LPMCs)

1. Cytomatrix 9 mm × 1.5 mm thick (Nohla Therapeutics, Cat. No. CY-903).
2. 50 ml centrifuge tube.
3. 90 mm polystyrene petri dish.
4. 40 μM sterile cell strainer.
5. Nunclon Delta Surface 24-well plate.
6. Surgical tweezers.
7. Hemocytometer plate.
8. 1× Phosphate-buffered saline (PBS).
9. Hank's Balanced Salt Solution (HBSS).
10. FACS buffer: 1× PBS, 10% fetal bovine serum (FBS), 2 mM EDTA.
11. R10: RPMI media 1640, 10% FBS.
12. Gut-T: RMPI media 1640, 10% FBS, L-glutamine 292 μg/ml, penicillin 100 IU/ml, streptomycin 100 μg/ml, 2-mercaptoethanol 3.5 μl/L, gentamicin 20 mg/ml, amphotericin B 2.5 μg/ml, metronidazole 1 μg/ml.
13. Collagen 1, Rat Tail 100 mg.
14. Human recombinant IL-2 1KIU/ml.
15. Trypan blue.

2.2 Isolation of Peripheral Blood Mononuclear Cells (PBMCs)

1. 50 ml centrifuge tube.
2. Hemocytometer plate.
3. Lymphoprep™.
4. 1× Phosphate-buffered saline (PBS).
5. Trypan blue.

2.3 Innate Lymphoid Cell (ILC) Activation

1. 75 mm flow cytometry (FACS) tubes.
2. 50 ml centrifuge tube.
3. R10: RPMI media 1640, 10% FBS.
4. Phorbol 12-myristate 13-acetate (PMA) 50 μg/ml.
5. Ionomycin from *Streptomyces conglobatus* 1 mM.
6. Monensin 3 mM.
7. Brefeldin A.

Table 1
Extracellular antibody master mix

Fluorochrome	Marker	μl/test	Company	Cat. No.	Dilution
Near-IR	Live/Dead	0.1	Life Technologies	L10119	1 in 1000
Biotin	TCRαβ	1	BioLegend	306704	1 in 100
Biotin	TCRγδ	1	BioLegend	331206	1 in 100
Biotin	CD1a	1	BioLegend	300112	1 in 100
Biotin	CD11c	1	BioLegend	301612	1 in 100
Biotin	CD14	1	BioLegend	325624	1 in 100
Biotin	CD19	1	BioLegend	302204	1 in 100
Biotin	CD34	1	BioLegend	343524	1 in 100
Biotin	CD94	1	Serotec	MCA2254B	1 in 100
Biotin	CD123	1	BioLegend	306004	1 in 100
Biotin	CD303 (BDCA2)	1	Miltenyi	130-090-691	1 in 100
Biotin	FcεRI	1	eBioscience	13-5899-82	1 in 100
FITC	CD3	0.5	BioLegend	300306	1 in 200
FITC	CD3	0.5	BioLegend	317306	1 in 200
V500	CD45	1	BD Horizon	560779	1 in 100
PE-Cy7	CD117	1	BioLegend	313212	1 in 100
BV605	CD127	1	BioLegend	351334	1 in 100
PerCP-Cy5.5	CD161	1	BioLegend	339908	1 in 100
BV421	CD294	2.5	BioLegend	350112	1 in 40
1× PBS		81.4	–	–	–

2.4 Extracellular Staining

1. 75 mm flow cytometry (FACS) tubes.
2. 1× Phosphate-buffered saline (PBS).
3. Extracellular antibody master mix (*see* Table 1).
4. FACS buffer: 1× PBS, 10% fetal bovine serum (FBS), 2 mM EDTA.
5. APC-Cy7 Streptavidin (BioLegend, Cat. No. 405208).
6. 4% Paraformaldehyde (PFA) solution.

2.5 Intracellular Staining

1. 75 mm flow cytometry (FACS) tubes.
2. FACS buffer: 1× PBS, 10% fetal bovine serum (FBS), 2 mM EDTA.
3. 0.2% Saponin buffer (100 mg Saponin, 50 ml 1× PBS).
4. Intracellular antibody master mix (*see* Table 3).

2.6 Data Acquisition and Analysis

1. BD LSRFortessa cell analyzer systems.
2. 75 mm flow cytometry (FACS) tubes.
3. UltraComp eBeads compensation beads.
4. FlowJo flow cytometry analysis software.

3 Methods

Functional analysis and phenotyping of ILCs can be carried out using colonic tissue resident lamina propria mononuclear cells (LPMCs) or circulating peripheral blood mononuclear cells (PBMCs) and both methods are outlined below. If ILC phenotyping is to be carried out using LPMCs, we still recommend isolating a small number of PBMCs (1×10^6) to make up control tubes. This is to avoid the use of valuable samples on unstained and fluorescence minus one (FMO) controls.

3.1 Isolation of Lamina Propria Mononuclear Cells (LPMCs)

1. Rectosigmoid colonic biopsies are obtained during colonoscopy and transferred directly into 10 ml of cold HBSS (*see* **Notes 1** and **2**). Biopsies are stored at 4 °C for a maximum of 2 h prior to processing to prevent sample degradation.
2. Dilute collagen to 100 μg/ml by adding 250 μl rat tail collagen stock solution to 10 ml 1× PBS. Incubate cytomatrices in the 100 μg/ml collagen solution for 30 min in a 37 °C water-bath. Use 1 cytomatrix per colonic biopsy. After 30 min, carefully pour out collagen and add 10 ml 1× PBS. The cytomatrices are now sufficiently coated in collagen and ready for use (*see* **Note 3**).
3. Prepare total volume of Gut-T required—this is 2 ml of Gut-T per cytomatrix/biopsy. Then add 1 μl of IL-2 per 10 ml of Gut-T, that is, for 20 colonic biopsies, prepare 40 ml of Gut-T with 4 μl of IL-2. Place the media in a 37 °C water-bath. When the media is sufficiently warmed, transfer 2 ml into each well of a 24-well plate.
4. Prepare the colonic biopsies by first rinsing them with fresh HBSS, then transferring to a petri dish (*see* **Note 4**). Working with surgical tweezers on the petri dish, place each biopsy on the surface of a cytomatrix. Invert the cytomatrix and apply pressure crushing the biopsy onto the surface of the cytomatrix. Turn the cytomatrix over so that the biopsy is facing upward and place into a well of the plate containing Gut-T and IL-2. Place in a cell culture incubator at 37 °C for 48 h.
5. After 48 h, carefully remove the cytomatrices/biopsies from the wells and transfer into a 50 ml tube containing 15 ml 1× PBS. Shake the tube vigorously to dislodge cells adherent to the cytomatrices. Transfer the resultant cell suspension with a

Pasteur pipette and pass through a 40 μm strainer into a new 50 ml tube labeled LPMCs. Rinse the cytomatrices thoroughly by adding a further 10 ml 1× PBS and pipetting up and down. Pass this volume through the 40 μm strainer into the tube labeled LPMCs. Discard the used 40 μm strainer and replace with a new one.

6. Transfer the supernatants from the 24-well plate into the 50 ml tube labeled LPMCs through the 40 μm strainer. Wash each well with 1 ml 1× PBS by pipetting up and down and transferring to tube labeled LPMCs. Depending on the number of biopsies obtained, it may be necessary to transfer this to a new 50 ml tube.
7. Centrifuge LPMCs at 688 × *g* f or 10 min at 4 °C. Discard the supernatant and resuspend pellet in 4 ml of Gut-T. Mix well by pipetting up and down, then take 10 μl of sample to perform cell count. Mix the 10 μl sample with an equal volume of trypan blue and transfer to hemocytometer plate for cell count. This will allow quantification of live lamina propria mononuclear cells. Typically, we isolate ~450,000 LPMCs per colonic biopsy.

3.2 Isolation of Peripheral Blood Mononuclear Cells (PBMCs)

1. Collect peripheral blood sample in a lavender top EDTA tube and store at room temperature. We recommend sampling within 6 h to avoid sample degradation.
2. Transfer 20 ml of LymphoPrep™ into a 50 ml centrifuge tube. Carefully layer 1:1 volume of blood on top of LymphoPrep™ using a pipette controller and centrifuge at 850 × *g*, Acc 1, Dec 1, for 30 min.
3. Carefully remove the PBMC layer which sits immediately below the serum using a Pasteur pipette and transfer into new 50 ml tube labeled PBMCs.
4. Centrifuge PBMCs at 688 × *g* or 10 min at 4 °C. Discard the supernatant and resuspend pellet in 20 ml of R10. Mix well by pipetting up and down, then take 10 μl of sample to perform cell count. Mix the 10 μl sample with an equal volume of trypan blue and transfer to hemocytometer plate for cell count. This will allow quantification of live peripheral blood mononuclear cells.
5. Keep PBMCs on ice ready for use. It is important to set aside approximately 1×10^6 cells which will be used as an unstained control and fluorescence minus one (FMO) controls for CD127, CD161, CD117 (c-kit), and CD294 (CRTH2) (*see* **Note 5**).

3.3 Activation of Cells

1. Prepare a "stimulated" (positive control) and "unstimulated" (negative control) master mix as follows. It is sufficient to make 2 ml of each.

(a) "Stimulated" master mix: R10 solution with 1 μl/ml of PMA, 1 μl/ml of Ionomycin, 1 μl/ml of Monensin and 1 μl/ml of Brefeldin A.

(b) "Unstimulated" master mix: R10 solution with 1 μl/ml of Monensin and 1 μl/ml Brefeldin A only.

2. Divide the sample of PBMCs or LPMCs equally into two FACS tubes labeled "stimulated" and "unstimulated" and centrifuge at 688 × *g* for 5 min at 4 °C. Discard the supernatants and resuspend one pellet in 500 μl of the "stimulated" master mix and the other pellet in 500 μl of the "unstimulated" master mix. Incubate for 3 h at 37 °C in a cell culture incubator.
3. Take the 1 × 10^6 PBMCs that were set aside and divide them equally amongst the control tubes. These are as follows: unstained, CD127 FMO, CD161 FMO, CD117 FMO, and CD294 FMO. Centrifuge the control tubes at 688 × *g* for 5 min at 4 °C, discard the supernatants and resuspend the pellets in 200 μl of the "unstimulated" master mix. Mix well by pipetting up and down and place in the cell culture incubator for 3 h with the other samples.
4. For an alternative method of ILC activation, *see* **Note 6**.

3.4 Cell Surface Staining

1. Following the 3 h incubation period, wash the samples and controls with 1 ml 1× PBS and centrifuge at 688 × *g* for 5 min at 4 °C. Discard the supernatants.
2. In the meantime, prepare the extracellular antibody master mix (*see* Table 1) (*see* **Note 7**). Each sample requires 100 μl of the master mix. Ensure a surplus of extracellular antibody master mix is made in case of loss during pipetting. Typically make up $n + 1$ volume of master mix and vortex well (*see* **Note 8**). Label this extracellular master mix 1. Add 100 μl of extracellular master mix 1 to each sample and mix well by pipetting up and down. Incubate for 30 min at room temperature protected from light.
3. Create a second extracellular antibody master mix specifically for the FMO controls. This should include all antibodies in Table 1 minus CD117 PE-Cy7, CD127 BV605, CD161 PerCP-Cy5.5 and CD294 BV421. Make up a sufficient volume for $n + 1$ and vortex well. Label this extracellular master mix 2. Add 13.1 μl of extracellular master mix 2 to each FMO control. Then add the additional antibodies to each corresponding FMO control (*see* Table 2) to make up total volume of 100 μl. Mix each FMO control well by pipetting up and down. Incubate FMO controls with the samples for 30 min at room temperature protected from light.

Table 2
Antibody volumes for FMO controls

Vol (μl)	CD117 FMO	CD127 FMO	CD161 FMO	CD294 FMO
Extracellular MM 2	13.1	13.1	13.1	13.1
CD117 PE-Cy7	–	1	1	1
CD127 BV605	1	–	1	1
CD161 PerCp-Cy5.5	1	1	–	1
CD294 BV421	2.5	2.5	2.5	–
1× PBS	82.4	82.4	82.4	83.9

4. The unstained control should be incubated in 100 μl of 1× PBS for 30 min at room temperature protected from light.
5. After 30 min, wash the samples and controls with 1 ml of FACS buffer and centrifuge at 688 × *g* for 5 min at 4 °C. Discard the supernatants.
6. Label the biotinylated antibodies with APC-Cy7 streptavidin by adding 2 μl of APC-Cy7 streptavidin in 100 μl of 1× PBS to the samples and FMO controls. Incubate for 15 min at room temperature protected from light. Then wash each tube with 1 ml of FACS buffer and centrifuge at 688 × *g* for 5 min at 4 °C. Discard the supernatants.
7. Fix the samples and controls by resuspending each pellet in 400 μl 4% PFA and mix well by pipetting up and down. Incubate for 15 min at room temperature protected from light (*see* **Note 9**). Then wash each tube with 1 ml of FACS buffer and centrifuge at 688 × *g* for 5 min at 4 °C. Discard the supernatants and resuspend the samples in 500 μl, and the FMO controls in 200 μl of FACS buffer. The FMO controls are now ready for use.
8. At this stage samples may be left overnight at 4 °C protected from light for intracellular cytokine staining typically on the next day (*see* **Note 10**).

3.5 Intracellular Cytokine Staining

1. Wash the samples with 1 ml of FACS buffer and centrifuge at 688 × *g* f or 5 min at 4 °C. Discard the supernatants.
2. In the meantime, prepare the intracellular antibody master mix (*see* Table 3) to make up 100 μl per sample. The cytokine antibodies are made up in a 1:1 volume of Brilliant Stain Buffer and 0.2% saponin buffer (*see* **Note 11**). Again, ensure a surplus of master mix is prepared by making up *n* + 1 volume and vortex well. Incubate the samples for 45 min at room temperature protected from light.

Table 3
Intracellular antibody master mix

Fluorochrome	Marker	μl/test	Company	Cat. No.	Dilution
AF700	IFN-γ	2	BioLegend	506515	1 in 50
BUV395	TNF-α	0.5	BioLegend	563996	1 in 200
BV710	IL-13	5	BD Horizon	564288	1 in 20
PE	IL-5	5	BioLegend	504303	1 in 20
BV786	IL-17A	3	BD Horizon	563745	1 in 33
APC	IL-22	1	BioLegend	366706	1 in 100
Brilliant stain buffer		46.75	BD Horizon	563794	–
02% Saponin buffer		46.75	–	–	–

3. Wash the samples with 1 ml of FACS buffer and centrifuge at 688 × *g* for 5 min at 4 °C. Discard the supernatants and resuspend in 1 ml FACS buffer ready for analysis.

3.6 Data Acquisition and Analysis

1. Prepare single color controls by labeling FACS tubes with corresponding fluorochromes (APC-Cy7, FITC, V500, PE-Cy7, BV605, PerCP-Cy5.5, BV421, AF700, BUV395, BV710, PE, BV786, and APC). Add one drop of UltraComp eBeads to each FACS tube, then add 1 μl of the corresponding fluorochrome antibody. Add 300 μl of FACS buffer to each tube to make up the single color control. Vortex well.
2. Data acquisition is performed using the multicolor BD LSRFortessa cell analyzer.
3. Calibrate the analyzer by initially running the unstained control and adjusting forward and side scatter ensuring good resolution of the lymphocyte population. Adjust photomultiplier tube (PMT) values for each fluorochrome to ensure there is no autofluorescence of unstained cells.
4. Next, individual single color control tubes should be analyzed making minor adjustments to the PMT values to minimize spectral overlap. Record this data setting positive and negative gated-populations. Calculate the compensation parameters of the cell analyzer. At this point set an appropriate threshold on forward scatter to eliminate debris while retaining the lymphocyte population.
5. As the ILC population is rare, it is important you acquire data from the whole sample. Do not set any specific collection criteria like time or cell count limits (*see* **Note 12**).

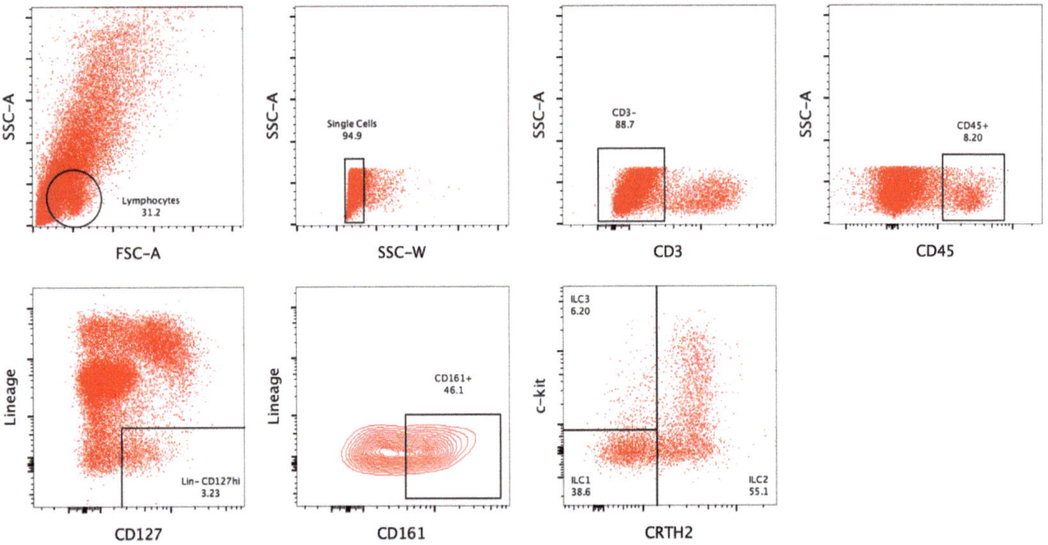

Fig. 1 Gating strategy to define the human ILC compartment. c-kit and CRTH2 determine ILC1, 2, and 3 subpopulations. FMOs are used to establish the position of positive gates

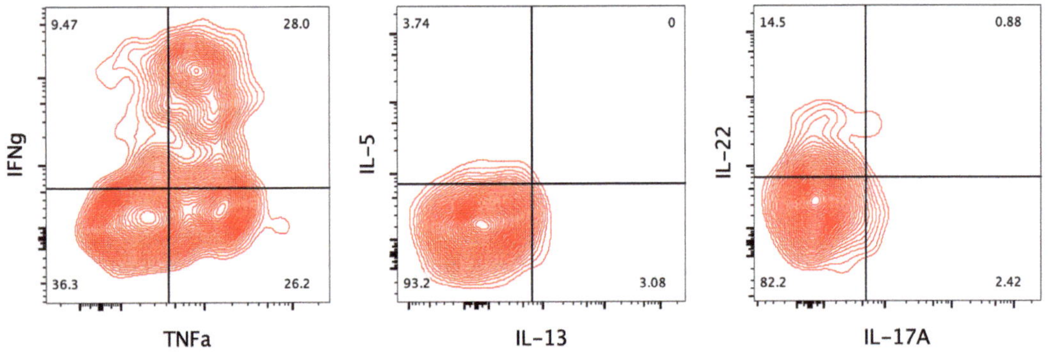

Fig. 2 Representative data from a patient with Crohn's disease looking at intracellular cytokine expression in ILCs following stimulation with PMA and ionomycin. There is an abundance of IFN-γ and TNF-α producing ILC1s and IL-22 producing ILC3s in keeping with the immuno-phenotype of Crohn's disease

6. Data analysis is performed using FlowJo flow cytometry analysis software. Define the lymphocyte population in forward vs. side scatter and follow the gating strategy outlined in Fig. 1. ILCs should be defined as follows—ILC1: lin^- $CD127^{hi}$ $CD161^+$ $c\text{-}kit^-$ $CRTH2^-$, ILC2: lin^- $CD127^{hi}$ $CD161^+$ $CRTH2^+$, ILC3: lin^- $CD127^{hi}$ $CD161^+$ $c\text{-}kit^+$ $CRTH2^-$. Once ILC populations are defined, it is possible to plot the ILC subsets with corresponding lineage specific cytokines, that is, IFN-γ and TNF-α for ILC1s, IL-5 and IL-13 for ILC2s, and IL-17A and IL-22 for ILC3s. This enables functional evaluation of ILC subtypes (*see* Figs. 2 and 3).

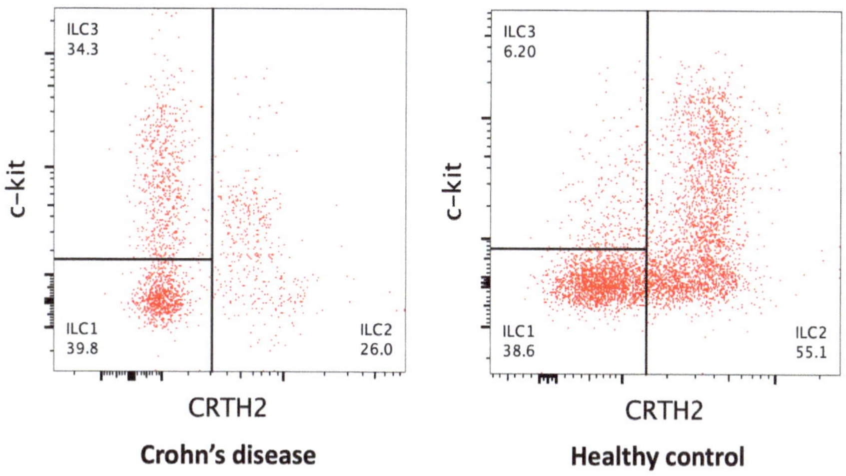

Fig. 3 Phenotypic differences in ILC subpopulations in a Crohn's disease patient vs. healthy control. A higher proportion of ILC1s and ILC3s are present in the inflamed mucosa of the patient with Crohn's disease compared to a healthy control

4 Notes

1. Biopsies are taken from the distal colon in view of data suggesting that the frequency of ILCs increases in the distal gastrointestinal tract [14]. This serves to maximize our yield of ILCs from colonic tissue.
2. Provided ethical approval is in place, we recommend the use of 16 or more biopsies to provide sufficient numbers of ILCs for meaningful analysis.
3. The use of cytomatrices to isolate human lymphocytes has been described in both skin and gastrointestinal mucosa [15, 16]. We use a modified protocol with significantly enhanced lymphocyte and ILC yields when directly compared to collagenase 1a digestion (unpublished data).
4. Transferring biopsies from HBSS to a petri dish can be performed with ease using a Pasteur pipette. It should be possible to suction all biopsies in one attempt in a very small volume of liquid for transfer to a petri dish. This allows easy manipulation of biopsies.
5. FMO controls help determine where the negative population of cells sit in order to accurately gate on what is positive. This is particularly useful for rare events or where cell populations do not clearly cluster together. For example, the CD117 FMO consists of cells stained with all fluorochrome antibodies in the extracellular master mix apart from CD117. By comparing data from the CD117 FMO against the fully stained sample you can clearly define the CD117 positive population.

6. An alternative method of activation has been described by Simoni et al. [17]. Cells are cultured in complete media with IL-12 50 ng/ml, IL-15 50 ng/ml and IL-18 50 ng/ml overnight and Brefeldin A added in the final 4 h prior to acquisition.
7. The extracellular master mix contains both biotinylated surface markers which will bind APC-Cy7 streptavidin and the live/dead discriminant near-IR. These both share a common emission wavelength of 750 nm and are detected by the red laser. This forms a "dumping" channel where by gating on the negative population you can easily remove all lineage specific markers and dead cells.
8. A common pitfall when staining cells is to not vortex the master mixes adequately. This can result in variability in the signal quality between samples making interpretation impossible.
9. It is possible to fix the cells by resuspending them in 400 μl 0.5% PFA and leaving them overnight at 4 °C protected from light. The samples should be washed the next day with 1 ml FACS buffer and centrifuged at 688 × *g* for 5 min at 4 °C prior to intracellular staining.
10. It is essential to fix cells prior to intracellular staining to ensure cytoskeletal stability and retain the cellular location of intracellular antigens.
11. Detecting intracellular antigens requires cell permeabilization before staining. Saponin interacts with membrane cholesterol, selectively removing it and leaving a porous cell surface membrane.
12. It may be necessary to dilute the sample if the acquisition rate is greater than 1000 events/s, as this risks blocking the analyzer.

Acknowledgments

Research enabling the optimization of this protocol was supported by the Medical Research Council and the National Institute for Health Research (NIHR) Biomedical Research Centre at Guy's and St Thomas' National Health Service (NHS) Foundation Trust and King's College London.

References

1. Parronchi P, Romagnani P, Annunziato F et al (1997) Type 1 T-helper cell predominance and interleukin-12 expression in the gut of patients with Crohn's disease. Am J Pathol 150:823–832
2. Rovedatti L, Kudo T, Biancheri P et al (2009) Differential regulation of interleukin 17 and interferon c production in inflammatory bowel disease. Gut 58:1629–1636. https://doi.org/10.1136/gut.2009.182170
3. Garrett WS, Lord GM, Punit S et al (2007) Communicable ulcerative colitis induced by T-bet deficiency in the innate immune system.

Cell 131:33–45. https://doi.org/10.1016/j.cell.2007.08.017

4. Powell N, Walker AW, Stolarczyk E et al (2012) The transcription factor T-bet regulates intestinal inflammation mediated by interleukin-7 receptor+ innate lymphoid cells. Immunity 37:674–684. https://doi.org/10.1016/j.immuni.2012.09.008
5. Buonocore S, Ahern PP, Uhlig HH et al (2010) Innate lymphoid cells drive interleukin-23-dependent innate intestinal pathology. Nature 464:1371–1375. https://doi.org/10.1038/nature08949
6. Cella M, Fuchs A, Vermi W et al (2009) A human natural killer cell subset provides an innate source of IL-22 for mucosal immunity. Nature 457:722–725. https://doi.org/10.1038/nature07537
7. Bernink JH, Peters CP, Munneke M et al (2013) Human type 1 innate lymphoid cells accumulate in inflamed mucosal tissues. Nat Immunol 14:221. https://doi.org/10.1038/ni.2534
8. Spits H, Cupedo T (2012) Innate lymphoid cells: emerging insights in development, lineage relationships, and function. Annu Rev Immunol 30:647–675. https://doi.org/10.1146/annurev-immunol-020711-075053
9. Constantinides MG, McDonald BD, Verhoef PA, Bendelac A (2014) A committed precursor to innate lymphoid cells. Nature 508:397–401. https://doi.org/10.1038/nature13047
10. Goldberg R, Prescott N, Lord GM et al (2015) The unusual suspects—innate lymphoid cells as novel therapeutic targets in IBD. Nat Rev Gastroenterol Hepatol 12:271–283. https://doi.org/10.1038/nrgastro.2015.52
11. Bernink JH, Mjösberg J, Spits H (2017) Human ILC1: to be or not to be. Immunity 46:756–757. https://doi.org/10.1016/j.immuni.2017.05.001
12. Caligiuri MA (2008) Human natural killer cells. Blood 112:461–469. https://doi.org/10.1182/blood-2007-09-077438
13. Crellin NK, Trifari S, Kaplan CD et al (2010) Human NKp44+IL-22+ cells and LTi-like cells constitute a stable RORC+ lineage distinct from conventional natural killer cells. J Exp Med 207:281–290. https://doi.org/10.1084/jem.20091509
14. Krämer B, Goeser F, Lutz P et al (2017) Compartment-specific distribution of human intestinal innate lymphoid cells is altered in HIV patients under effective therapy. PLoS Pathog 13:e1006373. https://doi.org/10.1371/journal.ppat.1006373
15. Di Marco Barros R, Roberts NA, Dart RJ et al (2016) Epithelia use butyrophilin-like molecules to shape organ-specific γδ T cell compartments. Cell 167:203–218. https://doi.org/10.1016/j.cell.2016.08.030
16. Clark RA, Chong BF, Mirchandani N et al (2006) A novel method for the isolation of skin resident T cells from normal and diseased human skin. J Invest Dermatol 126:1059–1070. https://doi.org/10.1038/sj.jid.5700199
17. Simoni Y, Fehlings M, Kløverpris HN et al (2017) Human innate lymphoid cell subsets possess tissue-type based heterogeneity in phenotype and frequency. Immunity 46:148–161. https://doi.org/10.1016/j.immuni.2016.11.005

Chapter 18

CRISPR/Cas9-Based Gene Engineering of Human Natural Killer Cells: Protocols for Knockout and Readouts to Evaluate Their Efficacy

Mélanie Lambert, Caroline Leijonhufvud, Filip Segerberg, J. Joseph Melenhorst, and Mattias Carlsten

Abstract

Natural killer (NK) cells are cytotoxic lymphocytes of our immune system with the ability to identify and kill certain virally infected and tumor-transformed cells. During the past 15 years, it has become increasingly clear that NK cells are involved in tumor immune surveillance and that they can be utilized to treat cancer patients. However, their ability to induce durable responses in settings of adoptive cell therapy needs to be further improved. One possible approach is to genetically engineer NK cells to augment their cytotoxicity per se, but also their ability to persist in vivo and home to the tumor-bearing tissue. In recent years, investigators have explored the potential of viral transduction and mRNA electroporation to modify NK cells. Although these methods have generated promising data, they are associated with certain limitations. With the increasing advances in the CRISPR/Cas9 technology, investigators have now turned their attention toward using this technology with NK cells as an alternative method. In this book chapter, we introduce NK cells and provide an historical overview of techniques to genetically engineer lymphocytes. Further, we elucidate protocols for inducing double-strand breaks in NK cells via CRISPR/Cas9 together with readouts to address its efficacy and functional outcome. We also discuss the pros and cons of the described readouts. The overall aim of this book chapter is to help introduce the CRISPR/Cas9 technology to the broader audience of NK cell researchers.

Key words NK cells, Gene editing, CRISPR/Cas9, Flow cytometry, RT-PCR, DNA sequencing, Digital droplet PCR

1 Introduction

1.1 Background to NK Cells

In the mid-1970s, a new immune cell was discovered that, in contrast to T cells, was able to lyse certain tumor cells without prior sensitization of the host [1–4]. This cell was given the name "Natural Killer (NK) cell." Based on its characteristics, including the mechanisms employed to rapidly destroy a target cell meanwhile releasing cytokines and chemokines to recruit and promote adaptive immunity, NK cells have traditionally been considered an

Shoba Amarnath (ed.), *Innate Lymphoid Cells: Methods and Protocols*, Methods in Molecular Biology, vol. 2121, https://doi.org/10.1007/978-1-0716-0338-3_18,

innate immune cell. However, this concept has recently been challenged based on data demonstrating that NK cells also can acquire memory-like properties following cytomegalovirus infection [5]. Today, we know that NK cells have several important functions in the body such as recognizing and destroying tumor cells, controlling bacterial and viral infections and regulating immune responses [6]. NK cells are also implicated in pregnancy by promoting the maternal–fetal interface and proper blood supply to the fetus while protecting it from infections [7, 8].

NK cell cytotoxicity exerted through the release of perforin- and granzyme-containing granule, a process known as "degranulation", is regulated by an intricate balance of signals from germ line-encoded activating and inhibitory cell surface receptors [9]. HLA class I-binding receptors, such as the Killer Immunoglobulin-like Receptors (KIRs), the Natural Killer Group 2A (NKG2A), and Leukocyte Immunoglobulin-like Receptor subfamily B member 1 (LILRB1; L IR-1), represent key molecules that prevent NK cell-mediated killing of normal cells [9]. In addition to this, the inhibitory KIRs and the NKG2A receptor also tune NK cell cytotoxicity to be strong against cells devoid of HLA class I, a process referred to as NK cell education [10]. Counterbalancing the inhibitory HLA class I-binding receptors, a large array of activation and coactivation receptors, such as the NKG2D, DNAM-1, NKp30, and NKp46 receptors, is required to trigger NK cell cytotoxicity [11]. Most of these receptors bind stress-induced or virally associated ligands such as MICA/B, ULBPs, PVR, Nectin-2, and B7-H6 [12]. In addition to these receptors, NK cells can also kill antibody-coated cells via ligation to their Fc-binding receptor CD16 [13]. This pathway is called antibody-dependent cellular cytotoxicity (ADCC). Beyond target lysis via degranulation mentioned above, NK cells can also eliminate target cells via stimulation of death receptors expressed on the target cell surface that trigger caspase-dependent apoptosis [14]. As mentioned above, NK cells are also involved in immune regulation. This is primarily mediated through the release of chemokines and cytokines that can attract and stimulate both innate and adaptive components of the immune system [15, 16], but also via elimination of highly activated immune cells [17].

One of the best studied aspects of NK cells is their ability to kill tumor-transformed cells. Over the past years, it has also become clear that NK cells are involved in tumor immune surveillance [18]. This is evidenced by indirect data from clinical cohorts as well as by more direct experimental data. The former has linked poor NK cell function early in life to increased risk of cancer [19], whereas the latter is exemplified by experimental animal studies showing that depletion of NK cells facilitates leukemia engraftment [20] and that mice deficient in activation receptors NKG2D or DNAM-1 develop more tumors compared to wild-type mice of the same strain [21, 22]. Another line of evidence comes from clinical studies on allogeneic stem cell transplantation (SCT) demonstrating that donor NK cells can be key to preventing post-transplantation leukemia relapse in certain settings. This was first published by Ruggeri et al. in 2002, where the authors demonstrated that KIR-ligand mismatching of donor NK cells in the graft-to-host direction prevented acute myeloid leukemia (AML) relapse following haploidentical SCT [23]. The role for NK cells in allogeneic SCT has thereafter been verified by several other studies [24, 25]. More recent data show that, in addition to these gene association studies, the expansion of adaptive memory-like donor NK cells following allogeneic SCT is linked to improved outcome in AML [26]. Hence, there are several lines of evidence for the central role of NK cells in tumor immune surveillance.

NK cells can also be utilized in settings of adoptive cell transfer to treat patients with cancer [27]. This was first demonstrated for AML in a study by Miller et al. published in 2005 [28]. Since this publication, there has been an burst of clinical trials demonstrating objective clinical responses in patients with refractory and/or relapsed myeloid malignancies [29–32]. In addition to the above, adoptive NK cell infusion as postconsolidation therapy for elderly AML patients or patients with high-risk disease has been shown to prolong survival [33, 34]. Recent data also indicate that the infusion of NK cells in conjunction to an allogeneic SCT can have beneficial effects for patients with myeloid malignancies [35]. Nevertheless, development of new and improved NK cell preparations for adoptive transfer to treat patients with cancer is needed in order to induce remission in a larger cohort of patients and achieve durable responses. Limiting factors such as poor in vivo persistence and inefficient localization to the tumor site need to be addressed. One approach that is currently being explored is genetic engineering of NK cells. In this book chapter we present an historical overview of gene engineering techniques followed by results, protocols, and readouts for the use of the CRISPR/Cas9 technology in NK cells.

1.2 Overview of Gene Manipulation and Gene Editing Approaches for Lymphocytes

Genetic manipulation of human lymphocytes has classically been based on the introduction of nucleic acid polymers using electroporation. However, the use of retroviral and lentiviral vectors has also been employed to effectively achieve stable ectopic expression of a given transgene. Newer technologies to edit endogenous genes and to introduce genes via homology-directed repair (HDR) have greatly augmented our ability to modulate the genome in ways that were unthinkable until only recently. Below we discuss the most commonly used approaches to manipulate human lymphocytes, from, in retrospect, fairly blunt tools to the current methodologies of precision genome engineering.

1.2.1 Transient Introduction of DNA and RNA

Nucleic acid polymers such as DNA and RNA have been introduced in eukaryotic cells via various means, most notably via calcium phosphate transfection [36]. While this method works with some adherent cell lines such as HEK- 293T cells, it is notoriously ineffective with most other cell types, including human lymphocytes. Electroporation has changed that, and various commercially available electroporators are currently on the market to efficiently achieve high-level transfection. Some of these are approved for Good Manufacturing Practice (GMP) cell engineering and achieve the same high-level gene engineering as with small-scale equipment, a feature that is utterly necessary for clinical translation [37].

Initially, DNA expression vectors were introduced into eukaryotic cells via the electroporation technique, but higher efficiencies while lower toxicities have been achieved using RNA [38, 39]. Although it is beyond the scope of this book chapter, suffice to say that eukaryotic cells have several mechanisms in place to respond to foreign and "misplaced" nucleic acids. The electroporation of cells itself elicits an interferon response, a "danger" response, that may affect the potency of the engineered cells [40]. Pathogen-associated molecular patterns play a major role in this response. Specific receptors expressed by eukaryotic cells include toll-like receptor (TLR) 4, which specifically bind and respond to bacterial unmethylated DNA [41, 42], and AIM2, which binds and responds to cytosolic double-stranded DNA [43–45], resulting in cytokine release, growth arrest, and apoptosis [46]. Unmodified RNA, similar to DNA, activates several TLRs and other damage- or pathogen-associated molecular pattern receptors [47–51], resulting in cytokine secretion and cell death [46, 47, 52]. Since electroporation with in vitro-transcribed RNA is less toxic while more efficient compared to DNA and not likely to yield stably engineered cells, we have used this method to introduce tumor targeting receptors into human cells [53, 54]. The gene expression levels can be further enhanced and toxicity minimized by introducing specific chemical modifications to the RNA during ex vivo synthesis [50, 55, 56].

1.2.2 Stable Gene Modification Strategies

Stable transgene expression can be generated using a number of approaches. Traditionally, transfection of the target cells with linearized plasmid DNA can yield low-level stable transfectants which were further selected based on a vector-encoded antibiotic resistance marker [57]. The transfection of plasmid DNA for clinical applications has largely been abandoned due to the development of novel, more efficient and safe methodologies such as retroviral and lentiviral vectors and transposon systems such as Sleeping Beauty and PiggyBac.

Retroviral vectors were developed for use in laboratory experiments and further improved by Dusty Miller and his team in the 1980s [58–61]. The safety of next-generation retroviral vectors was enhanced by the elimination of homologous sequences and a split-vector design. This vector design was used in one of the first gene therapy trials wherein patient hematopoietic stem/progenitor cells were engineered with the common-gamma chain gene to correct an inborn immunodeficiency [62, 63]. Unfortunately, this resulted in the oncogenic transformation in two patients due to the insertion-mediated activation of an endogenous proto-oncogene LMO2 [64, 65]. The vector was redesigned with a self-inactivating vector to improve safety [66] and has since been used in 186 clinical trials as of June 2019 (search of ClinicalTrials.gov, using the search terms "retrovirus AND vector"; and reviewed in [67]).

The retroviral gene transfer system offered several advantages over then existing gene modification strategies, including stable gene expression, the relative ease of handling, and potential to generate stable retrovirus producer cell lines. An important drawback of retroviral gene transfer is that only dividing cells could be transduced with this system [61], necessitating the induction of proliferation of the target population. Though this may not be an issue for T cells [68], some cell types such as hematopoietic stem and progenitor cells were less susceptible to this form of genetic engineering. In addition, forcing immature cells to cycle using cytokines might result in loss of potency. In the early 1990s, Didier Trono and colleagues developed the lentiviral vector system that would allow stable gene modification of nondividing cells, a point proven by the transduction of neurons [69, 70]. Lentiviral vectors are essentially human immunodeficiency virus-based vectors reengineered to only yield infectious particles if cotransfected with specific helper plasmids. This system was further developed with enhanced safety and efficiency in mind into what is now called the third generation vector [71]. The transduction of noncycling cells proved a major benefit, as was the integration site specificity: Gamma retroviral vectors typically integrate near a transcription start site [72], whereas lentiviral vectors preferentially integrate in intergenic and intronic regions of transcriptionally active genes [72, 73]. As with retroviral vectors, the manufacturing cost of lentiviral vectors is substantial and few facilities exist that can generate high-titer cGMP-grade vectors, overall leading to long wait times for clinical-grade lentiviral vector.

Another drawback of the use of retroviral and lentiviral vectors for gene therapy is that, despite the engineered safety, either virus type can theoretically yield replication-competent offspring in transduced cells, which has necessitated the inclusion of time-consuming and costly cell product release assays and postinfusion monitoring of recombined viral vectors. This issue has recently been revisited by several gene therapy laboratories, which, on summarizing hundreds of patient samples have not observed any such artifactual rearrangement, which is a testament to the safety of currently used vectors [74–77].

The high cost of virus production and technical expertise required to generate this critical reagent has contributed to the design of alternative, transposon-based gene modification strategies. Herein, a plasmid-encoded transposase facilitates the integration of a gene of interest into the genome of the target cells. Two main systems have been reported: The Sleeping Beauty system and the PiggyBac system. Both have been used in CAR T cell trials, wherein the sleeping beauty system proved safe [78]; early data with the PiggyBac transposon-engineered anti-BCMA CAR T cells showed promising results [79].

Virus- and transposon-mediated gene transfer methods have in common that the payload insertion into the genome is relatively random with the Sleeping Beauty system [78, 80] and lentiviral vectors, but near transcriptional start sites with the PiggyBac system [80], similar to retroviral vectors. The likelihood of malignant transformation with current retroviral and lentiviral vectors is extremely low, but increases with multiple integrations in the same cell when targeting immature hematopoietic cells [81, 82]. T cells, on the other hand, appear resistant to malignant transformation for reasons yet unclear [83]. If multiple integrations per cell increases the likelihood of malignant transformation, this can be managed by targeting a single integration per cell (*see* below).

The unpredictability of the vector integration site and the looming potential of oncogenic transformation with codisruption of growth-regulating sites have reinvigorated an interest in precision engineering: The targeted delivery of the transgene to prespecified sites. This has classically been achieved via homologous recombination between exogenous DNA carrying sequences homologous to the targeted sites and the cellular gene itself [84]. This method, though widely used to generate transgenic animals, is very inefficient [84, 85]. However, now, with breakthrough developments in genome engineering using clustered regularly interspaced short palindromic repeat (CRISPR)/CRISPR-associated protein 9 (CRISPR/Cas9) system [86], TALEN [87, 88], zinc finger nucleases [89], or megaTAL [90], all of which create double-stranded breaks (DSBs) at predetermined genomic sites [86, 91], the gene of interest can be inserted into those edited sites at very high efficiency using HDR. The HDR cassette design is very similar to those developed decades ago, but delivered using adeno-associated virus (AAV) -based vectors [92–95] or non–vector-based delivery via DNA electroporation [96, 97]. This method has successfully been employed to insert a tumor-targeting chimeric antigen receptor into the T cell receptor-alpha locus in T cells to generate universal donor CAR T cells [92, 93]. Most recently, the combination of CRISPR/Cas9 has gained popularity for directed gene insertions due to its simplicity of use [98]. In this system, the Cas9 nuclease uses a short CRISPR RNA (crRNA) plus a longer transactivating crRNA (tracrRNA) for the directed double-stranded disruption of the target site. Oftentimes the crRNA and tracrRNA are fused into one single guide RNA. The double-strand breaks are repaired either by nonhomologous end-joining (NHEJ), which, via the insertions or deletions of one or more nucleotides, disrupts the open reading frame and ablate gene function, or via HDR, using either an introduced or endogenous homologous template to modify the preexisting sequence. We here use the latter to insert exogenous sequences into the CRISPR/Cas9-disrupted site.

Taken together, several approaches have been explored to edit and/or introduce genes into human lymphocytes. Distinct disadvantages with the various methods, such as transient expression following RNA electroporation or the sometimes low efficacy and poor viability as well as the risk of insertion mutagenesis (malignant transformation) and viral reactivation following viral transduction, have hindered the broader use and safety of exploring gene engineered lymphocytes for immunotherapy in humans. The recent developments of CRISPR/Cas9 have added a new nonviral tool that allows for specific, yet permanent, gene editing of human lymphocytes and other cells.

1.3 Results from Gene Editing of Human NK Cells via CRISPR/Cas9 and the Purpose of This Book Chapter

In contrast to T cells, NK cells have historically been difficult to transduce and transfect. However, improved knowledge and technological advancements have more recently facilitated the exploration of gene engineered NK cells to improve treatment results when adoptively infused into patients. The most prominent examples of this is the ongoing clinical trial (NCT03579927) utilizing CAR-engineered cord blood-derived NK cells [99]. Nevertheless, although the efficacy of viral transduction approaches has been improved, it is far from that of mRNA transfection via the electroporation of NK cells [27]. With the latest advances and the use of mRNA, NK cells can be genetically modified to near 100%, although the expression is transient [100, 101]. In contrast to mRNA electroporation, the use of CRISPR/Cas9 induces permanent changes in NK cells (Fig. 1). As this technology is only recently being explored with primary human NK cells, only a few reports have been released today. The first clear study demonstrated that the CRISPR/Cas9 technology, when introduced via electroporation, can knock out the expression of the NKp46 and CD45 receptors as well as the CISH complex [102]. Another group of investigators have demonstrated that RT-PCR-verified knockout of the TGFBR2 gene in NK cells via CRISPR/Cas9 triggered the NK cells to become less sensitive to TGFβ-mediated suppression [103]. Notably, no publication has yet demonstrated the potential of combining CRISPR/Cas9 and HDR in NK cells.

One reason for the relatively few studies on CRISPR/Cas9 is the lack of knowledge, protocols, and experience of this

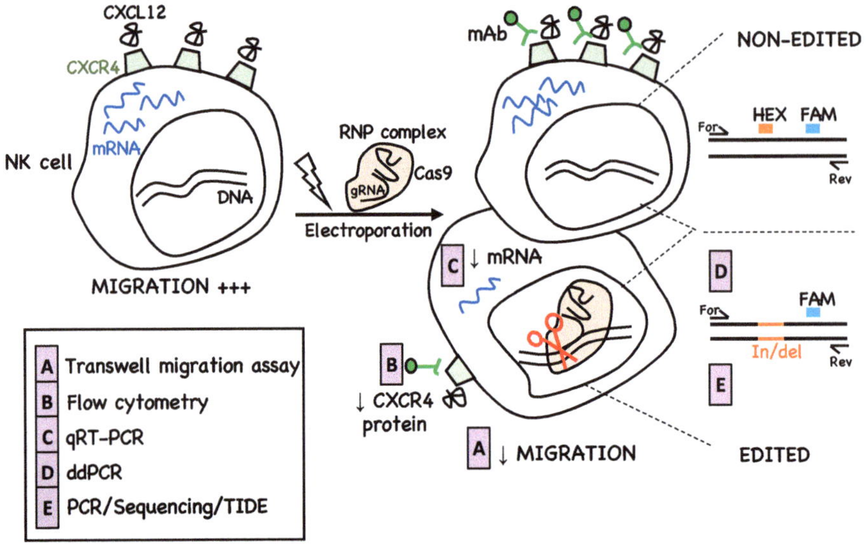

Fig. 1 Schematic overview of the CRISPR/Cas9 process and cellular level of action for the readouts presented to address efficacy in NK cells. Electroporation of NK cells with the RNP complex triggers DSBs of the gene of interest (here represented by CXCR4). The readouts for efficacy presented in this book chapter and its cellular localization are depicted. (**a**) In vitro transwell migration assay. Efficient knockout of the CXCR4 gene should trigger less NK cell migration toward the chemokine CXCL12. (**b**) Flow cytometry-based readout to address CXCR4 protein expression on the cell surface should detect lower levels following CRISPR-Cas9-mediated knockout. (**c**) mRNA expression levels of CXCR4 can be quantified by qRT-PCR and are expected to decrease following efficient knockout. (**d**) ddPCR addresses the gene editing efficacy of CXCR4 at the genomic level. (**e**) Sequencing of the region of editing and analysis by the TIDE software can be used to estimate the overall editing efficiency of the CXCR4 gene

technology. The purpose of this book chapter is to provide a protocol to induce CRISPR/Cas9-mediated DSBs in NK cells and describe the different readouts that are available to address the efficacy. We will exemplify this by the knockout of the chemokine receptor CXCR4 in NK cells, a receptor that is essential for NK cell development [104] and that was recently shown important to improve the bone marrow homing capacity of NK cells following upregulation via mRNA electroporation [101].

2 Materials

Except for the components of flow cytometry and the migration assay, prepare all solutions using ultrapure water and analytical grade reagents. Use filter barrier tips with compatible micropipettors to avoid contamination of reagents in sample preparation, including droplet generation. Prepare and store all reagents at room temperature (RT) (unless indicated otherwise). For the experiments presented below, we have used ex vivo expanded primary human NK cells from healthy donors. Please *see* Carlsten et al. for details on the expansion protocol [100].

2.1 Electroporation Components

1. Instrument: 4D-Nucleofector™ System (4D-Nucleofector™ Core Unit and 4D-Nucleofector™ X Unit).
2. P3 Primary Cell 4D-Nucleofector™ X Kit S (Lonza). Solution at RT.
3. NK cell culture medium: X-Vivo 20 (Lonza) supplemented with 10% heat-inactivated human AB serum (Sigma-Aldrich), 2 mM GlutaMAX-1 (Gibco), 500 IU/mL human IL-2 (PeproTech). Store at 4 °C, aliquot and prepare at 37 °C for use.
4. Ribonucleoprotein (RNP) complex (*see* "2.2 CRISPR/Cas9 components" below).
5. Electroporation 20 μL 16-well Nucleocuvette™ Strips.
6. 48-Well cell culture plate, growth area 0.95 cm^2 per well.
7. Plastic cell culture flask T25.

2.2 CRISPR/Cas9 Components

1. 2-component guide RNA: 200 μM Alt-R CRISPR-Cas9 crRNA + 200 μM Alt-R CRISPR-Cas9 tracrRNA (IDT).
2. 61 μM Alt-R Staphylococcus pyogenes (Sp) Cas9 Nuclease V3 (IDT).

2.3 In Vitro Migration Assay Components

1. Phosphate-buffered saline (PBS).
2. Fetal bovine serum (FBS).
3. CXCR4-specific Chemokine Ligand, CXCL12, 10 μg.
4. Corning® Costar® Transwell® cell culture plates and inserts with 5 μm pores size.
5. 96-Well, V-bottom, polypropylene tissue culture plate.
6. CyQUANT® solution: CyQUANT® proliferation kit (Thermofisher Scientific).
7. Corning® Costar® TC-Treated 96-Well Plates and 96-Well Black Polystyrene Clear Flat Bottom Microplates.
8. Plate reader Tecan Infinite F200Pro.

2.4 Phenotyping Components

1. LSR Fortessa II Flow cytometer.
2. 96-Well, V-bottom, polypropylene tissue culture plate.
3. BUV737 Mouse Anti-Human CD56 Antibody clone NCAM16.2. LIVE/DEAD™ Fixable Aqua Dead Cell Stain Kit. PE/Cy5 anti-human CD184 (CXCR4) Antibody clone 12G5.
4. FACS buffer (89.5% PBS×1, 10% FBS, 0.5% 0.5 M EDTA).
5. 1% paraformaldehyde solution in PBS.
6. FlowJo analysis software.

2.5 qRT-PCR Components

1. RNA extraction: RNeasy Mini Kit (Qiagen), 70% ethanol, 1.5 mL microcentrifuge capped tubes, NanoDrop instrument.
2. cDNA synthesis: 1 μg of mRNA is reverse transcribed into cDNA with the High-Capacity cDNA Reverse Transcription Kit following the manufacturer's instructions.

Components	Volume/Reaction (μL)
10× RT buffer	2.0
25× dNTP mix (100 mM)	0.8
10× RT random primers	2.0
MultiScribe reverse transcriptase	1.0
Nuclease-free water	4.2
mRNA (1 μg)	10
Total	20

Components are mixed all together in a PCR tube and the reaction is conducted following these four steps:

Step:	1	2	3	4
Temperature (°C)	25	37	85	4
Time	10 min	120 min	5 s	∞

3. qRT-qPCR: qPCR thermal cycler, MicroAmp Fast Optical 96-well reaction plate, MicroAmp Optical plate seals, Applied Biosystems™ Power™ SYBR™ Green Master Mix, Primers mix.

2.6 Sequencing Components

In order to determine the efficiency of the editing, we use TIDE software. Based on the quantitative sequence trace data from two standard capillary (Sanger) sequencing reactions, this web tool quantifies editing efficacy and identifies the predominant types of insertions and deletions (indels) in the DNA of a targeted cell pool.

1. Genomic DNA is usually isolated 3 days after transfection. DNA from a minimum of 1×10^6 cells should be isolated to get a comprehensive sampling of the complexity of the mutations that are introduced by the DNA repair system following the CRISPR/Cas9-induced DSB. Genomic DNA purification: DNeasy Blood & Tissue Kit (Qiagen), 100% ethanol, 1.5 mL microcentrifuge capped tubes, NanoDrop instrument.
2. PCR master mix (example makes 50 μL).

   PCR reactions are carried out in Platinum SuperFi PCR Master Mix (Thermo Scientific™) with primers (Eurofins

Genomics) flanking the expected disruption site. The projected break site should be located preferably ~200 bp downstream from the sequencing start site.

Volume	Component
10 μL	GC mix
25 μL	Mix PCR Platinum SuperFi
5 μL	Forward and reverse primers (5 μM)
10 μL	Genomic DNA (150 ng)

3. PCR is performed in a C1000 Touch™ Thermal Cycler using the following cycle parameters:

Step	Temperature	Time (min:s)	Number of cycles
Initial denaturation	95 °C	1:00	1
Denaturation Annealing Extension	98 °C 60 °C 72 °C	0:10 0:10 0:30	30
Final extension	72 °C	5:00	1
	4 °C	Hold	

4. PCR products are loaded onto 1–2% agarose gel. A sharp single band should be visible.
5. PCR products are purified using a kit according to manufacturer's instructions (GeneJET Gel Extraction and DNA Cleanup Micro Kit, Thermo Scientific™).
6. Purified PCR samples are prepared for commercial Sanger sequencing.
7. The TIDE web tool is used to examine the target gene disruption efficiency and is available at https://tide.nki.nl.

2.7 Digital Droplet PCR Components

ddPCR nonhomologous end joining genome edit detection (NHEJ GED) assays are designed specifically for NHEJ detection using proprietary computational algorithms. ddPCR NHEJ GED assays contain a pair of primers, a FAM probe, and a HEX probe. The FAM probe binds distant from the site of potential NHEJ, and is designed to count all copies of the amplicon. A HEX-labeled drop-off probe is designed to bind WT DNA and targets the predicted cut site. The HEX probe will not bind, or will "drop off," when the allele has an insertion or deletion at this site. It is the loss of this HEX signal (while maintaining FAM signal) that indicates presence of an NHEJ allele. An example of the data is illustrated in Fig. 2d.

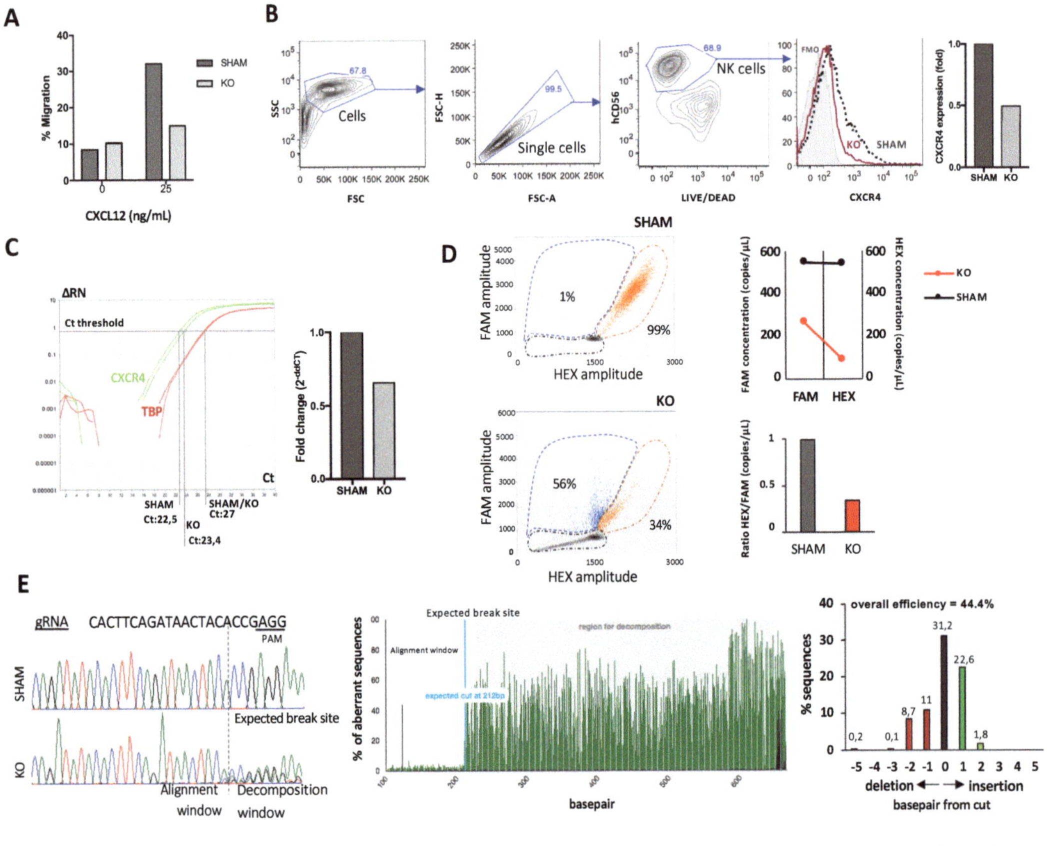

Fig. 2 Data from gene editing of NK cells from one healthy donor determined by the proposed readouts. (**a**) In vitro transwell migration of expanded NK cells toward CXCL12 (also called SDF-1α) 3 days after RNP electroporation. (**b**) Cell surface expression of CXCR4 on expanded NK cells as determined by flow cytometry 3 days after RNP electroporation. (**c**) CXCR4 gene expression quantified by qRT-PCR 3 days after RNP electroporation. dCt values are calculated for CXCR4 expression relative to TATA-box binding protein (TBP) expression, which is used as an internal housekeeping gene control. (**d**) ddPCR assay 3 days after RNP electroporation. 2D droplet plots with droplet group definition for analysis. FAM and HEX concentration (copies/μL) are shown for each condition. Quantification of the plots are represented by the FAM and HEX concentration (copies/μL) and the ratio HEX/FAM is calculated. (**e**) Sequencing data as analyzed by TIDE 3 days after RNP electroporation. TIDE requires as input a guide RNA sequence as well as DNA from your control and the edited sample. Sequencing chromatograms are viewed with Chromas free trace software. The quality control sequence is visualized in black and treated sample in green. A constant composite sequence is yielded after the break site. Trace decomposition yields the spectrum of indels with their frequencies. *KO* knockout, *SHAM* wild-type control

1. 20× FAM and HEX hydrolysis probe and primer assay mixtures in water.
2. Genomic DNA in water adjusted to 20 ng/μL.
3. Restriction enzyme that does not cut within the amplicon. 2–4 U restriction enzyme can be directly added to a 20 μL ddPCR reaction without additional incubation.
4. ddPCR™ Supermix for Probes (No dUTP).

5. Droplet Generation Oil for Probes (stored at RT).
6. DG8™ Gaskets for QX200™ Droplet Generator.
7. DG8™ Cartridges for Droplet Generator.
8. Pierceable Foil Heat Seal.
9. ddPCR™ 96-Well Plates.
10. QX200™ Droplet DigitalTM PCR system.
11. PX1™ PCR Plate Sealer.
12. DG8™ Cartridge Holder.
13. 96-Well C1000 Touch™ Thermal Cycler.

3 Methods

3.1 Protocol for CRISPR/Cas9 of NK Cells

1. Prepare the guide RNA (gRNA) by mixing Alt-R CRISPR-Cas9 crRNA and Alt-R CRISPR-Cas9 tracrRNA in a sterile PCR tube at 1:1 ratio (1.25 μL of each) so the final concentration of the gRNA is 100 μM. Heat at 95 °C for 5 min in a PCR thermal cycler. Cool down for 5–10 min to RT. Prepare the RNP complex by adding 1.5 μL of Cas9 and 1 μL of PBS to the gRNA and leave at RT for 10 min. Keep the RNP complex on ice (*see* **Notes 1–3**).
2. Turn on the 4D-Nucleofector™ System. Select the electroporation protocol CM-137.
3. Determine the number of cells needed for the study, and collect the desired number of cells from the culture flask into a 50 mL tube.
4. Centrifuge the sample at 300 × *g* for 7 min.
5. Aspirate the supernatant completely.
6. Wash the sample with 5 mL PBS; centrifuge at 300 × *g* for 7 min.
7. Determine the density and total volume of cell suspension needed for transfection. Carefully suspend the cells in P3 buffer in 15 μL in a sterile PCR tube.
8. Without waiting, carefully add 5 μL of the RNP complex to the cells and mix carefully three times with the same pipette tip.
9. Transfer the contents of the PCR tube to a sterile cuvette in the Nucleocuvette Strip. Sweep each of the four corners of the chamber with cell suspension to make sure there are no air bubbles left in the cuvette.
10. Put the lid on the strip and plug into the 4D-Nucleofector™ System.
11. Press "Start" and wait until the program indicates "Complete."

12. Carefully take the strip back to the biosafety cabinet and transfer all of the contents into the middle of one empty well of a tissue culture-treated 48-Well flat bottom plate.
13. Add between 0.5 and 1.0 mL of fresh media to the cells without pipetting up and down. Incubate the plate at 37 °C, 6.5% CO_2 (these conditions may vary for different cell type) for 1 h. Transfer the cells into a T25 or T75 flasks. Optimal cell density is 0.5–2 × 10^6 cells/mL of NK cell medium.

3.2 Protocols for Readouts of CRISPR/Cas9 Efficacy

Different readouts to address the knockout efficacy can be used. Below, five methods to address the CXCR4 knock-out efficacy are presented. They all have different pros and cons (*see* **Note 4**).

3.2.1 Migration-Based Readout

The transwell migration assay is a commonly used test to study the migratory response of cells to a ligand. For this assay, ex vivo expanded NK cells are resuspended in PBS and placed on the upper layer of a cell culture insert with permeable membrane (5.0 nm). A solution of PBS supplemented with 10% FBS and the CXCR4 ligand, CXCL12, is placed below the cell-permeable membrane. Following an incubation period (1 h), the cells that have migrated through the membrane are lysed and labeled using the CyQUANT Cell Proliferation Assay kit which is a highly sensitive fluorescence-based method for quantifying cells. The content is then measured and compared to control wells using an available plate reader that can detect the emission wavelength of the dye within the CyQuant kit. We have used an Infinite F200Pro instrument (Tecan).

1. 2–3 days after electroporation, create an even suspension of cells by flushing the bottom of the culture flask using a 5 mL serological pipette. Collect the appropriate number of cells for the assay.
2. Wash and resuspend the cells in PBS for a suspension of 0.5 × 10^6 cells/mL media. Let the cells rest in 37 °C until ready to use.
3. Dilute the chemokine ligand in warm PBS supplemented with 10% FBS, to the maximum desired ligand concentration. For CXCL12, 25 ng/mL is sufficient for significant migration of expanded NK cells. The ligand solution can be further diluted out to serial concentrations.
4. Remove transwell inserts from the plate and transfer 650 μL of the chemokine solution into the appropriate wells. For the wells mentioned below under **step 5**, less volume should be added (so that the final volume is 650 μL after the control cells have been added). Carefully place the transwell inserts back into the wells.
5. In triplicate, add 5 × 10^4, 2.5 × 10^4, 1.25 × 10^4 NK cells, respectively, directly into the wells without transwell inserts. These wells will be used to calculate the percentage of

migration (control wells). This can preferably be done through serial dilution.

6. Add 100 μL of the cell suspension on top of each transwell insert (5×10^4 NK cells total per well).
7. Incubate for 1 h at 37 °C.
8. After the assay is complete, carefully lift the transwell inserts out of the wells. Collect the migrated cells from each well and transfer 200 μL into a 96-Well, V-bottom plate in triplicate.
9. Centrifuge the plate at 300 × *g* for 3 min. Discard the liquid from the plate into an appropriate waste container.
10. Isolated pellets are frozen at −80 °C for a minimum of 1 h before the analysis.
11. Resuspend all samples in 250 μL CyQUANT® solution and plate 200 μL in a 96-Well flat bottom black assay plate (200 μL per well).
12. Make sure to have at least one well that only contains the CyQUANT solution to exclude background fluorescence when analyzing the data.
13. Protect the plate from light and place on ice until the samples are ready to be run.
14. Using the Tecan Infinite F200Pro microplate reader, measure fluorescence of each well (485/535 nm). Data can be exported and analyzed in Microsoft Excel.
15. Percent migration can be calculated using the correlation coefficient determined by calculating a correlation curve based on the control wells mentioned above.

3.2.2 Flow Cytometry-Based Readout

1. Define your time point(s) for measuring the surface receptor expression of your protein of interest. For most receptors that we have analyzed, expression decrease within 72 h after electroporation.
2. At the selected time point(s), flush the bottom of the cell culture flask using a sterile 5 mL serological pipette to create an even cell suspension. Count the cells and take approximately 0.1×10^6–0.5×10^6 cells per sample for the staining. Transfer the cells in a 96-Well, V-bottom plate.
3. Centrifuge the plate at 300 × *g* for 3 min. Discard the liquid from the plate into an appropriate discard container.
4. Suspend each sample in 50 μL of antibody and viability marker master mix (anti-CXCR4-PeCy5, anti-CD56-BUV737, and LIVE/DEAD™ Fixable Aqua resuspended in FACS buffer).
5. Incubate the samples in the dark, at 4 °C, for 15–30 min. Add 150 μL FACS buffer to each well and gently suspend the cells.

Centrifuge the plate at 300 × *g* for 3 min. Discard the supernatant in an appropriate waste container.

6. Wash once more with 200 μL FACS buffer, centrifuge the plate at the same speed and time as above and discard the supernatant as previously outlined.
7. Resuspend the samples with 50 μL 1% paraformaldehyde solution and wait for 15 min at RT in the dark.
8. Add 150 μL FACS buffer and gently suspend the cells. Centrifuge the plate at 300 × *g* for 3 min. Discard the supernatant in an appropriate waste container. Resuspend samples in 250 μL FACS buffer and run on the flow cytometer.
9. Analyze receptor expression by quantifying geometric mean fluorescence intensity (GMFI) using the FlowJo analysis software.

3.2.3 RT-PCR-Based Readout

1. At the selected time point, flush the bottom of the cell culture flask using a sterile 5 mL serological pipette to create an even cell suspension. Count the cells and take approximately 1–5 × 10^6 cells per sample. Transfer the cells in a 1.5 mL Eppendorf tube.
2. Centrifuge the sample at 300 × *g* for 7 min. Aspirate the supernatant completely.
3. Extract mRNA from the pellet with the RNeasy Mini Kit by following the manufacturer's instructions. Measure RNA concentration using the NanoDrop instrument.
4. 1 μg of mRNA is reversely transcribed into cDNA with the High-Capacity cDNA Reverse Transcription Kit following the manufacturer's instructions in 20 μL final volume.
5. 10 ng of cDNA is mixed with 12.5 μL of Power SYBR® Green PCR Master Mix, 1 μL of a mix of forward and reverse primers (5 μM) and water up to 25 μL per reaction.
6. Exponential amplification is performed according to the following program: 20 s at 95 °C, 1 min at 95 °C, 20 s at 60 °C (×40) within StepOne™ Real-Time PCR systems. Data can be exported and analyzed in Microsoft Excel.

3.2.4 Sequencing-Based Readout

1. At the selected time point(s), flush the bottom of the cell culture flask using a sterile 5 mL serological pipette to create an even cell suspension. Count the cells and take approximately 1–5 × 10^6 cells per sample. Transfer the cells in a 1.5 mL Eppendorf tube. Genomic DNA is isolated from the cell pool that was transfected with PBS (control) and from cells exposed to RNP complex (experimental sample).
2. Centrifuge the sample at 300 × *g* for 7 min. Aspirate the supernatant completely.

3. Extract gDNA from the pellet with the DNeasy Blood & Tissue Kit, following the manufacturer's instructions. Measure gDNA concentration using the NanoDrop instrument.
4. PCR product are loaded onto 1–2% agarose gel and extracted on a UV ramp to be purified using GeneJET Gel Extraction and DNA Cleanup Micro Kit.
5. Purified PCR products are prepared for commercial Sanger sequencing.
6. The resulting sequence trace files (.ab1 or .scf format) are then uploaded into the TIDE web tool (http://tide.nki.nl). In addition, a character string representing the gRNA sequence (20 nt) is required as input. Then, the guide RNA sequence is aligned to the control sequence in order to determine the position of the expected Cas9 break site. Next, in all Sanger sequence traces an alignment window is automatically selected that runs from 100 to 15 bp upstream of the break site. The sequence segment in this window of the experimental sample is aligned to the control sequence in order to determine any offset between the sequence reads. All parameters in TIDE have default settings but can be adjusted if necessary.
7. An R^2 value is calculated as a measure of the goodness of fit, and the statistical significance of the detection of each indel is calculated.

3.2.5 ddPCR-Based Gene Disruption Assessment

Two different probes are designed within a single amplicon. The first is a FAM probe that is nonoverlapping with the cut site and which always binds to the genomic DNA. It provides a positive FAM amplitude for counting total genome copies in the sample. The second is a HEX (NHEJ) probe that binds at the cut or nick site of the genomic locus. It has a wild type sequence, so in the case where NHEJ occurs and results in either insertion or deletion of sequences at the cut site, the probe will no longer be able to bind resulting in a loss of HEX signal.

1. Dilute genomic DNA samples (*see* Subheading 3.2.3) to 100–150 ng/μL in distilled water.
2. Assemble the master mix below on ice by mixing the reagents. Multiply these by the number of samples. Distilled water is used as a negative control.
3. Aliquot 19 μL of master mix into eight-tube PCR strips.
4. Add 3 μL (for a total of 100–150 ng) of genomic DNA per sample. Also, add controls to tubes or wells designated for the controls.
5. If the total number of samples to be analyzed is not a multiple of 8, the remaining empty tubes of the PCR strip must be filled

with 25 μL of ddPCR Buffer Control Kit diluted to 1× by distilled water.

6. Gently pipet up and down 7–8 times to mix reactions by a 20 μL eight-channel pipette and carefully apply the mixtures into "sample" wells. Add 70 μL of oil into the "oil wells."
7. Generate droplets and transfer in plate.
8. Seal the plate with foil using the PX1™ PCR Plate Sealer set to 180 °C, 5 s.
9. Perform a thermal cycling using 96-Well C1000 Touch™ Thermal Cycler.
10. Analyze the droplets by the ddPCR analyzer.
11. Go to "Analyze" and then "2-D Amplitude" in the QuantaSoft software.

4 Notes

1. The protocol provided in this chapter to induce DSBs in NK cells via CRISPR/Cas9 needs to be optimized for each gene, cell preparation, and purpose. Factors that may be of importance for high efficacy is the crRNA–tracrRNA ratio, the gRNA–Cas9 ratio, and the concentration of RNP complex. Also, alternative electroporation programs may need to be explored to obtain the most optimal result in a given cell in a given state. Activated and expanded NK cells are likely easier to induce DSBs in, compared to resting NK cells.
2. If new target genes will be explored using the CRISPR/Cas9 approach, we recommend that at least three gRNAs are evaluated initially. The gRNA can be designed and verified using, for instance, CRISPR-Cas9 guide RNA design checker (IDT), or tools provided by MIT (http://crispr.mit.edu/), Chop Chop (http://chopchop.cbu.uib.no/), and Benchling (https://benchling.com/crispr). Initial screens of the gRNAs in easy-to-transfect cell lines can be done but can, in some cases, be misleading. If screens can be done in other lymphocytes, we recommend primary T cells as these cells may be somewhat easier to induce DSBs in compared to NK cells. For the initial screens and the following optimization, we recommend either sequencing or ddPCR as readout as these are the most detailed readouts for this purpose. Once the gRNA and the protocol are established, one can select other readouts that may serve the overall purpose better. Importantly, always address viability of the cells regardless of readout.
3. Although the purpose of this book chapter is to provide a protocol for CRISPR/Cas9 and examples of readouts to help

research groups to get started with gene editing of NK cells using the CRISPR/Cas9 system, it is important to emphasize that we are just beginning to explore this technology and that there may be issues associated with its use. One such issue is the risk of off-target effects. Even though the gRNAs are designed to guide the Cas9 enzyme to specific and predefined sites within the genome, there is a potential risk that the enzyme induces DSBs at other genomic locations, which in worst-case scenario can disrupt the function of other important genes or create oncogenes with the capability of generating immortalized cells. The extent of off-target effects and the possible consequences related to this process is currently not fully understood, which is demonstrated by a recent Editorial in Nature Biotechnology [105]. However, as reviewed by Lino et al. [106], precautions to limit potential off-target effects, for example through the development of mutant Cas9 systems, are currently being explored. Even though the extent and severeness of off-target effects following CRISPR-Cas9 gene editing is not completely understood, especially with regard to NK cells, we recommend using the gRNA design tools mentioned above for choosing an optimal guide. Further, we highly recommend to test the CRISPR/Cas9 edited NK cells for intact basic functionality, for example by running proliferation assays and to test their ability to degranulate and/or kill target cells as well as their ability to produce cytokines.

4. What readout to select for addressing the efficacy may vary depending on the purpose of inducing DSBs. Given below are a couple of opinions on the readouts presented in the chapter. Of note, additional readouts not presented in this chapter may be available.
   - The functional readout, here exemplified by in vitro migration following knockout of CXCR4, covers the more global changes that may be induced in the cell. The purpose of knocking out a given gene can be reached although the efficacy of inducing DSBs is not 100%. This is exemplified in this chapter.
   - The flow cytometry-based readout can be sensitive and is a good readout on the protein level, but does not fully reveal the true efficacy on a genetic level. Also, as some proteins are very stable, flow cytometry may be useful to address this aspect but can also be misleading if the half-life of a given protein is unknown (i.e., long). This is, however, a standard method available in most laboratories working with lymphocytes. An alternative would be Western blotting.
   - The qRT-PCR readout is more sensitive compared to the protein level-based readouts mentioned above, and is a

good surrogate marker for protein expression while not being dependent on the half-life of the protein. However, this approach can be biased by the design of the primers and will never cover the potential misfolding and degradation of proteins as well as potential loss of function despite protein expression and vice versa. This is a relatively cheap and quick readout that, if designed correct, would represent a good surrogate marker for DSB efficacy.

- The sequence-based readout provides more details than the qRT-PCR-based readout. Here you get the number of insertions, deletions, and frameshifts along with the degree of unmodified sequences. This approach is a bit more cumbersome and requires the instrumentation and knowledge to do DNA sequencing or the cost to out-source the sequencing along with the ability to identify the most optimal primers to amply the DNA fragment for sequencing.
- The digital-droplet PCR-based readout provides a clearer view on the efficacy on a DNA level. Primer and probes design may be the key element in avoiding rainy reactions. In some case it may be more straight forward to design new primers for the target sequence. This method, once set up and optimized for your gene of interest is quick and reliable. However, the method is more expensive than the abovementioned ones.
- Please note that for functional readout, flow cytometry and qRT-PCR never will provide an understanding of the efficacy on a DNA level. This is in contrast to the readout done by sequencing and ddPCR. However, we recommend that a functional readout or at least a readout on the protein level is used to complement the readouts on a genetic level as this will cover two important aspects of the gene editing via CRISPR/Cas9.

Acknowledgements

We would like to acknowledge Marios Dimitriou and Teresa Montera Blanco for support with establishing the ddPCR technology and the MedH Flow Cytometry core facility (Karolinska Institutet), supported by KI/SLL, for providing cell analysis services. This work has been supported by funding from Swedish Society for Medicine, Wallenberg Clinical Fellow, Swedish Cancer Foundation, Swedish Childhood Cancer Association, and the Swedish Research Council.

References

1. Herberman RB, Nunn ME, Lavrin DH (1975) Natural cytotoxic reactivity of mouse lymphoid cells against syngeneic and allogeneic tumors. I. Distribution of reactivity and specificity. Int J Cancer 16:216–229. https://doi.org/10.1002/ijc.2910160204
2. Herberman RB, Nunn ME, Holden HT, Lavrin DH (1975) Natural cytotoxic reactivity of mouse lymphoid cells against syngeneic and allogeneic tumors. II. Characterization of effector cells. Int J Cancer 16:230–239. https://doi.org/10.1002/ijc.2910160205
3. Kiessling R, Klein E, Wigzell H (1975) "Natural" killer cells in the mouse. I. Cytotoxic cells with specificity for mouse Moloney leukemia cells. Specificity and distribution according to genotype. Eur J Immunol 5:112–117. https://doi.org/10.1002/eji.1830050208
4. Kiessling R, Klein E, Pross H, Wigzell H (1975) "Natural" killer cells in the mouse. II. Cytotoxic cells with specificity for mouse Moloney leukemia cells. Characteristics of the killer cell. Eur J Immunol 5:117–121. https://doi.org/10.1002/eji.1830050209
5. O'Sullivan TE, Sun JC, Lanier LL (2015) Natural killer cell memory. Immunity 43:634–645. https://doi.org/10.1016/j.immuni.2015.09.013
6. Vivier E, Tomasello E, Baratin M, Walzer T, Ugolini S (2008) Functions of natural killer cells. Nat Immunol 9:503–510. https://doi.org/10.1038/ni1582
7. Dosiou C, Giudice LC (2005) Natural killer cells in pregnancy and recurrent pregnancy loss: endocrine and immunologic perspectives. Endocr Rev 26:44–62. https://doi.org/10.1210/er.2003-0021
8. Moffett-King A (2002) Natural killer cells and pregnancy. Nat Rev Immunol 2:656–663. https://doi.org/10.1038/nri886
9. Long EO, Kim HS, Liu D, Peterson ME, Rajagopalan S (2013) Controlling natural killer cell responses: integration of signals for activation and inhibition. Annu Rev Immunol 31:227–258. https://doi.org/10.1146/annurev-immunol-020711-075005
10. Anfossi N, André P, Guia S, Falk CS, Roetynck S, Stewart CA, Breso V, Frassati C, Reviron D, Middleton D et al (2006) Human NK cell education by inhibitory receptors for MHC class I. Immunity 25:331–342. https://doi.org/10.1016/j.immuni.2006.06.013
11. Bryceson YT, March ME, Ljunggren HG, Long EO (2006) Activation, coactivation, and costimulation of resting human natural killer cells. Immunol Rev 214:73–91. https://doi.org/10.1111/j.1600-065X.2006.00457.x
12. Chester C, Fritsch K, Kohrt HE (2015) Natural killer cell immunomodulation: targeting activating, inhibitory, and co-stimulatory receptor signaling for cancer immunotherapy. Front Immunol 6:601. https://doi.org/10.3389/fimmu.2015.00601
13. Bryceson YT, March ME, Barber DF, Ljunggren H-G, Long EO (2005) Cytolytic granule polarization and degranulation controlled by different receptors in resting NK cells. J Exp Med 202:1001–1012. https://doi.org/10.1084/jem.20051143
14. Screpanti V, Wallin RP, Ljunggren HG, Grandien A (2001) A central role for death receptor-mediated apoptosis in the rejection of tumors by NK cells. J Immunol 167:2068–2073. https://doi.org/10.4049/jimmunol.167.4.2068
15. Raulet DH (2004) Interplay of natural killer cells and their receptors with the adaptive immune response. Nat Immunol 5:996–1002. https://doi.org/10.1038/ni1114
16. Salagianni M, Lekka E, Moustaki A, Iliopoulou EG, Baxevanis CN, Papamichail M, Perez SA (2011) NK cell adoptive transfer combined with Ontak-mediated regulatory T cell elimination induces effective adaptive antitumor immune responses. J Immunol 186:3327–3335. https://doi.org/10.4049/jimmunol.1000652
17. Pallmer K, Oxenius A (2016) Recognition and regulation of T cells by NK cells. Front Immunol 7:251. https://doi.org/10.3389/fimmu.2016.00251
18. Malmberg KJ, Carlsten M, Björklund A, Sohlberg E, Bryceson YT, Ljunggren HG (2017) Natural killer cell-mediated immunosurveillance of human cancer. Semin Immunol 31:20–29. https://doi.org/10.1016/j.smim.2017.08.002
19. Imai K, Matsuyama S, Miyake S, Suga K, Nakachi K (2000) Natural cytotoxic activity of peripheral-blood lymphocytes and cancer incidence: an 11-year follow-up study of a general population. Lancet 356:1795–1799. https://doi.org/10.1016/S0140-6736(00)03231-1

20. Nakata J, Nakano K, Okumura A, Mizutani Y, Kinoshita H, Iwai M, Hasegawa K, Morimoto S, Fujiki F, Tatsumi N et al (2014) In vivo eradication of MLL/ENL leukemia cells by NK cells in the absence of adaptive immunity. Leukemia 28:1316–1325. https://doi.org/10.1038/leu.2013.374
21. Guerra N, Tan YX, Joncker NT, Choy A, Gallardo F, Xiong N, Knoblaugh S, Cado D, Greenberg NR, Raulet DH (2008) NKG2D-deficient mice are defective in tumor surveillance in models of spontaneous malignancy. Immunity 28:571–580. https://doi.org/10.1016/j.immuni.2008.02.016
22. Iguchi-Manaka A, Kai H, Yamashita Y, Shibata K, Tahara-Hanaoka S, Honda S, Yasui T, Kikutani H, Shibuya K, Shibuya A (2008) Accelerated tumor growth in mice deficient in DNAM-1 receptor. J Exp Med 205:2959–2964. https://doi.org/10.1084/jem.20081611
23. Ruggeri L, Capanni M, Urbani E, Perruccio K, Shlomchik WD, Tosti A, Posati S, Rogaia D, Frassoni F, Aversa F et al (2002) Effectiveness of donor natural killer cell aloreactivity in mismatched hematopoietic transplants. Science 295:2097–2100. https://doi.org/10.1126/science.1068440
24. Hsu KC, Keever-Taylor CA, Wilton A, Pinto C, Heller G, Arkun K, O'Reilly RJ, Horowitz MM, Dupont B (2005) Improved outcome in HLA-identical sibling hematopoietic stem-cell transplantation for acute myelogenous leukemia predicted by KIR and HLA genotypes. Blood 105:4878–4884. https://doi.org/10.1182/blood-2004-12-4825
25. Cooley S, Weisdorf DJ, Guethlein LA, Klein JP, Wang T, Le CT, Marsh SGE, Geraghty D, Spellman S, Haagenson MD et al (2010) Donor selection for natural killer cell receptor genes leads to superior survival after unrelated transplantation for acute myelogenous leukemia. Blood 116:2411–2419. https://doi.org/10.1182/blood-2010-05-283051
26. Cichocki F, Taras E, Chiuppesi F, Wagner JE, Blazar BR, Brunstein C, Luo X, Diamond DJ, Cooley S, Weisdorf DJ et al (2019) Adaptive NK cell reconstitution is associated with better clinical outcomes. JCI Insight 4:2. https://doi.org/10.1172/jci.insight.125553
27. Childs RW, Carlsten M (2015) Therapeutic approaches to enhance natural killer cell cytotoxicity against cancer: the force awakens. Nat Rev Drug Discov 14:487–498. https://doi.org/10.1038/nrd4506
28. Miller JS, Soignier Y, Panoskaltsis-Mortari A, McNearney SA, Yun GH, Fautsch SK, McKenna D, Le C, Defor TE, Burns LJ et al (2005) Successful adoptive transfer and in vivo expansion of human haploidentical NK cells in patients with cancer. Blood 105:3051–3057. https://doi.org/10.1182/blood-2004-07-2974
29. Rubnitz JE, Inaba H, Ribeiro RC, Pounds S, Rooney B, Bell T, Pui CH, Leung W (2010) NKAML: a pilot study to determine the safety and feasibility of haploidentical natural killer cell transplantation in childhood acute myeloid leukemia. J Clin Oncol 28:955–959. https://doi.org/10.1200/JCO.2009.24.4590
30. Kottaridis PD, North J, Tsirogianni M, Marden C, Samuel ER, Jide-Banwo S, Grace S, Lowdell MW (2015) Two-stage priming of allogeneic natural killer cells for the treatment of patients with acute myeloid leukemia: a phase I trial. PLoS One 10:6. https://doi.org/10.1371/journal.pone.0123416
31. Romee R, Rosario M, Berrien-Elliott MM, Wagner JA, Jewell BA, Schappe T, Leong JW, Abdel-Latif S, Schneider SE, Willey S et al (2016) Cytokine-induced memory-like natural killer cells exhibit enhanced responses against myeloid leukemia. Sci Transl Med 8:357. https://doi.org/10.1126/scitranslmed.aaf2341
32. Bjorklund AT, Carlsten M, Sohlberg E, Liu LL, Clancy T, Karimi M, Cooley S, Miller JS, Klimkowska M, Schaffer M et al (2018) Complete remission with reduction of high-risk clones following haploidentical NK-cell therapy against MDS and AML. Clin Cancer Res 24:1834–1844. https://doi.org/10.1158/1078-0432.CCR-17-3196
33. Curti A, Ruggeri L, Parisi S, Bontadini A, Dan E, Motta MR, Rizzi S, Trabanelli S, Ocadlikova D, Lecciso M et al (2016) Larger size of donor alloreactive NK cell repertoire correlates with better response to NK cell immunotherapy in elderly acute myeloid leukemia patients. Clin Cancer Res 22:1914–1921. https://doi.org/10.1158/1078-0432.CCR-15-1604
34. Fehniger TA, Miller JS, Stuart RK, Cooley S, Salhotra A, Curtsinger J, Westervelt P, DiPersio JF, Hillman TM, Silver N et al (2018) A phase 1 trial of CNDO-109-activated natural killer cells in patients with high-risk acute myeloid leukemia. Biol Blood Marrow Transplant 24:1581–1589. https://doi.org/10.1016/j.bbmt.2018.03.019
35. Lee DA, Denman CJ, Rondon G, Woodworth G, Chen J, Fisher T, Kaur I, Fernandez-Vina M, Cao K, Ciurea S et al

(2016) Haploidentical natural killer cells infused before allogeneic stem cell transplantation for myeloid malignancies: a phase I trial. Biol Blood Marrow Transplant 22:1290–1298. https://doi.org/10.1016/j.bbmt.2016.04.009

36. Graham FL, van der Eb AJ (1973) A new technique for the assay of infectivity of human adenovirus 5 DNA. Virology 52:456–467. https://doi.org/10.1016/0042-6822(73)90341-3

37. Neumann E, Schaefer-Ridder M, Wang Y, Hofschneider PH (1982) Gene transfer into mouse lyoma cells by electroporation in high electric fields. EMBO J 1:841–845. Available from: http://www.ncbi.nlm.nih.gov/pubmed/6329708

38. Zhao Y, Zheng Z, Cohen CJ, Gattinoni L, Palmer DC, Restifo NP, Rosenberg SA, Morgan RA (2006) High-efficiency transfection of primary human and mouse T lymphocytes using RNA electroporation. Mol Ther 13:151–159. https://doi.org/10.1016/j.ymthe.2005.07.688

39. Van Tendeloo VFI, Ponsaerts P, Lardon F, Nijs G, Lenjou M, Van Broeckhoven C, Van Bockstaele DR, Berneman ZN (2001) Highly efficient gene delivery by mRNA electroporation in human hematopoietic cells: superiority to lipofection and passive pulsing of mRNA and to electroporation of plasmid cDNA for tumor antigen loading of dendritic cells. Blood 98:49–56. https://doi.org/10.1182/blood.V98.1.49

40. Huerfano S, Ryabchenko B, Forstová J (2013) Nucleofection of expression vectors induces a robust interferon response and inhibition of cell proliferation. DNA Cell Biol 32:467–479. https://doi.org/10.1089/dna.2012.1950

41. Hemmi H, Takeuchi O, Kawai T, Kaisho T, Sato S, Sanjo H, Matsumoto M, Hoshino K, Wagner H, Takeda K et al (2000) A Toll-like receptor recognizes bacterial DNA. Nature 408:740–745. https://doi.org/10.1038/35047123

42. Chuang T-H, Lee J, Kline L, Mathison JC, Ulevitch RJ (2002) Toll-like receptor 9 mediates CpG-DNA signaling. J Leukoc Biol 71:538–544. Available from: http://www.ncbi.nlm.nih.gov/pubmed/11867692

43. Bürckstümmer T, Baumann C, Blüml S, Dixit E, Dürnberger G, Jahn H, Planyavsky M, Bilban M, Colinge J, Bennett KL et al (2009) An orthogonal proteomic-genomic screen identifies AIM2 as a cytoplasmic DNA sensor for the inflammasome. Nat Immunol 10:266–272. https://doi.org/10.1038/ni.1702

44. Fernandes-Alnemri T, Yu J-W, Datta P, Wu J, Alnemri ES (2009) {AIM2} activates the inflammasome and cell death in response to cytoplasmic {DNA}. Nature 458:509–513. https://doi.org/10.1038/nature07710

45. Hornung V, Ablasser A, Charrel-Dennis M, Bauernfeind F, Horvath G, Caffrey DR, Latz E, Fitzgerald KA (2009) {AIM2} recognizes cytosolic {dsDNA} and forms a caspase-1-activating inflammasome with {ASC}. Nature 458:514–518. https://doi.org/10.1038/nature07725

46. Dou Z, Ghosh K, Vizioli MG, Zhu J, Sen P, Wangensteen KJ, Simithy J, Lan Y, Lin Y, Zhou Z et al (2017) Cytoplasmic chromatin triggers inflammation in senescence and cancer. Nature 550:402–406. https://doi.org/10.1038/nature24050

47. Karikó K, Ni H, Capodici J, Lamphier M, Weissman D (2004) mRNA is an endogenous ligand for Toll-like receptor 3. J Biol Chem 279:12542–12550. https://doi.org/10.1074/jbc.M310175200

48. Diebold SS, Kaisho T, Hemmi H, Akira S, Reis E, Sousa C (2004) Innate antiviral responses by means of TLR7-mediated recognition of single-stranded RNA. Science 303:1529–1531. https://doi.org/10.1126/science.1093616

49. Hornung V, Ellegast J, Kim S, Brzózka K, Jung A, Kato H, Poeck H, Akira S, Conzelmann KK, Schlee M et al (2006) 5-′-Triphosphate RNA is the ligand for RIG-I. Science 314:994–997. https://doi.org/10.1126/science.1132505

50. Nallagatla SR, Hwang J, Toroney R, Zheng X, Cameron CE, Bevilacqua PC (2007) 5-′-Triphosphate-dependent activation of PKR by RNAs with short stem-loops. Science 318:1455–1458. https://doi.org/10.1126/science.1147347

51. Sun L, Wu J, Du F, Chen X, Chen ZJ (2013) Cyclic GMP-AMP synthase is a cytosolic DNA sensor that activates the type I interferon pathway. Science 339:786–791. https://doi.org/10.1126/science.1232458

52. Karikó K, Muramatsu H, Keller JM, Weissman D (2012) Increased erythropoiesis in mice injected with submicrogram quantities of pseudouridine-containing mRNA encoding erythropoietin. Mol Ther 20:948–953. https://doi.org/10.1038/mt.2012.7

53. Zhao Y, Moon E, Carpenito C, Paulos CM, Liu X, Brennan AL, Chew A, Carroll RG, Scholler J, Levine BL et al (2010) Multiple

injections of electroporated autologous T cells expressing a chimeric antigen receptor mediate regression of human disseminated tumor. Cancer Res 70:9053–9061. https://doi.org/10.1158/0008-5472.CAN-10-2880

54. Beatty GL, O'Hara MH, Lacey SF, Torigian DA, Nazimuddin F, Chen F, Kulikovskaya IM, Soulen MC, McGarvey M, Nelson AM et al (2018) Activity of mesothelin-specific chimeric antigen receptor T cells against pancreatic carcinoma metastases in a phase 1 trial. Gastroenterology 155:29–32. https://doi.org/10.1053/j.gastro.2018.03.029
55. Hendel A, Bak RO, Clark JT, Kennedy AB, Ryan DE, Roy S, Steinfeld I, Lunstad BD, Kaiser RJ, Wilkens AB et al (2015) Chemically modified guide RNAs enhance CRISPR-Cas genome editing in human primary cells. Nat Biotechnol 33:985–989. https://doi.org/10.1038/nbt.3290
56. Anderson BR, Muramatsu H, Nallagatla SR, Bevilacqua PC, Sansing LH, Weissman D, Karikó K (2010) Incorporation of pseudouridine into mRNA enhances translation by diminishing PKR activation. Nucleic Acids Res 38:5884–5892. https://doi.org/10.1093/nar/gkq347
57. Jensen MC, Popplewell L, Cooper LJ, DiGiusto D, Kalos M, Ostberg JR, Forman SJ (2010) Antitransgene rejection responses contribute to attenuated persistence of adoptively transferred CD20/CD19-specific chimeric antigen receptor redirected T cells in humans. Biol Blood Marrow Transplant 16:1245–1256. https://doi.org/10.1016/j.bbmt.2010.03.014
58. Miller AD, Rosman GJ (1989) Improved retroviral vectors for gene transfer and expression. Biotechniques 7:980–990
59. Miller AD, Buttimore C (1986) Redesign of retrovirus packaging cell lines to avoid recombination leading to helper virus production. Mol Cell Biol 6:2895–2902. Available from: http://www.ncbi.nlm.nih.gov/pubmed/3785217
60. Danos O, Mulligan RC (1988) Safe and efficient generation of recombinant retroviruses with amphotropic and ecotropic host ranges. Proc Natl Acad Sci U S A 85:6460–6464. Available from: http://www.ncbi.nlm.nih.gov/pubmed/3413107
61. Miller DG, Adam MA, Miller AD (1990) Gene transfer by retrovirus vectors occurs only in cells that are actively replicating at the time of infection. Mol Cell Biol 10:4239–4242. Available from: http://www.ncbi.nlm.nih.gov/pubmed/2370865
62. Hacein-Bey-Abina S, Le Deist F, Carlier F, Bouneaud C, Hue C, De Villartay J-P, Thrasher AJ, Wulffraat N, Sorensen R, Dupuis-Girod S et al (2002) Sustained correction of X-linked severe combined immunodeficiency by ex vivo gene therapy. N Engl J Med 346:1185–1193. https://doi.org/10.1056/nejmoa012616
63. Cavazzana-Calvo M, Hacein-Bey S, De Saint Basile G, Gross F, Yvon E, Nusbaum P, Selz F, Hue C, Certain S, Casanova JL et al (2000) Gene therapy of human severe combined immunodeficiency (SCID)-X1 disease. Science 288:669–672. https://doi.org/10.1126/science.288.5466.669
64. Hacein-Bey-Abina S, von Kalle C, Schmidt M, Le Deist F, Wulffraat N, McIntyre E, Radford I, Villeval J-L, Fraser CC, Cavazzana-Calvo M et al (2003) A serious adverse event after successful gene therapy for X-linked severe combined immunodeficiency. N Engl J Med 348:255–256. https://doi.org/10.1056/nejm200301163480314
65. Hacein-Bey-Abina S, Von Kalle C, Schmidt M, McCormack MP, Wulffraat N, Leboulch P, Lim A, Osborne CS, Pawliuk R, Morillon E et al (2003) LMO2-associated clonal T cell proliferation in two patients after gene therapy for SCID-X1. Science 302:415–419. https://doi.org/10.1126/science.1088547
66. Modlich U, Bohne J, Schmidt M, Von Kalle C, Knöss S, Schambach A, Baum C (2006) Cell-culture assays reveal the importance of retroviral vector design for insertional genotoxicity. Blood 108:2545–2553. https://doi.org/10.1182/blood-2005-08-024976
67. Cavazzana M, Bushman FD, Miccio A, André-Schmutz I, Six E (2019) Gene therapy targeting haematopoietic stem cells for inherited diseases: progress and challenges. Nat Rev Drug Discov 18:447. https://doi.org/10.1038/s41573-019-0020-9
68. Rosenberg SA, Aebersold P, Cornetta K, Kasid A, Morgan RA, Moen R, Karson EM, Lotze MT, Yang JC, Topalian SL et al (2010) Gene transfer into humans—immunotherapy of patients with advanced melanoma, using tumor-infiltrating lymphocytes modified by retroviral gene transduction. N Engl J Med 323:570–578. https://doi.org/10.1056/nejm199008303230904
69. Naldini L, Blomer U, Gage FH, Trono D, Verma IM (1996) Efficient transfer, integration, and sustained long-term expression of the transgene in adult rat brains injected with a lentiviral vector. Proc Natl Acad Sci U

S A 93:11382–11388. https://doi.org/10.1073/pnas.93.21.11382

70. Naldini L, Blömer U, Gallay P, Ory D, Mulligan R, Gage FH, Verma IM, Trono D (1996) In vivo gene delivery and stable transduction of nondividing cells by a lentiviral vector. Science 272:263–267. https://doi.org/10.1126/science.272.5259.263
71. Zufferey R, Dull T, Mandel RJ, Bukovsky A, Quiroz D, Naldini L, Trono D (1998) Self-inactivating lentivirus vector for safe and efficient in vivo gene delivery. J Virol 72:9873–9880. Available from: http://www.ncbi.nlm.nih.gov/pubmed/9811723
72. Mitchell RS, Beitzel BF, Schroder ARW, Shinn P, Chen H, Berry CC, Ecker JR, Bushman FD (2004) Retroviral DNA integration: ASLV, HIV, and MLV show distinct target site preferences. PLoS Biol 2:8. https://doi.org/10.1371/journal.pbio.0020234
73. Schröder ARW, Shinn P, Chen H, Berry C, Ecker JR (2002) Bushman F. HIV-1 integration in the human genome favors active genes and local hotspots. Cell 110:521–529. Available from: http://www.ncbi.nlm.nih.gov/pubmed/12202041
74. Marcucci KT, Jadlowsky JK, Hwang WT, Suhoski-Davis M, Gonzalez VE, Kulikovskaya I, Gupta M, Lacey SF, Plesa G, Chew A et al (2018) Retroviral and lentiviral safety analysis of gene-modified T cell products and infused HIV and oncology patients. Mol Ther 26:269–279. https://doi.org/10.1016/j.ymthe.2017.10.012
75. Cesana D, Volpin M, Serina Secanechia YN, Montini E Safety and efficacy of retroviral and lentiviral vectors for gene therapy. In: Safety and efficacy of gene-based therapeutics for inherited disorders. Springer, New York, pp 9–35. https://doi.org/10.1007/978-3-319-53457-2_2
76. Cornetta K, Duffy L, Turtle CJ, Jensen M, Forman S, Binder-Scholl G, Fry T, Chew A, Maloney DG, June CH (2018) Absence of replication-competent lentivirus in the clinic: analysis of infused T cell products. Mol Ther 26:280–288. https://doi.org/10.1016/j.ymthe.2017.09.008
77. Lyon D, Lapteva N, Gee AP (2018) Absence of replication-competent retrovirus in vectors, T cell products, and patient follow-up samples. Mol Ther 26:6–7. https://doi.org/10.1016/j.ymthe.2017.12.003
78. Kebriaei P, Singh H, Huls MH, Figliola MJ, Bassett R, Olivares S, Jena B, Dawson MJ, Kumaresan PR, Su S et al (2016) Phase I trials using sleeping beauty to generate {CD19}-specific {CAR} T cells. J Clin Invest 126:3363–3376. https://doi.org/10.1172/JCI86721
79. Gregory T, Cohen A, Costello C (2018) Efficacy and safety of P-Bcma-101 CAR-T cells in patients with relapsed/refractory (r/r) multiple myeloma (MM). Blood 132:1012. https://doi.org/10.1182/blood-2018-99-111419
80. Huang X, Guo H, Tammana S, Jung YC, Mellgren E, Bassi P, Cao Q, Tu ZJ, Kim YC, Ekker SC et al (2010) Gene transfer efficiency and genome-wide integration profiling of sleeping beauty, Tol2, and PiggyBac transposons in human primary t cells. Mol Ther 18:1803–1813. https://doi.org/10.1038/mt.2010.141
81. Woods NB, Muessig A, Schmidt M, Flygare J, Olsson K, Salmon P, Trono D, Von Kalle C, Karlsson S (2003) Lentiviral vector transduction of NOD/SCID repopulating cells results in multiple vector integrations per transduced cell: risk of insertional mutagenesis. Blood 101:1284–1289. https://doi.org/10.1182/blood-2002-07-2238
82. Espinoza DA, Fan X, Yang D, Cordes SF, Truitt LL, Calvo KR, Yabe IM, Demirci S, Hope KJ, Hong SG et al (2019) Aberrant clonal hematopoiesis following lentiviral vector transduction of HSPCs in a rhesus macaque. Mol Ther 27:1074–1086. https://doi.org/10.1016/j.ymthe.2019.04.003
83. Newrzela S, Cornils K, Li Z, Baum C, Brugman MH, Hartmann M, Meyer J, Hartmann S, Hansmann ML, Fehse B et al (2008) Resistance of mature T cells to oncogene transformation. Blood 112:2278–2286. https://doi.org/10.1182/blood-2007-12-128751
84. Smithies O, Gregg RG, Boggs SS, Koralewski MA, Kucherlapati RS (1985) Insertion of DNA sequences into the human chromosomal β-globin locus by homologous recombination. Nature 317:230–234. https://doi.org/10.1038/317230a0
85. Thomas KR, Folger KR, Capecchi MR (1986) High frequency targeting of genes to specific sites in the mammalian genome. Cell 44:419–428. https://doi.org/10.1016/0092-8674(86)90463-0
86. Simeonov DR, Marson A (2019) CRISPR-based tools in immunity. Annu Rev Immunol 37:571–597. https://doi.org/10.1146/annurev-immunol-042718-041522
87. Moscou MJ, Bogdanove AJ (2009) A simple cipher governs DNA recognition by TAL effectors. Science 326:1501. https://doi.org/10.1126/science.1178817

88. Christian M, Cermak T, Doyle EL, Schmidt C, Zhang F, Hummel A, Bogdanove AJ, Voytas DF (2010) Targeting DNA double-strand breaks with TAL effector nucleases. Genetics 186:756–761. https://doi.org/10.1534/genetics.110.120717
89. Urnov FD, Miller JC, Lee YL, Beausejour CM, Rock JM, Augustus S, Jamieson AC, Porteus MH, Gregory PD, Holmes MC (2005) Highly efficient endogenous human gene correction using designed zinc-finger nucleases. Nature 435:646–651. https://doi.org/10.1038/nature03556
90. Boissel S, Jarjour J, Astrakhan A, Adey A, Gouble A, Duchateau P, Shendure J, Stoddard BL, Certo MT, Baker D et al (2014) MegaTALs: a rare-cleaving nuclease architecture for therapeutic genome engineering. Nucleic Acids Res 42:2591–2601. https://doi.org/10.1093/nar/gkt1224
91. Porteus M (2015) Genome editing: a new approach to human therapeutics. Annu Rev Pharmacol Toxicol 56:163–190. https://doi.org/10.1146/annurev-pharmtox-010814-124454
92. Eyquem J, Mansilla-Soto J, Giavridis T, van der Stegen SJC, Hamieh M, Cunanan KM, Odak A, Gönen M, Sadelain M (2017) Targeting a CAR to the TRAC locus with CRISPR/Cas9 enhances tumour rejection. Nature 543:113–117. https://doi.org/10.1038/nature21405
93. MacLeod DT, Antony J, Martin AJ, Moser RJ, Hekele A, Wetzel KJ, Brown AE, Triggiano MA, Hux JA, Pham CD et al (2017) Integration of a CD19 CAR into the TCR alpha chain locus streamlines production of allogeneic gene-edited CAR T cells. Mol Ther 25:949–961. https://doi.org/10.1016/j.ymthe.2017.02.005
94. Hale M, Lee B, Honaker Y, Leung WH, Grier AE, Jacobs HM, Sommer K, Sahni J, Jackson SW, Scharenberg AM et al (2017) Homology-directed recombination for enhanced engineering of chimeric antigen receptor T cells. Mol Ther Methods Clin Dev 4:192–203. https://doi.org/10.1016/j.omtm.2016.12.008
95. Porteus MH (2019) A new class of medicines through DNA editing. N Engl J Med 380:947–959. https://doi.org/10.1056/NEJMra1800729
96. Roth TL, Puig-Saus C, Yu R, Shifrut E, Carnevale J, Li PJ, Hiatt J, Saco J, Krystofinski P, Li H et al (2018) Reprogramming human T cell function and specificity with non-viral genome targeting. Nature 559:405–409. https://doi.org/10.1038/s41586-018-0326-5
97. Nguyen DN, Roth TL, Li J, Chen PA, Mamedov MR, Vo LT, Tobin V, Apathy R, Goodman D, Shifrut E, et al (2019) A Cas9 nanoparticle system with truncated Cas9 target sequences on DNA repair templates enhances genome targeting in diverse human immune cell types. bioRxiv 591719. https://doi.org/10.1101/591719
98. Mali P, Yang L, Esvelt KM, Aach J, Guell M, DiCarlo JE, Norville JE, Church GM (2013) RNA-guided human genome engineering via Cas9. Science 339:823–826. https://doi.org/10.1126/science.1232033
99. Liu E, Tong Y, Dotti G, Shaim H, Savoldo B, Mukherjee M, Orange J, Wan X, Lu X, Reynolds A et al (2018) Cord blood NK cells engineered to express IL-15 and a CD19-targeted CAR show long-term persistence and potent antitumor activity. Leukemia 32:520–531. https://doi.org/10.1038/leu.2017.226
100. Carlsten M, Levy E, Karambelkar A, Li L, Reger R, Berg M, Peshwa MV, Childs RW (2016) Efficient mRNA-based genetic engineering of human NK cells with high-affinity CD16 and CCR7 augments rituximab-induced ADCC against lymphoma and targets NK cell migration toward the lymph node-associated chemokine CCL19. Front Immunol 7:105. https://doi.org/10.3389/fimmu.2016.00105
101. Levy E, Reger R, Segerberg F, Lambert M, Leijonhufvud C, Baumer Y, Carlsten M, Childs R (2019) Enhanced bone marrow homing of natural killer cells following mRNA transfection with gain-of-function variant CXCR4R334X. Front Immunol 10:1262. https://doi.org/10.3389/fimmu.2019.01262. eCollection 2019
102. Rautela J, Surgenor E, Huntington ND (2018) Efficient genome editing of human natural killer cells by CRISPR RNP. bioRxiv 406934. https://doi.org/10.1101/406934
103. Naeimi Kararoudi M, Dolatshad H, Trikha P, Hussain S-RA, Elmas E, Foltz JA, Moseman JE, Thakkar A, Nakkula RJ, Lamb M et al (2018) Generation of knock-out primary and expanded human NK cells using Cas9 ribonucleoproteins. J Vis Exp 14:136. https://doi.org/10.3791/58237
104. Noda M, Omatsu Y, Sugiyama T, Oishi S, Fujii N, Nagasawa T (2011) CXCL12-CXCR4 chemokine signaling is essential for NK-cell development in adult mice. Blood 117:451–458. https://doi.org/10.1182/blood-2010-04-277897

105. (2018) Keep calm and edit on. Nat Biotechnol 36:667–667. https://doi.org/10.1038/nbt.4221

106. Lino CA, Harper JC, Carney JP, Timlin JA (2018) Delivering crispr: a review of the challenges and approaches. Drug Deliv 25:1234–1257. https://doi.org/10.1080/10717544.2018.1474964

Index

A

Adoptive transfer 11, 13–19, 116, 126, 130–134, 137, 215
Adults 3, 9, 16, 19, 38, 46, 102, 150, 185–198
Allergic airway inflammation (AAI) 99–112
Asthma 8, 99, 100

B

B16 .... 155, 156, 161

C

Cell culture 84, 126, 155, 156, 203, 205, 221, 226–228
Cell isolation 73, 74, 78, 155
Coculture 5, 71–82
Common lymphoid progenitors (CLP) 1, 7, 8, 19, 200
CRISPR/Cas9 .... 79, 213–231
Crypts 186, 187, 189, 190, 194, 195
Cytokines 1, 23, 38, 59, 72, 84, 93, 99, 118, 129, 142, 153, 185, 200, 213
Cytotoxic granules exocytosis 85

D

Digital-droplet PCR 223, 225, 232
DNA sequencing 232

E

Enteroid 5, 185–198

F

Fetal mLN ILC 55
Flow cytometry 9, 10, 13, 14, 18, 20, 23–35, 38–40, 42, 43, 51, 61, 64, 66, 68, 71–82, 84–86, 88, 90, 96, 100, 103, 111–113, 124, 125, 131–135, 143, 146–149, 155–162, 201, 202, 208, 220, 224, 227, 232

G

Gating 10, 13, 14, 16, 27, 29, 30, 32, 42, 43, 61, 76, 88, 90, 91, 103, 104, 112, 121, 123, 126, 135, 149, 151, 208, 210
Gene editing 214–220, 224, 231, 232
Glucocorticoids 83, 84
G-protein coupled receptors (GPCRs) 93–98
Group 2 innate lymphoid cell (ILC2) 2–4, 7, 8, 24, 27–30, 32, 33, 35, 37, 42, 43, 47, 51, 57, 59, 60, 71–82, 93, 97, 99–112, 115–126, 129, 154, 161, 200, 208
Group 3 innate lymphoid cell (ILC3) 4, 7, 8, 24, 27–30, 32, 33, 35, 37–48, 51, 53, 55, 59–61, 63, 93, 94, 97, 115, 116, 129–135, 137, 138, 154, 200, 208, 209

I

IFN-γ 85, 88, 90, 93, 122, 125, 130, 199, 200, 208
IL-22 .... 3, 4, 24, 38, 39, 46, 47, 60, 93, 130, 154, 199, 200, 208
Imaging 104, 105, 108, 109, 112
Immunofluorescence 18, 51–58
Infant 185–198
Inflammatory bowel disease (IBD) 116, 185, 199–210
Innate immunity 130, 214
Innate-like lymphocytes 3, 19, 199
Innate lymphoid cell (ILC) 1–5, 7–21, 23–35, 37, 39, 42, 43, 45–48, 51–69, 83–91, 93–98, 115, 120, 121, 126, 129–139, 153–162, 165, 185, 199–210
Innate lymphoid cell (ILC) characterization 1–5, 153–162
Innate lymphoid cell (ILC) protocols 1–5, 7–21, 66, 116
Intestine 26–28, 38, 40, 42, 46–48, 57, 65, 67, 68, 72, 88, 94, 95, 130, 132, 135, 139, 141–144, 146, 147, 150, 154, 185–187
Intracellular staining 12, 13, 43, 103, 112, 121, 125, 158, 202, 210
Intraepithelial lymphocytes (IEL) 141–151

L

Lung 29–31, 35, 57, 72, 75, 78, 99, 100, 102, 105, 107–109, 112, 115, 116, 119–121

Shoba Amarnath (ed.), *Innate Lymphoid Cells: Methods and Protocols*, Methods in Molecular Biology, vol. 2121, https://doi.org/10.1007/978-1-0716-0338-3, © Springer Science+Business Media, LLC, part of Springer Nature 2020

M

Media 25, 27, 29, 30, 35, 39, 42, 62, 63, 66, 77, 79, 81, 84–87, 94–96, 100, 102, 107, 108, 120, 126, 131, 134, 139, 143–146, 150, 155–158, 162, 187–189, 191–196, 198, 201, 203, 210, 226
Melanoma 154, 156, 161
Mesenteric lymph nodes (mLN) 27, 29, 30, 35, 40, 43, 45, 48, 53, 102, 112
MHC2 ....3, 4, 24, 43, 47, 71
Microbiota 48, 169, 173, 180, 183, 199, BNF–184
Microscopy25, 27, 29, 52, 57, 105, 108, 109, 118, 119, 191
Migration 2, 94, 96–98, 220, 221, 224, 226, 227, 231
MiSeq 166, 168, 177, 179–183
Mucosal barrier 38

N

Natural killer (NK) cells 1, 2, 7, 8, 17, 19, 23, 43, 47, 59–62, 66, 84, 85, 88, 90, 129, 135, 139, 153, 154, 200, 213–231

O

Organoid 185–198

P

PCR 166–169, 172–174, 177, 180, 181, 222–225, 228–230
Peyers patches 4, 26, 40, 42, 65, 67, 95, 130, 138, 144, 150, 200
Phosphorylation 59–69
Preterm 185–198
Programmed death-1 (PD-1) 2, 3, 10, 20, 72, 90, 162

R

Regulation 9, 24, 60, 83, 186–188, 190, 214
RT-PCR 222, 231, 232

S

Signal transducer and activator of transcription (STAT) 59–69
16S rRNA 165, 166
Small intestine (SI) 26–28, 34, 38, 40, 46, 48, 65, 67, 68, 88, 94, 95, 130, 132, 135, 141–144, 147, 150
Small intestine lamina propria 26–28
Stem cell 7, 9, 78, 186–191, 195, 200, 215
Stimulation 4, 39, 42, 46, 47, 61–63, 66, 72, 77, 81, 84–90, 100, 102, 110, 112, 124, 125, 148, 154, 155, 157, 158, 208, 214

T

γδ T cells 142
T helper cell 71, 153, 200
Transcription factor 1–4, 7, 8, 10, 15, 19, 23, 24, 27–33, 35, 37, 38, 43, 45, 51, 59–69, 74, 80, 83, 93, 97, 104, 116, 124, 129, 153, 154, 158, 199, 200
Type 2 immune responses 71, 99

U

Unconventional T cells 142

Y

Yersinia enterocolitica 130–132, 134–139

MIX
Papier aus verantwortungsvollen Quellen
Paper from responsible sources
FSC® C105338

If you have any concerns about our products,
you can contact us on
ProductSafety@springernature.com

In case Publisher is established outside the EU,
the EU authorized representative is:
Springer Nature Customer Service Center GmbH
Europaplatz 3, 69115 Heidelberg, Germany

Printed by Libri Plureos GmbH
in Hamburg, Germany